PRACTICAL DESIGN OF REINFORCED CONCRETE

PRACTICAL DESIGN OF REINFORCED CONCRETE

RUSSELL S. FLING
Consulting Engineer

JOHN WILEY & SONS
New York • Chichester • Brisbane • Toronto • Singapore

Library of Congress Cataloging in Publication Data:

Fling, Russell S.
    Practical design of reinforced concrete.

    Includes bibliographies and index.
        1. Reinforced concrete construction.   2. Structural
design.   I. Title.

TA683.2.F54   1987   624.1'8341   86-11158
ISBN 0-471-80827-X

Printed in the United States of America
10 9 8 7 6 5 4 3 2 1

This book is dedicated to
Engineers who Strive to Protect the Public
by Designing Safe, Serviceable Structures
While Earning a Living.

PREFACE

The purpose of this text is to help the student and structural engineer* design reinforced concrete buildings efficiently, correctly, and accurately. Insofar as possible, material is presented in logical order as the structural design would be prepared in a design office. Necessary deviations are made to explain basic concepts before they are used in design. Students and practicing engineers should thoroughly understand these basic concepts before applying design procedures in practice.

Structural investigation consists of predicting the behavior, such as load-carrying capacity or deflection, of a structural member, frame, or system when all necessary information is given. Presumably there is only one correct answer. *Structural design* consists of selecting the size, shape, materials, and details of a structural member, frame or system when the required behavior, such as load capacity and other conditions, are given. Sometimes it is necessary to employ an iterative procedure in which a design or certain conditions are assumed and then investigated. The design or conditions are modified based on the first investigation and are then investigated again. Iterations continue until convergence, that is, until the investigation indicates that the assumed design or conditions are satisfactory but not oversized. However, for efficiency, design should proceed without iterations.

Design is the opposite of investigation. In design, there are many correct solutions. Perhaps no one solution is the "best." Design is by far the most common activity of practicing structural engineers. Some people consider design more difficult than investigation because of the many options involved, but it need not be so. Description of investigation procedures sometimes must precede description of design procedures, but the latter is emphasized in this text.

For efficiency, design aids should be compact, pertinent, and few in number. Most preferable are aids that can be committed to memory for conditions most commonly encountered. It is the intent of this text to present graphs and tables that will illustrate visually the effect parameters have on final design. Some graphs and tables in Appendix A will also serve as design aids by giving values useful in design without further computation for situations most commonly encountered in design practice.

---

*I implicitly assume that students become designers and that those who design concrete structures present evidence of their competence to state authorities who then grant them the privilege of using the title of engineer.

Likewise, simplified or short procedures should be significantly shorter than complete or long procedures, without sacrificing safety or construction economy. The design should move quickly and directly to determine required information while promoting the engineer's understanding of structural behavior.

To facilitate reaching these goals, all chapters of this text list objectives for the subjects included in the chapter. A student should refer to the objectives again after having studied all material in the chapter. Also included are suggestions for short procedures, where appropriate, and an analysis of the premium to be paid in construction cost, if any, for use of such procedures.

The structural engineer's objective in designing a structure is to prepare contract documents that describe the reinforced concrete structure to be constructed. These documents should be completed with a minimum of time, cost, and effort while meeting all other design objectives. The engineer's objective is *not* to make calculations or prepare a design, per se. However, some authorities (e.g., the engineer's supervisor, the owner, or the building department) will require proof that this was done in a satisfactory manner, so it behooves the engineer to prepare design computations in a competent manner for such an eventuality.

Only cast-in-place reinforced concrete is covered in this text. It does not include precast, prestressed, or composite construction or related subjects such as determination of loads and rigorous methods of moment distribution.

English units are used throughout the text because it is unlikely that metric or SI units will be adopted or be used to a significant degree in the United States in the near future. Conversion tables are included in Appendix D.

Although engineering principles are the same all over the world, this text is directed toward achieving the maximum economy of design and construction of reinforced concrete structures of moderate size or smaller in the United States. Design procedures appropriate only for tall buildings (such as the P-delta method of evaluating column slenderness), or appropriate only for advanced design using computers (such as the equivalent frame method for analyzing two-way slabs), or used only by those preparing design aids (such as balanced conditions in columns), are not discussed. Instead, reference is made to other publications. Also, calculation of cutoff points for positive moment bars in flexural members is not covered because I am opposed to the use of short bars. Much calculation effort is required to determine length of short bars, and their use risks misplacement with potentially serious consequences.

In this text, I have introduced some subjective opinions distilled from many years of practice as a structural engineer engaged in the design of a wide variety of structures in reinforced concrete and other materials. Some of these opinions are found in the footnotes, which are intended to facilitate the task of understanding the engineering principles involved, as well as to present practical observations on construction of reinforced concrete. I have tried to distinguish between established theory on which there is general agreement, and personal opinion, which is still open to debate.

References given at the end of each chapter direct the reader to detailed information on important subjects. Reference lists are not complete but provide only an initial supplementary reading for those interested in pursuing a subject further.

Some references are classic papers, which, though 10 to 20 years old, still provide the most definitive discussion of their subject.

Extensive reference is made to American Concrete Institute (ACI) Building Code Requirements for Reinforced Concrete (ACI 318-83). Hereafter, it is referred to as the ACI Code, or simply, the Code. Practicing structural engineers should obtain a copy of the Code and the accompanying commentary for use in design practice [1].* The ACI Code has been widely adopted by many governmental jurisdictions throughout the United States, as well as by many foreign countries.

In the text, those equations used most often in design (prime equations) are emphasized by being enclosed in a box. Prime equations and their derivations should be memorized by practicing engineers. Certain other equations are as important as prime equations but are not used as often in design and therefore need not be memorized. Some equations are secondary equations used in deriving or supporting prime equations.

I welcome any suggestions for improving the wording, correcting any errors, and including or omitting any material in future revisions. I may be contacted through the publisher, John Wiley & Sons.

Russell S. Fling

Selected References

1. ACI Committee 318. *Building Code Requirements for Reinforced Concrete, (ACI 318-83)*; *Commentary on Building Code Requirements for Reinforced Concrete*. American Concrete Institute. Detroit, Michigan, 1983.
   (ACI Code and Commentary are available from the American Concrete Institute, Box 19150, Redford Station, Detroit, Michigan 48219. Because these documents are mentioned throughout the text, they are not listed as references for each chapter.)

*Number in brackets refers to reference at the end of this preface.

ACKNOWLEDGMENTS

The inspiration for this book had its genesis with my early employment by Raymond C. Reese, a vigorous and perceptive engineer who later became chairman of the ACI Code Committee and President of ACI. The inspiration was nurtured and developed, unbeknown to me at the time, by ACI committees on which I served for many years. If you are pleased with this book, give credit to ACI and its committees. I take full blame for any failure to fulfill the aims and objectives in its preparation.

Credit should also be given to clients and contractors who challenged me in my years of active practice.

During preparation of the book, I received much encouragement from friends, Mary K. Hurd, Paul F. Rice, and others, and audiences before which I spoke. Without their support this book would not have been written. The offices of Paul J. Ford and Lantz, Jones & Nebraska reviewed each chapter during preparation and contributed many useful suggestions, ideas, and criticisms. I am grateful for their help.

I would also like to thank the following reviewers: Gilman Barker, Wentworth Institute of Technology; David W. Fowler, University of Texas; Jacob Grossman, Robert Rosenwasser Associates, P.C.; Grant Halversen, West Virginia University; Edward S. Hoffman, The Frank Klein Company; Harry L. Jones, Texas A&M University; Arthur H. Nilson, Cornell University; Paul F. Rice, Concrete Reinforcing Steel Institute; and Paul Zia, North Carolina State University.

Finally, the indulgence of my wife Dona and the assistance of my daughter Karen in the final stages of preparation were most welcome when my own enthusiasm was lagging.

CONTENTS

APPENDIXES

PRACTICAL DESIGN OF REINFORCED CONCRETE

CHAPTER 1

INTRODUCTION

1.1 OBJECTIVES

The twin goals of every engineer should be to perform his or her work competently and efficiently. These goals can only be approached if the engineer has a clear mental picture of the results to be expected from the design. With a firm, well-defined view of the objectives, the engineer can begin to strive toward perfection in performance and to make intelligent decisions on how best to spend his or her time and effort most effectively. The following objectives of the design process for reinforced concrete buildings are listed approximately in descending order of priority. Structural safety is clearly the most important objective and must take precedence over all other objectives.

1. Assure Structural Safety Structural safety refers to freedom from collapse or failure of the structure to support the loads imposed upon it. Such failure could mean that the structure and everything supported by it falls to the ground or that the structure simply deforms excessively. Because failure results in loss of property and could also result in injury or death, it must be prevented.

Structural safety, in some cases, also includes safety during construction. For example, long span bridges usually must be designed for the various stages of construction when only a part of the completed structure is in place. Designers of building structures should be alert to similar circumstances. For example, composite construction (not covered in this book) requires consideration of construction stages. Shoring and reshoring are normally the responsibility of the contractor, but if these cannot be done in the usual manner, the structural engineer should change the design or give special instructions to the contractor.

The term *factor of safety* usually means the ratio of the load that would cause collapse to the service load. Service loads are the actual weight of materials, such as concrete, masonry, partitions, windows, ceilings, and so forth (dead loads), and an assumed actual weight of people, furniture, stored material, and other building contents (live loads). Service loads also include the actual forces exerted by wind, soils, and liquids on the structure.

Factor of safety can also apply to component parts of the construction or to serviceability conditions. Then the term should be modified appropriately as, for example, "factor of safety against shear failure" or "factor of safety against excessive deflection."

Due to their importance, structural safety concepts are explained more fully in Section 1.2.

2. *Meet Functional Requirements*   The owner or architect gives the structural engineer certain requirements that must be met by the completed structure. Typically, the structure must meet the following requirements:

- Provide a certain floor area and volume.
- Meet certain dimensions and locations of walls, columns, and other elements.
- Meet building code requirements, other than structural design, such as fire resistance, inclusion of stairs, elevators, and so forth.
- Allow installation of under-floor duct work in the concrete for electrical, computer, and telecommunication wiring or allow drilling holes through the floor to reach conduit below floor construction.
- Meet special requirements for equipment that the owner will install in the building, such as concrete foundations for heavy machinery or deflection limits for sensitive laboratory equipment.
- Facilitate attaching or constructing other portions of the building.
- Allow completion within a certain time period.
- Present aesthetically pleasing surfaces where exposed to view or allow installation of finishing materials.
- Provide security against intrusion, especially for buildings with valuable contents, such as museums and financial institutions.
- Provide extra security against tornadoes and other severe weather conditions for buildings housing critical equipment, such as computers for centralized data processing.

3. *Meet Serviceability Requirements*   It is normally not sufficient that the structure avoid collapse and meet the functional requirements of the owner. It must also behave in a satisfactory manner during its service life. Following are the most important requirements.

- Deflection must be limited for a variety of reasons. Because this is a complex subject, it is covered in detail in Chapter 8.
- Premature or excessive cracking must be limited in certain circumstances.
- Weathering, abrasion, erosion, and corrosion must be avoided. These are important subjects, but they are beyond the scope of this text and will not be discussed in detail.

4. Reduce Construction Cost   Thousands of years ago, mankind proved that it was possible to construct durable, permanent, attractive structures that met the needs of society. The real challenge to today's engineer is to continue to do so at the lowest possible cost. Changing market conditions, materials, and labor, as well as new ideas and techniques, make this an on-going challenge. Suggestions for reducing construction cost will be found throughout this book, especially in Chapter 14, Preliminary Design.

5. Reduce Design Cost   Clients have limits to the amount that they will pay for preparation of a structural design, so it is in the engineers' best interest to complete their work as efficiently as possible. In some cases, the success of this effort will determine whether engineers have a job, and in other cases, it will determine the level of their compensation. Suggestions for saving design time will be found throughout this text.

6. Reduce Time of Construction   At the time this text was written, interest on construction loans was 1% per month or more. It follows that saving a month during construction will save 1% of the costs expended in construction to date of the time saving. The owner may realize additional savings by occupying the building earlier.

7. Coordination with Other Construction Activities and Building Services   Structural frames are rarely built in isolation. Plumbing lines, and heating, ventilating, and air-conditioning ducts pass through and below floor framing. Electrical conduit is buried in concrete. Walls, doors, and windows attach to the concrete frame. Owners will insist that the structural frame accommodate other parts of the building with ease.

8. Personal Preferences   Practicing engineers quickly learn that owners, architects, and contractors frequently have personal preferences for a type of structural frame, method of construction, and materials. Successful engineers will at least consider such preferences.

Practicing engineers must comply with local building codes, which include provisions for design of electrical and mechanical systems, architectural layouts, fire safety, site work, and foundations, as well as structural design. The ACI Building Code Requirements for Reinforced Concrete (ACI 318-83), referred to in this text as the ACI Code, is usually a part of local building codes.

The ACI Code is written by a committee (ACI 318) organized by the American Concrete Institute, a private, nonprofit organization. ACI Committee 318 is composed of a carefully balanced group of practicing engineers, college professors, concrete researchers, contractors, government representatives, and industry representatives. The first ACI 318 Code was published in the early part of this century. Since then, it has gained widespread recognition as the most complete, authoritative set of rules, written in English, guiding the design of reinforced concrete. However, as the product of a private organization, the ACI Code has no legal authority, even though ACI membership is open to everyone (even worldwide), and its standards are adopted by a lengthy, strict procedure.

In the United States, most cities, countries, and states having jurisdiction over the design and construction of buildings have chosen to adopt the ACI Code, usually without modification but sometimes with minor alterations. When the ACI Code has been adopted by a unit of government, the Code becomes law and must be followed by engineers designing structures to be built within that government's jurisdiction. If the government with jurisdiction has no laws or regulations on design of concrete, most engineers will comply with the ACI Code as a matter of good practice and to deflect potential future criticism of their design.

1.2 STRUCTURAL SAFETY

There are two approaches to design, the working stress method and the strength method. Both were first proposed before 1900, but shortly afterward, the working stress method became universally accepted and used until 1956. In that year, a joint ASCE-ACI committee published a report on current knowledge of strength-design methods [1]. The report included specific design recommendations that were adopted by the ACI Code in 1956. Today's ACI Code (adopted in 1983) is devoted almost exclusively to the strength method. The working stress method is permitted with qualifications outlined in an appendix in the ACI Code. This text discusses only the strength method.

The working stress method uses actual service loads and allowable stresses that include a factor of safety. Analysis of a section assumes that concrete as well as steel behaves elastically. Major drawbacks to the working stress method are (1) no recognition is given to different degrees of uncertainty about various kinds of loads, and (2) concrete stress is not proportional to its strain up to its ultimate crushing strength. As a result, the factor of safety is quite variable and does not recognize the true strength of reinforced concrete.

The strength method uses loads greater than service loads by factors (called load factors) that represent an assessment of the variability of each load. The structure is then designed to provide the strength necessary to carry the factored loads. Computation of strength takes into account the nonlinear stress-strain behavior of concrete. The computation also considers the dependability of different aspects of strength, such as flexure, shear, torsion, and axial load, by using strength reduction factors ϕ. The result is a structure that has a more uniform factor of safety considering the risk of higher-than-expected loads and lower-than-expected strengths. The strength predicted by ACI Code procedures is somewhat less than the true ultimate strength because of conservative estimates of material strength and design procedures. Load factors and strength reduction factors ϕ are discussed later in this section.

The factor of safety could be slightly more than unity if our knowledge of the actual loading and our methods of analysis were perfect, if actual material properties always met or exceeded specified properties, and if construction were executed perfectly. However, this is not a perfect world, so the factor of safety must always exceed unity by an appropriate margin.

Establishing quantitative values for the factor of safety has defied rational analysis so far. Instead, values are set by precedent and by the accumulated wisdom of many experienced engineers, scientists, and technicians.

The ACI Code assures safety in three ways. The first way is to increase assumed loads by an amount that depends on the nature of the loading; this is called a "load factor." The second way is to reduce the specified strength of materials by an amount that depends on the material and its use in the structure; this is called a "strength reduction factor." The third way is to incorporate Code limits that are more conservative than test results. In addition, the general building code, of which the ACI Code is a part, will require design loadings that usually (but not always) are more conservative than actual loadings that the structure will experience.

Following are some of the uncertainties about loads and design assumptions:

1. Actual loads may be larger than those assumed in design due to underestimation or to changing use of the structure.

2. Actual load distribution may be more severe than that assumed in the design.

3. Loads, forces, moments, and shears in the structure may be larger than those calculated using normal assumptions and procedures.

4. Structural behavior may be more severe than that assumed in design if the engineer overlooks a critical condition or the structure reacts to loading in an unforeseeable way.

5. Loads from several sources usually do not occur at their maximum values simultaneously.

Following are some of the uncertainties about materials and construction:

1. Material strengths may be lower than specified.

2. Failure may occur in a brittle mode rather than a ductile mode. A brittle failure is characterized by a sudden and total collapse of a member with little or no advance warning, such as excessive deflection or cracking. Glass and pottery fail in a brittle mode; steel and timber usually do not. If a structural frame or member must fail for any reason, ACI Code philosophy is to have it occur with large deflection and much cracking so as to warn everyone nearby of the risk of impending collapse. Ideally, collapse would not occur because the structure would deflect so much it would be unusable. Overload causing the deflection would be removed, the member would be shored, or other steps would be taken to prevent failure. Ductility is an important design consideration, and its effect on design procedures is mentioned throughout this text. Lack of ductility is the principal reason glass is not used as a structural material, even though it is stronger, stiffer, and less expensive than steel or concrete.

3. Concrete outlines or arrangement may be smaller or more severe than those specified in the design.

4. Reinforcing steel may be misplaced to a less effective position.

5. Construction methods and sequences may stress the structure more highly than those contemplated in the design.

In addition, the ACI Code Committee has considered the following:

1. The worst conditions outlined above probably do not occur simultaneously.

2. Failure of a member carrying modest loads is less harsh than failure of a large important member.

3. Redundancy is desirable to mitigate the consequences of failure of a member and limit progressive collapse. Progressive collapse is the sequential failure of an entire structure or large portions of a structure due to the initial failure of one member.

In general, the ACI Code Committee has incorporated uncertainties about loads and design assumptions into load factors. Likewise, uncertainties about materials and construction have been incorporated into strength reduction factors. In some cases, uncertainties in both categories have affected both factors, as well as the degree of conservatism implicit in the ACI Code equations and procedures. Practicing engineers may want to consider these same uncertainties and increase the factor of safety for certain situations, either explicitly or implicitly, by increasing member sizes or the amount of reinforcing steel used. Due to legal and moral responsibilities, practicing engineers rarely reduce the factor of safety.

In two recent papers, J.G. MacGregor presents a current overview of the methodology used and the assumptions made in deriving load factors and strength reduction factors for the ACI Code [2, 3]. He also makes suggestions for future revisions to the Code.

Load Factors

In Section 9.2, the ACI Code specifies the load factors to be used to determine U, the required strength to resist factored loads or related internal moments and forces. Regardless of whether other loads such as wind, earthquake, earth pressure, or the structural effects of volume changes are present, the required strength U must always be at least equal to the following:

$$U = 1.4D + 1.7L \tag{1.1}$$

where D = dead loads, or related internal moments and forces
and L = live loads, or related internal moments and forces.

When wind loads are present, one-third higher stresses are permitted. This is equivalent to reducing the load factors by one-fourth; thus

$$U = 0.75(1.4D + 1.7L + 1.7W) \tag{1.2}$$

To obtain the most severe condition, load combinations that include both full value and zero value for L must be investigated. In addition, dead load may be overstated to be conservative for the usual case. When dead loads counteract the effects of wind loads, load combinations with reduced dead load must be considered

$$U = 0.9D + 1.3W \tag{1.3}$$

If resistance to earthquake loads or forces or the load effects of earthquake, or the related internal moments or forces E are included in design, $1.1E$ must be substituted for W in Eqs. 1.2 and 1.3. Wind and earthquake need not be considered simultaneously.

If resistance to earth pressure H is included in design, $1.7H$ must be added to the right-hand side of Eq. 1.1 and within the parentheses of Eq. 1.2. If dead or live loads reduce the effect of earth pressure, $0.9D$ must be substituted for $1.4D$, and a zero value for L must be considered

$$U = 0.9D + 1.7H \qquad (1.4)$$

If resistance to loadings due to weight and pressure of fluids with well-defined densities and controllable maximum heights F is included in design, such loading must have a load factor of 1.4 and be added to all loading combinations that include live load.

The effect of impact must be included with live load L.

Where structural effects of differential settlement, creep, shrinkage, or temperature change T may be significant, higher stresses are permitted as for wind, and the required strength U shall be at least equal to

$$U = 0.75(1.4D + 1.4T + 1.7L) \qquad (1.5)$$

Because live load may not always be present, required strength U must not be less than

$$U = 1.4D + 1.4T \qquad (1.6)$$

The ACI Code intends that loads from various sources be considered in the most critical combinations, except that wind and earthquake need not be combined because of the very low probability that both would occur at maximum values at the same time.

Strength reduction factors are given in Table A1.1. (Code 9.3).

An overall factor of safety for a structure is somewhat ambiguous. It is more precise to state a factor of safety for a particular loading condition with respect to a specified failure mode. Nevertheless, engineers must sometimes estimate the factor of safety for nonengineers. The Code favors a ductile failure in flexural members (if a failure must occur at all) so an engineer might explain that the overall factor of safety ranges from 1.56 ($U/\phi = 1.4/0.9$) to 1.89 (1.7/0.9) for flexural members and 2.0 ($U/\phi = 1.4/0.7$) to 2.43 (1.7/0.7) for tied columns.

When reference is made to "loads," "moments," "torsional moments," and "shears," actual service loads and the moments, forces, and shears they produce are meant. Service moments, torsional moments, and shears are designated by the symbols M, T, and V.

When the loads, moments, and shears have been multiplied by the appropriate load factors, as discussed above, they are referred to as "factored load," "factored moment," "factored torsional moment," and "factored shear." The latter three are designated by the symbols M_u, T_u, and V_u.

The "nominal moment strength," "nominal torsional moment strength," and "nominal shear strength" of a concrete section are the calculated strengths before they have been reduced by the strength reduction factor ϕ. They are designated by the symbols M_n, T_n, and V_n. These strengths are not necessarily the true ultimate strengths. True ultimate strengths might be somewhat higher than nominal strengths due to conservative assumptions, overstrength of materials, and construction, as discussed in Section 1.4.

Table 1.1 summarizes these loads and their meanings.

1.3 STEPS IN DESIGN PROCESS

If engineers are to reduce their own time and cost and thereby increase their value to their client or employer, they must understand the design process, how much time is allocated to each step, and which steps offer the most potential for cost and time saving. Following is a brief discussion of steps in the design process.

Step 1. Preliminary Design   In this step, the framing scheme, the general arrangement of structural elements, and their relationship to other parts of the project are established. Also, concrete outlines are determined and possibly the quantities of reinforcing steel. If performed well, there should be no need to change concrete outlines after completion of this step. (See Chapter 14 for a discussion of the preparation of a preliminary design.)

The engineer has far greater opportunity to effect construction cost savings* in this design step than in any other. Furthermore, less time is normally spent in this step than in most others; therefore, it is unwise to try to save design effort in this step.

TABLE 1.1   Symbols and their Meanings

Symbol	Name	Meaning
D	Service dead load	Actual dead load
L	Service live load	Actual live load
M	Service moment	Actual moment
P	Service axial load	Actual axial load
T	Service torsional moment	Actual torsional moment
V	Service shear	Actual shear
M_u	Factored moment	Actual moment times load factor
P_u	Factored axial load	Actual axial load times load factor
T_u	Factored torsional moment	Actual torsional moment times load factor
V_u	Factored shear	Actual shear times load factor
M_n	Nominal moment strength	Strength of section before reduction by strength reduction factor (ϕ)
P_n	Nominal axial load strength at given eccentricity	
T_n	Nominal torsional moment strength	
V_n	Nominal shear strength	

*Also, great opportunity exists to save time spent later in design and preparation of construction documents—a matter of no small interest to engineers trying to support a family or a modest life-style.

Step 2. ***Final Design Calculations*** This is the most important step in determining the safety of the completed design. Work performed in this step has a much smaller effect on the construction cost than commonly supposed and certainly less than the cost-saving possibilities in preliminary design. This step in the design process is the one most affected by design code complexities. There are significant possibilities to save design time and cost in this step, and these will be discussed throughout this text.

Step 3. ***Contract Documents*** As a structural design is completed, it is necessary to transfer the concept and details from the mind of the engineer to the mind of those persons who will actually build the structure. Without the proper execution of contract documents, a structural design is useless (if the builder doesn't know what to build) or dangerous (if the builder misinterprets the intention of the engineer). Preparation of clear documents is an essential part of the design.

Contract documents usually consist of drawings (dimensioned pictures) and specifications (written descriptions). Together, they show the contractor what structure to build as contemplated in the design. In recent years there has been considerable work done in defining the scope of these documents, what information should be contained in them, and how it should be presented. The engineer can refer to documents prepared by the American Concrete Institute [4, 5], Concrete Reinforcing Steel Institute [6], and the Construction Specifications Institute [7].

Simplifying the contract documents will frequently reduce the construction cost, as well as save design time and cost. Engineers who want to reduce the time and cost spent in design should pursue simplification as a goal.

Step 4. ***Coordination*** The structural engineer is part of a team for the design and construction of a facility desired by the owner. As such, it is essential that the engineer coordinate his efforts with those of other members of the team. The time spent in coordination may be largely out of control by the structural engineer, but there are some things the engineer can do to minimize coordination effort. The engineer can anticipate the needs of other parties and provide for them in the preliminary design and as the final design progresses. For example, building structures have ventilating duct work and plumbing pipes that penetrate floor slabs. Horizontal runs of ducts and pipes are usually located just below the floor and roof slabs. If the structural design accommodates ducts and pipes, coordination effort is reduced.

Step 5. ***Services During Construction*** Fabrication shop detail drawings, concrete mix designs, and laboratory reports should be checked for conformity with the design concepts by the structural engineer who designed the project. Frequently, the engineer will be called upon to explain the contract documents to workers in the field. Sometimes trips to the field during construction will be necessary to observe construction, answer questions, and verify that work is progressing in a manner contemplated by the design. Time spent in answering questions and in field trips, especially, can be minimized by preparation of unambiguous contract documents.

Explanation of contract documents and occasional field trips will be required of all design engineers. A higher level of professional service is to provide periodic or full-time field inspection, especially at the start of construction. It is easier to instruct workers on proper construction than it is to remove and rebuild, or modify work incorrectly built.

Step 6. Services After Construction   If a project is well designed and well built, it should perform in a satisfactory manner so that no calls will be made to the engineer after completion. Obviously then, the more competently that a structure is designed and the better that contract documents are prepared, the less time an engineer will spend on a project after construction is completed. Of course, changed conditions may require structural modifications and more design work as a new project.

A summary of the steps in the design process and the typical range of time allocated to each step is shown in Table 1.2.

Design Philosophy

Structural design involves more than computations of moments, shears, forces, and member sizes. Successful engineers develop a personal philosophy or approach to design that results in safe, serviceable, and functional buildings even though the engineers, being human, make minor mathematical errors or deviations from Code requirements. The next few paragraphs should help students and beginning engineers to develop their own philosophy. As their careers advance, engineers should set aside some time every year or so to refine their own approach to design.

A 10% to 20% overstress has rarely, if ever, been the primary cause of structural failure,* although overstress would make a dangerous situation worse. On the other hand, overlooking some aspect of design has frequently (some say usually) been the primary cause of failure when the design is responsible for failure. In designing a structure, it is important for the engineer to be sure that the structure has been analyzed correctly for anticipated loads, the kind of stresses they cause (e.g., moment, shear, torsion, axial load, bearing), the direction of stresses (e.g., positive moment vs. negative moment, axial tension vs. compression, etc.), and the approximate magnitude of the stresses.

*Engineers must never use this statement as an excuse for sloppy or careless calculations, nor as justification for intentional overstress of this magnitude.

TABLE 1.2   Steps in the Design Process

Steps	Time Allocation (%)	Opportunity for Saving Design Time
1. Preliminary design	5 to 10	Little
2. Final design calculations	5 to 20	Much
3. Contract documents	30 to 50	Much
4. Coordination	20 to 30	Some
5. Services during construction	8 to 15	Some
6. Services after construction	1 to 2	Little
Total	100	

Calculations should be made quickly and simply so that the engineer's attention can be focused on design objectives and more time can be spent studying the structure and its lifetime environment, anticipating potential problems, and developing strategies for minimizing or avoiding them.

Questions the engineer should consider during final design include the following:

- Will loads cause high torsion? Which members will be most highly stressed by torsional moments? Will flexure and transverse shear be increased in some members by the effects of torsion?

- Will the building be constructed on such a short time schedule that the contractor might overstress the structure by reshoring procedures? What design changes or specification provisions will reduce the risk of damage?

- Will the location of plumbing, air ducts, or electrical conduit impair the strength of the structural frame? Should the structural engineer change the frame or request changes in pipes, ducts, or conduit?

- Will volume changes due to temperature, shrinkage, or creep cause damaging stresses or other difficulties?

- Could serviceability be compromised by excessive deflection or cracking?

- Is floor construction, especially in flat plates, adequate in shear around columns? Consider pipes, ducts, and other floor penetrations near columns, including those that may be unknown to the engineer at the time of design.

- Will the structural design concept require unusual construction procedures that might be misinterpreted or might lead to construction errors?

- Does the safety or serviceability of the structure depend on unusually tight tolerances?

- Could nonstructural materials or elements of construction interact with the frame to the detriment of the project?

- Have all loads been carried to supporting soils?

- Have all joints been properly reinforced for moments, axial forces, and shears?

- Have all important members been designed and scheduled or delineated on the contract documents? Do members with apparently similar loading have similar designs?

- After the final design is complete and the engineer surveys the work, does the structure appear "in balance," well designed, accurately and well delineated? Have all members been designed with a complete set of consistent assumptions and do all members have consistent factors of safety?

1.4 CALCULATION ACCURACY

It is far more important for an engineer to understand how a structure reacts to loading, how the structure interacts with nonstructural elements, how to arrange structural members, how to proportion concrete outlines, and where to place reinforcing steel to resist tensile stresses than it is for the engineer to compute the exact concrete outlines or the "right" amount of reinforcing steel to the second or third decimal place. Structural theory may seem reasonably precise, but the data from which designs are prepared are not. Most design procedures are based on tests with wide variations in data. Many assumptions are made for the sake of simplicity and ease of computation.

Live loads specified in building codes are generally rounded to 5 or 10 psf (see Appendix C, Abbreviations) or more for incremental variations of 20% or more. Actual live loads in some structures are 20% or 30% of specified live loads and sometimes never reach the specified levels. In other structures, the live load may be far exceeded, especially during construction.

Dead load of concrete is normally computed by using an assumed unit weight (150 pcf for normal weight concrete); however, the unit weight of plain concrete may vary by 5% or 10%, depending on the aggregates used and the amount of entrapped and entrained air. Allowable tolerances in concrete outlines permit variations in weight of up to 10% or so. Reinforcing steel adds 2% to 20% to the unit weight of plain concrete.

Dead load of other building materials and components have similar variations in weight—in some cases, even larger.

Even a cursory examination of test data on wind and seismic loads indicates that they are more variable than live and dead loads. Newspapers and magazines report numerous instances of structures damaged or collapsed by wind and seismic loads much higher than that assumed in the design.

In the distribution of moments in concrete frames, it is commonly assumed that concrete is homogeneous, isotropic, monolithic, and elastic, but it cannot be said that concrete has any of those properties precisely. Concrete is made from aggregate, cement, and water—three dissimilar materials. Steel is then added to the mixture. Concrete and steel have widely varying properties, thus casting doubt on the homogeneity assumption. Concrete cracks and the steel component is linear, thus destroying the isotropy assumption. None but the smallest reinforced concrete structure is cast monolithically before it is loaded. Concrete approaches elastic behavior for short-term, light loads but shows considerable inelastic deformations under all, even light, long-term loads. The common assumptions are necessary for engineers to complete designs in a reasonable amount of time. So far, no one has made a thorough study to determine how much the moments and shears in real structures vary from those computed using these assumptions, but it seems likely that variations are at least 10% or more.

ACI Code equations and procedures are based on extensive test data. Even though the research is carefully performed, these data are usually variable in a range of 50% to 100% or more.

The strength of materials is by no means perfect. Concrete and steel strengths will occasionally fall below the specified strengths, especially for individual bars and for small batches of concrete. Normally the yield strength of steel is about 10% higher than specified, and the strength of laboratory-cured concrete averages 20% higher than that specified.

Concrete members varying from specified dimensions by the allowable tolerances will vary in strength as much as 8% lower to 12% higher. Allowable variations in location of reinforcing steel will affect the strength as much as ±10% or 12%, and actual steel placement could be even more variable [8].

Finally, selection of concrete outlines and bar sizes and spacing can only be made with an accuracy of ±5% to 30% or more.

Table 1.3 summarizes these variations.

In spite of all the variations and uncertainties outlined above, it is quite possible to design safe structures that are economical yet not overly conservative. The engineer must not waste time and effort on excessively accurate calculations or yield to the temptation of making sloppy calculations simply because so much is unknown.

To strike a proper balance between accuracy and efficiency, recommendations on the accuracy of design calculations are summarized in Table 1.4.

Until the advent of computers, engineers worked very satisfactorily with "slide rule" accuracy, which is about three significant figures at the center of the rule where the analog of 0.5 is 3.16. Thus, overall accuracy is 1/316 $\simeq$ 0.3% per operation. Someone with good eyes and a steady hand can read a slide rule to, say, 0.1% accuracy, but the average person will read it to only 0.3% accuracy.

TABLE 1.3 Summary of Variations in Design and Construction[a]

	Not Conservative[b] (%)	Conservative[b] (%)
Loading		
Live	[c]	NA[d]
Dead, concrete	30	10
Dead, other	30	20
Wind	50 to 100	NA
Seismic	50 to 100	NA
Design assumptions	10	25 to 50
Construction		
Material strengths	10	20
Sizes and placement	20	20
Member selection		
Bar areas	3	5 to 30
Concrete strengths	3	10 to 30
Concrete outlines	3	20 to 40

[a] Values given are intended to give an approximate range that typically might be expected under extreme conditions of structures designed and built to meet the ACI Code. Most structures will have less variation than the values given. Structures not meeting the Code may have larger variations.

[b] In real structures, the extreme variations will not all occur at the same time or affect the structure in the same manner, that is, either on the conservative or on the unconservative side.

[c] One approach to estimating the possible overload from live loads is to assume the conversion of the structure to another occupancy with higher live load requirements. The limit to this approach is a building with goods stored to the ceiling.

[d] NA = not applicable.

TABLE 1.4 Recommendations on Calculation Accuracy[a,b]

Loading	
Record to the nearest	1 psf for slabs
	10 plf for joists
	100 plf for beams
	100 lbs for concentrated loads
	100 lbs for total loads and reactions on flexural members
	1.0 k for column and footing loads
Dimensions	0.1 ft span length and location of load
	0.1 in. effective beam and slab depth
Computations	0.1 ft-kip moments
	0.01 sq. in. bar areas
Design selection	½ in. slab thickness
	1 in. other concrete outlines[c]
	1000 psi concrete strength
	½ in. bar spacings in slabs and walls (1 in. is preferable)

[a] In no case need loadings and dimensions be recorded to an accuracy greater than three significant figures.

[b] See Appendix C for abbreviations.

[c] Larger incremental variations in concrete outlines are usually more economical (see Chapter 14, Preliminary Design).

If 10 consecutive operations of multiplication or division are performed on a slide rule, the error will still be less than about 3% ($1.003^{10} = 1.0304$) even if the maximum error occurs on each operation and in the same direction. This is far less than the error introduced by any one of the factors listed in Table 1.3. Countless structures have been designed with "slide rule" accuracy with satisfactory results.

Computations should be carried out with an accuracy of three significant figures because three figures are easy to handle by manual calculation as well as by hand calculator, and more than three figures are time consuming to record in writing and lead to a false sense of accuracy. One exception to this rule is a calculation involving the difference between two large numbers. In this case, the number of significant figures should include three in the expected final answer. For example, the difference between 738,567 and 738,455 is 112. Rounded to 739,000 and 738,000, the difference is 1000, about nine times as large a the true answer.

Due to rounding, accuracy demands one more decimal place in the calculations than in the final answer. For example, three significant figures in the calculations allow two significant figures in the answer.

Selection of concrete outlines, concrete strength, and reinforcing steel should normally exceed requirements shown by calculation. However, most practicing engineers allow some overstress under certain conditions. Engineers will consider these factors.

1. *Redundancy.* Can the load be carried by another member or system if one member fails? Less or no overstress is tolerated if there is no alternate load path.

2. *Importance of member.* Less or no overstress is permitted in a major structural member where the consequences of failure are severe. Smaller members with low failure severity are permitted some overstress occasionally. Engineers are usually more conservative in design of columns than in design of flexural members.

3. *Size of member.* Thin slabs are affected more by placing tolerances than are deeper members. For example, a ⅜-in. tolerance is 9% of an effective depth of 4 in., whereas a ½-in. tolerance is 2% of an effective depth of 24 inches. Also, larger members with many bars are less affected by an occasional low strength bar.

4. *Moment redistribution.* When a flexural member creeps and relaxes, the moment will usually migrate from negative to positive moment regions. Hence, engineers may allow some overstress in steel over the support but will be generous with bottom reinforcement at mid-span, especially in end spans.

5. *Step size.* If there is a big step or increment to the next larger convenient size, an engineer is more likely to allow some overstress than if the next larger size is only slightly more than that required by calculation. For example, if the calculated required area of a column is 145 sq. in., the engineer might use a 12-in. square column rather than a larger column. If 5.13 sq. in of steel is required, the engineer might use 4 #10 bars (5.08 sq. in.) rather than 4 #11 bars (6.24 sq. in.) in a column.

6. *Material properties.* In the United States, reinforcing steel usually has a yield strength 5% to 10% more than specified. Recognizing this, engineers may allow some overstress in reinforcement. If engineers know that concrete will be dependably well controlled at an average strength in excess of that specified, they may also allow some overstress on concrete. When concrete will have questionable quality control and unreliable strength, engineers are more conservative on concrete stresses.

7. *Design review.* If an engineer's calculations will be reviewed critically, less or no overstress will be permitted.

8. *Design versus investigation.* Most engineers will allow significantly higher overstress when investigating an existing structure than when designing a new structure. An existing structure is a known condition that cannot be changed, whereas a new structure under design still faces the uncertainties of construction.

9. *Personal bias.* If an engineer's personal bias is to be generous with loads and design procedures, the final bar selection or selection of concrete outlines may be comparably overstressed.

Summing up all the above factors, experienced engineers will first try to provide designs required by their calculations without overstress but will permit overstress of up to 3% or 5% in some members in order to make advantageous selections or to simplify construction. When investigating existing structures (whether designed by themselves or by others), overstresses of up to 5% to 15% are permitted by most engineers.

In rounding numbers to three significant figures, round to the nearest number. For example, round 1234 to 1230 and round 1236 to 1240. If the fourth digit is 5 followed by zeros, round to the nearest even number; for example, round 23,450 to 23,400 and round 23,350 to 23,400.

To minimize errors in arithmetic and logic, engineers should prepare neat, systematic calculations in consistent format. Such calculations can be more easily reviewed by someone else and even by the engineer who prepared the calculations. Calculations should include the units psf, klf, psi, in., ft, etc., on all calculations, especially on extensive or unusual computations. In this text, units are indicated in the answers to computations (as in the solution to an equation) but not within the computation. In practice, engineers should also indicate the units within the computation, especially for long computations infrequently performed.

1.5 SHORT PROCEDURES

Throughout this text, simplified or short procedures will be suggested. The engineer must make frequent decisions on when to use short procedures and when to use long procedures. Long procedures are those that take full advantage of all Code provisions, research results, and theories to reduce the amount of labor and materials used in construction. Such reductions usually come at the expense of additional calculation labor.

Following are situations in which the use of short design procedures is appropriate:

1. Preliminary design of almost all structures. Only on large, complicated, intricate, and/or sensitive structures need long procedures be used. Even then, short procedures would be the appropriate first step.
2. Final design of small structures.
3. Final design of individual members of larger structures where the quantity of such individual members is small.
4. Checking the design prepared by others or prepared by oneself using long procedures.

On the other hand, long procedures should be used in the following situations:

1. When repetition of identical units makes potential construction cost savings by using long procedures much larger than design costs.
2. When building height or an expensive foundation makes it desirable to reduce concrete outlines and therefore dead load as much as possible in order to reduce cost of columns and foundations.
3. When unique or innovative construction does not clearly fall within the limitations prescribed for use of short procedures.
4. When, for aesthetic or other reasons, it is necessary to design structures to the limit of dimensions, material strengths, and reinforcement.
5. When materials are used that do not fit the standards used in short procedures (e.g., concrete strength, $f_c' = 2$ or 7 ksi or steel yield strength, $f_y = 33$ or 70 ksi).

Following are key considerations in deciding when to use a short procedure instead of a long procedure in the final design:

1. Will the short procedure result in a safe structure?
2. How much will the short procedure cost in additional construction labor and material, if any?
3. How much design time will the short procedure save, if any?

Engineers should satisfy themselves that all simplified procedures used in design result in safe structures. Procedures suggested in this text are believed to be always conservative.

To evaluate considerations 2 and 3 (above), it is necessary to compare each case individually. At first, the evaluation can be rather formal and in writing. With a little practice, in most cases, the evaluation will be informal, unwritten, and will be performed quickly and automatically.

The evaluation consists of comparing the cost of design to the cost of construction. The evaluation should favor short procedures over long procedures by a factor of 3 or more (depending on structural fee arrangements and other conditions) because most owners would rather make the structure a little safer than spend the same amount of money on more structural calculations. Design cost analysis should take into account real time (almost always longer than first estimated) and the chances that long procedures will not save construction cost after all.

The additional time limit T in minutes, to use a long procedure is

$$T = \frac{C_c nP}{C_d K_1 K_2}$$
(1.7)

Where: C_c = potential construction cost saving per member
C_d = unit design cost per minute
n = number of similar members
P = probability that long procedures will result in the potential construction cost savings
K_1 = a factor favoring simplified procedures, always greater than one
K_2 = the average ratio of actual design time to estimated design time, usually greater than one.*

Example 1.1 A column requires four #9 vertical bars by a short procedure. The engineer suspects that four #8 bars might be justified by a long procedure. How much time can the engineer spend on the long procedure?

Solution The difference in weight of vertical bars is 44 lbs for a column of average height. Assume

C_c = 44 lbs × $0.25/lb = $11.00
C_d = $30/hr = $0.50/min
P = 50%
K_1 = 3
K_2 = 2

*Almost everyone I know, including me, will take longer to do something than originally anticipated.

$$\text{Then } T = \frac{11 \times 1 \times 0.50}{0.50 \times 3 \times 2} = \frac{\$11}{\$1} \times \frac{1}{6} = 1.83 \text{ min}$$

In this example, the time limit for use of the long procedure is less than 2 minutes for just one column or about 18 minutes for 10 columns alike. If the design cannot be completed within these time limits using the long procedure (including the time to make this analysis), the short procedure should be used. It is essential that the practicing engineer make this type of cost analysis, using current cost figures and appropriate factors, continually during preparation of the design.

In this example, the engineer might simplify Eq. 1.7 for other situations on the same or similar projects by dividing material cost ($11) by 6 to obtain the design time limit in minutes (1.83).

In a variation of the 80-20 rule (80% of the work is done by 20% of the people), 80% of most projects is routine and conventional. Using short procedures, engineers should strive to complete the design of this 80% as quickly as possible in 20% of the time. With the time thus saved, the engineer can spend the remaining 80% of design time on the 20% of the project that is unusual, difficult, or most in need of careful consideration.

1.6 PROPERTIES OF CONCRETE

Reinforced Concrete

Reinforced concrete is unique among common structural materials because it is composed of two dissimilar materials: plain concrete and steel. Concrete is relatively inexpensive and strong in compression but weak in tension. It provides strong, durable, attractive floor and wall surfaces and protects steel from corrosion and heat from fires. Reinforcing steel is relatively more expensive than concrete and strong in tension but less effective in compression because of buckling. Reinforced concrete takes advantage of the best properties of both materials while compensating for their less desirable features.

Plain concrete and steel work well together because chemical bond and mechanical anchorage prevent slippage of the bars relative to the concrete and because their coefficients of thermal expansion are nearly the same. For steel the coefficient is 6.5×10^{-6}, and for concrete the coefficient is 5.5 to 7.5×10^{-6} per degree Fahrenheit. Thus, wide variations in temperature do not produce damaging internal secondary stresses. For reinforced concrete, the coefficient of thermal expansion for plain concrete is used because concrete is the dominant material.

The coefficient of thermal expansion has other implications. For example, if an unrestrained reinforced concrete frame 200 ft long changes temperature by 65°F (say from 20 to 85°F), it will change in length by 1 in., but stress in the structure will remain unchanged. If fully restrained, the length will not change, but the change in axial stress on longitudinal steel will be about 10,000 psi and the change in axial stress on concrete will be about 1300 psi unless relieved by creep. In actual service, conditions will be between these two extremes but usually closer to the unrestrained frame.

Reinforced concrete should have reinforcing steel so placed that every potential crack in the concrete in any direction is crossed by reinforcing steel adequate to resist the tensile forces. Skill in reinforced concrete design consists in anticipating significant tensile forces and providing steel to resist them. Exceptions to this rule will be discussed later.

Concrete

Concrete is composed of inert aggregate particles held together by a paste of portland cement and water. The paste may contain admixtures to enhance or modify its properties or the properties of the concrete. In most areas of the United States, materials are available at a reasonable price to make concrete with strength and other properties suitable for most uses. When very high-strength concrete or concrete with special properties is desired, materials not available locally and techniques not familiar to local suppliers may be required. In such cases, the engineer should carefully balance desired properties with reasonable demands on concrete suppliers.

Compressive Strength

Compressive strength is the most important mechanical property of concrete. It is measured by breaking 6-in.-diam. by 12-in.-long cylinders in a standardized procedure established by the American Society for Testing and Materials (ASTM) [9]. Even though the test is standardized, when a large number of cylinders are tested, the results will be quite variable. Therefore, the average of all cylinder tests on a given project for one class of concrete must be higher than the specified strength, f'_c* by a substantial margin, generally 15% or more. Even then, a few individual cylinders will break at strengths below f'_c.

The average strengths of test cylinders for each pour may be plotted on a quality control chart similar to that in Fig. 1.1. Each sample number usually represents concrete cast on one day. When the average of the five previous groups of test cylinders falls below the required average strength, remedial measures should be taken to raise the average strength of test cylinders.

The acceptability of concrete based on cylinder tests is determined by statistical means outlined in an ACI Standard [10]. Engineers must become familiar with ACI and ASTM standards on concrete testing in order to properly assess the acceptability of concrete furnished to the project during construction and to answer the owner's legitimate concerns when an occasional cylinder test falls below the specified strength.[†]

The most important factor affecting the strength of concrete is the ratio of water to cement by weight (commonly called the water-cement ratio or W/C ratio), as shown in Fig. 1.2. Concrete must also be workable to enable it to be placed and consolidated or densified in forms and around reinforcement and to achieve its full

---

*In this text, concrete strength, f'_c, will normally be given in ksi (kips per sq. in.) to avoid repetitious listing of surplus ciphers.

[†]Owners sometimes become very emotional about one low cylinder test, even though the low test resulted from abuse of the cylinder after casting. Engineers must remain objective to determine when low tests warrant replacing concrete in the structure. Large sums of money are frequently involved in that decision.

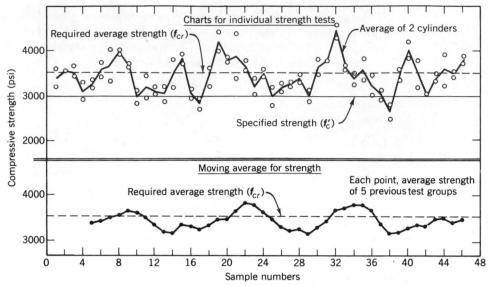

FIGURE 1.1 Quality control charts for concrete. (Courtesy of American Concrete Institute.)

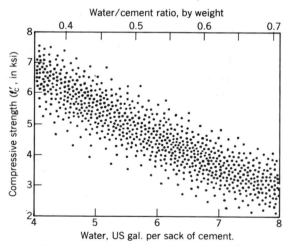

FIGURE 1.2 Effect of Water−Cement ratio on 28-day strength of non-air-entrained concrete (Courtesy of the American Concrete Institute.)

potential strength. To maintain workability at the low W/C ratios requires for high-strength concrete, more cement is required, but more cement without a concurrent reduction in the W/C ratio will not automatically lead to higher strength.

When concrete strengths of 6 ksi or greater are used, there are important considerations for design that are not discussed in this text nor covered by the ACI Code. These considerations are discussed in Reference 11.

Stress-Strain Behavior

Stress-strain behavior of concrete depends on strength of the concrete, materials used in the concrete, rate of loading, and type and size of test specimen. Typical curves are shown in Fig. 1.3 [12]. Note that stress and strain are not proportional near the ultimate strength for all concretes, but low-strength concrete exhibits more ductility, that is, it deforms more than high-strength concrete before failure. The shape of the compression zone in compression members also affects the ultimate strain of concrete. The descending portion of the curves is important in structural behavior. Even though maximum stress is reached at strains of about 0.015 to 0.002, the ACI Code conservatively states, "Maximum usable strain at extreme concrete compression fiber shall be assumed equal to 0.003″ (Code 10.2.3).

Modulus of Elasticity

Modulus of elasticity* for concrete usually refers to the secant modulus because, strictly speaking, stress and strain are never proportional. The secant is taken at $0.5f_c'$ because this approximates stress levels under working loads (see Fig. 1.4). Unlike steel and other materials, the modulus of elasticity for concrete depends on its strength and its unit weight. The modulus can be computed by the empirical equation

$$\boldsymbol{E_c} = 0.033\boldsymbol{w}^{1.5}\sqrt{\boldsymbol{f_c'}} \qquad (1.8)$$

where $\boldsymbol{w}$ is the unit weight of the hardened concrete in pcf and $\boldsymbol{f_c'}$ is its cylinder strength (Code 8.5.1). Both $\boldsymbol{f_c'}$ and $\sqrt{\boldsymbol{f_c'}}$, as well as $\boldsymbol{E_c}$, are in units of psi. Equation 1.8

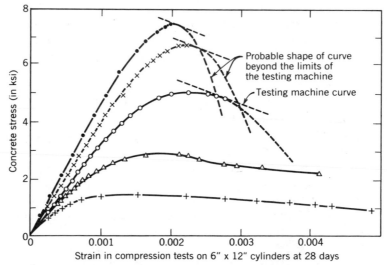

FIGURE 1.3 Typical concrete stress-strain curves under short-time loading (from Ref. 12). (courtesy of the American Concrete Institute.)

*The modulus of elasticity of concrete will normally be given in ksi (kips per sq. in.) in this text. The engineer can easily supply the extra ciphers if the calculation so requires for consistency of units.

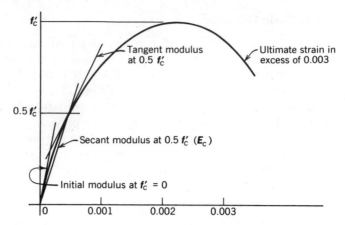

FIGURE 1.4   Modulus of elasticity of concrete.

was developed for concretes weighing 90 to 155 pcf [13]. For concretes using natural aggregates, $w \simeq 145$ pcf and $E_c = 57{,}000\sqrt{f_c'}$.

Expressing E_c in units of ksi but $\sqrt{f_c'}$ in units of psi gives

$$E_c = 57\sqrt{f_c'} \tag{1.9}$$

Creep

Creep is a time-dependent phenomenon. On first loading at less than $0.5f_c'$, the strain in concrete is nearly elastic but continues to increase immediately even though the load remains constant. The rate of increase decreases with time. Figure 1.5 illustrates the case of a concrete member without load for a short time after casting, then loaded initially with a constant load. Later, the load is removed entirely. Later still, the initial load is placed on the member again. The curve would have the same shape (but not the same dimensions) if only part of the load were removed and later replaced. If additional load were placed on the member after the initial load (instead of load being removed), the curve would take the shape of the dotted line above the solid line in Fig. 1.5.

Creep is increased by more mixing water, lower ambient humidity, less thickness of the concrete member or greater surface/volume ratio, loading at an earlier age, temperatures over 120°F, more cement, more fine aggregate, and more air in the concrete. Creep can be approximated in the "elastic" range (below $0.50f_c'$ by

$$C_t = \frac{t^{0.60}}{10 + t^{0.60}}\, C_u \tag{1.10}$$

where t = the duration of loading in days,
   C_t = the ratio of creep strain at time t to the initial (elastic) strain
and C_u = the ratio ultimate creep strain to initial strain. For average conditions

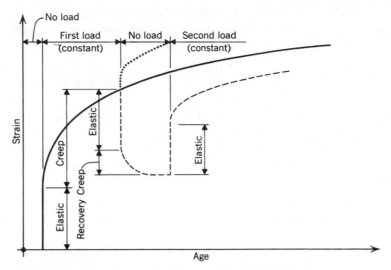

FIGURE 1.5 Typical creep strain history of loaded concrete.

(4-in. slump, 40% relative humidity, moist cured and loaded at 7 days), C_u is 2.35 [14].

In addition to calculating deflection and loss of prestress, creep is important in other applications. For example, if columns have an average stress on the concrete of 1000 psi, they can be expected to shorten due to creep about ⅛ in. per story height, ultimately. Creep will tend to soften the effect of stress concentrations as, for example, when one column settles more than others.

Note that creep is related to initial strain so that less creep occurs with concrete with a higher modulus of elasticity. This is true whether the higher modulus comes about because of loading at a later time, higher strength concrete, or materials yielding a higher modulus. Because mild steel does not creep at normal stress levels, concrete under axial compression load tends to shift its load to embedded steel.

Shrinkage

Shrinkage, like creep, is a time-dependent phenomenon and is affected by many of the same factors, although not in exactly the same degree. Any workable concrete mix contains more water than needed for hydration of the cement. The excess water evaporates from the surface when exposed to air, and the concrete shrinks in volume. More excess water, drier air, windy conditions, and thinner concrete sections all lead to more rapid evaporation and, hence, greater shrinkage. Concrete immersed in water tends to regain most of the volume lost in prior shrinkage.

Shrinkage can be approximated by

$$\epsilon_{(sh)t} = \frac{t}{35 + t} \times \epsilon_{(sh)u} \tag{1.11}$$

where $\epsilon_{(sh)u}$ = the ultimate shrinkage strain, and

$\epsilon_{(sh)t}$ = the shrinkage strain at time, t, in days after age 7 days for moist cured concrete.

For the average conditions listed above for creep, the ultimate shrinkage strain can be taken as 800×10^{-6}. For more precise estimates of shrinkage, see Ref. 14. Equations 1.10 and 1.11 are plotted in Fig. 1.6.

If embedded steel is asymmetrical about the neutral axis of the concrete, shrinkage warping will occur. Warping is usually in the same direction as elastic deflection and creep due to external loads. One exception occurs when the centroid of steel is below the neutral axis in a T-beam in regions of negative moment (see Chapter 8).

Tensile Strength

Tensile strength is much more variable and more difficult to measure than compressive strength. It varies from 10% to 15% of f'_c. There are three methods of testing, but none give entirely satisfactory results. The first method, direct tension tests, gives erratic results due to problems in the testing equipment and setup.

The second test method, flexural tests of plain concrete specimens, gives a modulus of rupture that is somewhat higher than the tensile strength because the maximum tensile stress occurs only at the extreme fiber. The modulus of rupture is calculated by dividing the bending moment at rupture by the section modulus.

The third test method, the split-cylinder test, is most commonly accepted today because it is simple to perform and gives results as useful as the other test methods. It consists of breaking a standard 6-in.-diam. by 12-in.-long cylinder in a compression testing machine while the cylinder is lying on its side. The split-cylinder tensile strength, f_{ct}, is computed from

$$f_{ct} = 2P/\pi dL \tag{1.12}$$

where P = test load at failure,

d = diameter of the cylinder and

L = length of the cylinder.

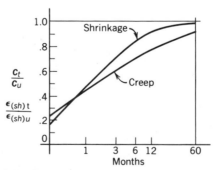

FIGURE 1.6 Specific creep and shrinkage strains.

Some tensile strength in concrete is essential for proper functioning of a structure, even in compression members. In reinforced concrete, the concrete is assumed to resist no tensile stresses in flexure or direct tension. However, the concrete tensile strength contributes to the strength of members in shear, torsion, and anchorage. Minimal deflection depends on the tensile strength of concrete. It also keeps the face shell of concrete on column and beam cages and prevents splitting of concrete along the plane of reinforcement.

Concrete using lightweight aggregate normally has a lower tensile strength than normal-weight concrete. The ACI Code assumes the tensile splitting strength $f_{ct} = 6.7\sqrt{f'_c}$ for normal-weight concrete. In the absence of specific testing on concrete, the tensile splitting strength f_{ct} and modulus of rupture f_r for "all-lightweight" and "sand-lightweight" concrete must be taken as 75% and 85% of the value for normal weight concrete, respectively. Linear interpolation may be used when partial sand replacement is used. This gives values of $f_{ct} = 5.7\sqrt{f'_c}$ for sand-lightweight concrete, and $f_{ct} = 5.0\sqrt{f'_c}$ for all-lightweight concrete (Code 9.5.2.3 and 11.2). The units for f_{ct}, f'_c, and $\sqrt{f'_c}$ are in psi.

Shear Strength

Shear strength of concrete is very difficult to measure but is probably almost as high as its compressive strength. "Pure" shear stresses are rarely, if ever, a concern in designing concrete. Instead, the diagonal tension stresses associated with shearing forces are high with respect to the tensile strength of concrete. Even though the phenomenon is called "shear," engineers should always remember that diagonal tension is the real concern.

For over 75 years, the American Concrete Institute (ACI) has been the leading authority in the United States on the use of concrete. The ACI is also highly respected throughout the world. For concise, up-to-date information on a wide variety of subjects regarding concrete, engineers can refer to ACI Standards and committee reports. On normal building projects, engineers will be especially interested in the following subjects and documents:

Proportioning	ACI211
Use of Admixtures	ACI212
Evaluation of Test Results	ACI214
Measuring, Mixing, Transporting, and Placing	ACI304
Curing	ACI308
Consolidation	ACI309
Inspection	ACI311

These documents and many others are available from ACI in the five-volume set of the Manual of Concrete Practice [15] or as individual documents. Engineers should study and become familiar with all pertinent codes, specifications, and recommended practices by ACI and ASTM.

1.7 PROPERTIES OF REINFORCING STEEL

All reinforcing bars have lugs or deformations of a minimum height and spacing to assure a mechanical anchorage of steel to concrete rather than relying on the weak chemical bond or adhesion between smooth bars and concrete. The deformations also include markings to indicate the grade of steel, the size of bar and the manufacturer. Bar sizes are designated by numbers that indicate the number of eighths of an inch in the nominal diameter. Thus, a #10 bar is about 1¼ in. in diameter. The actual diameter of bars measured across the deformations is about ⅛ in. larger than the nominal diameter. Figure 1.7 illustrates some common styles of bar deformations, and Appendix E gives properties of bars.

Welded wire fabric (WWF) consists of smooth or deformed wires fabricated into sheets or rolls. Each intersection of wires is welded to provide a mechanical anchorage to concrete. If wire is deformed, the deformations also anchor the wires.

Reinforcing steel is normally provided in grades 40, 50, and 60 corresponding to yield points of 40, 50, and 60 ksi*; grade 60 is the most common. In these grades, steel exhibits a long plateau in the stress-strain curve in which the steel elongates as much as 2% without much increase in stress. On further straining, the steel reaches its ultimate strength ranging from 70 to 90 ksi and an elongation of 7% to 9% or more. This large elongation is essential to assure sufficient ductility in the steel to permit bending to the desired shapes, as well as to provide ductility to reinforced concrete. Regardless of the yield strength or ultimate strength of steel, the maximum useful steel stress is taken as the stress at a strain of 0.35% (Code 3.5.3.2). This is somewhat higher than the maximum strain assumed in concrete of 0.30%.

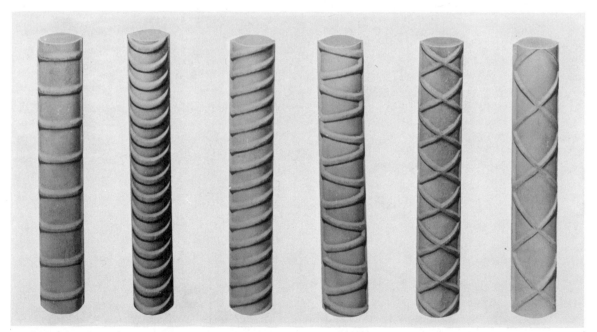

FIGURE 1.7 Deformed reinforcing bars (courtesy of Concrete Reinforcing Steel Institute).

Because the modulus of elasticity of steel is assumed to be 29,000 ksi* (Code 8.5.2), and because inelastic behavior is more likely to occur at high stresses, the highest usable stress in reinforcing steel is 80 ksi, using currently available materials and theory (Code 9.4). At the time this was written, no reinforcing steel was commonly available with a yield strength above 60 ksi, except for WWF, which has a specified yield strength of 65 ksi.

Steel in compression is assumed to have the same yield strength and ultimate strength as in tension.

1.8 COVER AND SPACING OF REINFORCEMENT

The ACI Code has minimum spacing requirements for reinforcing bars to permit concrete to flow easily, without voids or honeycomb, around and through a layer of bars. Minimum spacing also helps ensure proper anchorage of the bars in concrete. If bars were spaced too closely in one layer, the concrete might split along the layer of reinforcement. A limit on maximum spacing of bars helps assure good distribution of cracks and ensures that no portion of the structure is without reinforcement. For slabs and walls, maximum bar spacing helps prevent a concentrated load from punching through the slab or wall between bars.

Clear distance between parallel bars in a layer must not be less than the bar diameter d_b nor 1 in., nor one and one-third times the nominal maximum size of coarse aggregate in the concrete. Where parallel reinforcement is placed in two or more layers, bars in the upper layer must be placed directly above bars in the bottom layer with clear distance between layers of at least 1 in. In columns, the clear distance between parallel bars must be at least $1.5d_b$ and at least $1\frac{1}{2}$ in. In all cases, the clear distance limitation between bars also applies to the clear distance between a contact splice and adjacent splices or bars. Figure 1.8 summarizes these requirements (Code 7.6 and 3.3.3).

Two, three, or four parallel bars may be bundled in contact to act as a unit if they are enclosed within stirrups or ties. (Stirrups in beams and ties in columns are closely spaced small bars, usually #3 to #5 size, placed perpendicular to the longitudinal axis of the member. Stirrups are discussed in Chapter 6 and ties in Chapter 9.) In beams, bars larger than #11 bars must not be bundled because of the difficulty in anchoring to develop full stress in the bars; however, bundled bars in compression of #14 and #18 size are acceptable in columns. In flexural members, individual bars in a bundle may be terminated within a span if the ends are staggered at least 40 bar diameters d_b. Where spacing limitations and concrete cover are based on bar diameter d_b, a unit of bundled bars must be treated as a single bar of a diameter derived from the total area of bars in the bundle. For example, a bundle of 1 #10 plus 1 #11 bar would have a total area of 2.83 sq. in. (1.56 + 1.27). The equivalent bar diameter would be 1.90 in. ($\sqrt{4 \times 2.83/\pi}$), and the clear distance to the next bar or bundle must be 2.9 in. (1.9 × 1.5) in a column. Requirements for bundled bars are summarized in Fig. 1.9 (Code 7.6.6). In general, bundled bars should be used sparingly. Bundled bars are useful in members with congested reinforcement to

---

*Steel modulus of elasticity E_c and yield strength f_y will normally be given in ksi (kips per sq. in.) in this text. Engineers should supply the extra ciphers if the calculation so requires for consistency of units.

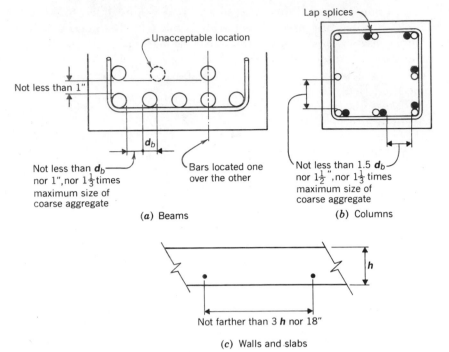

FIGURE 1.8   Spacing limits for reinforcement.

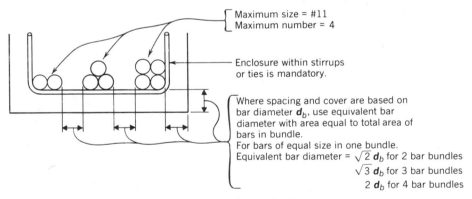

FIGURE 1.9   Spacing and cover requirements for bundled bars.

allow clearance between perpendicular intersecting bars from other members (as in a beam-column joint) and to provide space between bars to permit easy placement of concrete.

Requirements for minimum concrete cover over reinforcing bars permits concrete to flow easily around bars, provides protection for steel from heat of fires, and provides protection of steel from corrosion. Minimum concrete cover requirements are tabulated in Table A1.2 (Code 7.7).

Where concrete is exposed to deicing salts or other corrosive environment, as in a parking garage, cover on all reinforcing steel and other ferrous metals should be at least 2 in.

PROBLEMS

1. Write a 200-word essay on the public perception of structural safety in buildings.
2. What is the factored load on an 8-in. slab supporting 5 ft of water in a swimming pool?
3. What are the maximum factored moments on an isolated 18-in. × 30-in. simple beam spanning 30 ft clear and supporting a mid-span crane load of 40 kips? Under certain conditions, the crane can exert a 25-kip uplift on the beam. Assume a 10% impact factor.
4. How much will a 110-ft-long, unrestrained concrete frame contract when its temperature drops from 72 to 15°F?
5. What is the modulus of elasticity of concrete weighing 105 lbs/cu. ft. with f'_c = 4.5 ksi, using ACI Code procedures?
6. How much will a 200-ft-high concrete column shorten elastically, neglecting the reinforcement, if the average service stress on the concrete is 500 psi? Assume normal weight concrete and f'_c = 5 ksi.
7. What will be the creep shortening in the column in Problem 6 after 5 years, assuming average conditions?
8. What will be the shrinkage shortening in the column in Problem 6 after 5 years, assuming average conditions and no restraint from the reinforcement?
9. What will be the total shortening of the column in Problem 6 due to elastic shortening, creep, and shrinkage after 5 years?

Selected References

1. ACI-ASCE Committee 327. "Ultimate Strength Design." *ACI Journal, Proc.* V.52, pp 505−524, January 1956.
2. J.G. MacGregor, S.A. Mirza, and B. Ellingwood. "Statistical Analysis of Resistance of Reinforced and Prestressed Concrete Members." *ACI Journal, Proc.* V.80, No.3, pp 167−176, May−June 1983.
3. James G. MacGregor. "Load and Resistance Factors for Concrete Design," *ACI Journal, Proc.* V.80, No.4, pp 279−287, July−August 1983.
4. ACI Committee 315. *Details and Detailing of Concrete Reinforcement (ACI 315-80).* American Concrete Institute, Detroit, 1980.
5. ACI Committee 315. *Manual of Engineering and Placing Drawings for Reinforced Concrete Structures (ACI 315R-80).* American Concrete Institute, Detroit, 1980.
6. *Manual of Standard Practice.* Concrete Reinforcing Steel Institute, Chicago, 1980, and 1983 supplement.
7. *Master Format CSI Document MP-2-1.* Construction Specifications Institute, C51, Alexandria, Virginia, March 10, 1983. See also *Section Format CSI Document MP-2-2.* Construction Specifications Institute, Alexandria, Va., May 1980.
8. P.R. Morgan, T.E. Ng, N.H.M. Smith, and G.D. Base. "How Accurately Can Reinforcing Steel Be Placed? Field Tolerance Measurement Compared to Codes," *ACI Concrete International Design and Construction,* V.4, No.10, pp 54−65, October 1982.
9. *Standard Test Method for Compressive Strength of Cylindrical Concrete Specimens (ASTM C39-72),*" (reapproved 1979), American Society for Testing and Materials, Philadelphia, 1979.

10. ACI Committee 214. "Recommended Practice for Evaluation of Strength Test Results of Concrete (ACI 214-77)," American Concrete Institute, Detroit, 1977.

11. ACI Committee 363. "State-of-the-Art Report on High-Strength Concrete," *ACI Journal*, V.81, No. 4, pp 364–411, July–August 1984.

12. Eivind Hognestad, N.W. Hanson, and Douglas McHenry. "Concrete Stress Distribution in Ultimate Strength Design," *ACI Journal, Proc.* V.52, pp 455–479, December 1955.

13. Adrian Pauw. "Static Modulus of Elasticity of Concrete as Affected by Density," *ACI Journal, Proc.* V.57, pp 679–687, December 1960.

14. Dan E. Branson and M.L. Christiason. "Time-Dependent Concrete Properties Relating to Design—Strength and Elastic Properties, Creep and Shrinkage," In *Designing for Effects of Creep, Shrinkage, Temperature in Concrete Structures (SP-27).* American Concrete Institute, Detroit, 1971, pp. 257–277.

15. *ACI Manual of Concrete Practice, Parts 1, 2, 3, 4, and 5.* American Concrete Institute, Detroit, 1985.

CHAPTER 2

BEAM FLEXURE

2.1 OBJECTIVES

Design procedures discussed here are based on the assumption that concrete out-lines have been established in preliminary design (Chapter 14) and that moments acting on critical sections have been determined (Chapter 13). Following are the objectives of this chapter:

1. Select the number and size of longitudinal tension bars required at critical sections.
2. Verify the adequacy of the cross section in compression at critical sections.
3. Select the number and size of longitudinal compression bars required at critical sections, if such reinforcement is necessary.

The critical sections mentioned above are usually the points of maximum moment. Engineers must be alert for exceptions. For example, when the cross section becomes smaller a short distance from the point of maximum moment where the reduction in moment is less than the reduction in capacity, a critical section would be at the section having the maximum moment in the area of re-duced cross section.

2.2 RECTANGULAR BEAM ANALYSIS

Before the first crack, flexural members of reinforced concrete behave almost like plain concrete members and follow the ordinary rules for strength of materials. Stress in the steel is very low, being proportional to the stress in the adjacent concrete fibers. Ignoring creep, stress in steel is n times the stress in concrete at the same level or height in the beam, where n is the modular ratio = E_s/E_c. For normal weight concrete, n is in the range of 7 to 10. After first cracking, a reinforced concrete beam can never return to the uncracked state. Behavior of the cracked beam is of primary interest to engineers in the design of reinforced concrete.

31

A reinforced concrete beam resists external moment by an equal and opposite internal moment provided by a couple between internal forces **T** and **C** separated by the internal lever arm. **T** is the tensile force provided by the tension steel, and **C** is the compressive force provided by concrete on the compression face of the beam plus any longitudinal steel in the beam on the compression side of the neutral axis (see Fig. 2.1).

Design efforts are directed toward the following:

1. Determining the size of the internal lever arm (**d** − **a**/2), where **d** is the effective depth of the beam or the distance from the centroid of the tension steel to the compression face of the beam and **a** is the height of the compression stress block in the concrete. The internal lever arm varies from about 75% of **d** to nearly 100% of **d** [see Fig. 2.1(d)].

2. Determining the required magnitude of the coupling forces. These forces are equal in magnitude. Their maximum value is limited to the value of **T** for reasons that will be discussed later and because steel is the more expensive material.

3. Ensuring that **C** can be sufficiently larger than **T** to assure ductile behavior and prevent a brittle failure.

Rectangular Beams

Consider the rectangular beam shown in Fig. 2.1(a). It is assumed that a plane section before bending remains plane after bending. After the section has been subjected to moments sufficiently large to cause cracks on the tensile side, the resulting strain diagram is shown in Fig. 2.1(b). The neutral axis is approximately at mid-height of the section before cracking. After cracking, it moves toward the compression face of the beam.

Because steel is a ductile material with large extensibility, it is safe and convenient to assume that the tensile force **T** is concentrated at the centroid of the steel whether the steel is in one layer or in two or more layers.

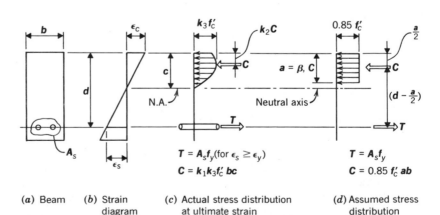

$T = A_s f_y \text{(for } \epsilon_s \geq \epsilon_y)$
$C = k_1 k_3 f_c' \, bc$

$T = A_s f_y$
$C = 0.85 \, f_c' \, ab$

(a) Beam (b) Strain diagram (c) Actual stress distribution at ultimate strain (d) Assumed stress distribution

FIGURE 2.1 Singly reinforced concrete beam. N.A. = neutral axis.

As noted in Section 1.6, when concrete strain ϵ_c is larger than about 50% of ultimate strain, ϵ_{cu}, stress is not proportional to strain. Stress distribution in concrete is shown in Fig. 2.1(c). The exact curve defining the stress magnitude is unknown, but it approximates a parabola. The maximum flexural strength is reached when ϵ_c equals the crushing strain of the concrete. At that point, the steel strain may or may not have reached strain equal to f_y/E_s, but assume for now that it has. (The case in which steel strain has not reached yield before the concrete strain reaches ultimate strain is considered later under the discussion of balanced conditions.)

Even though the compressive stress distribution in a beam has the same general shape as in a test cylinder, the maximum stress in a beam is different from concrete cylinder strength, f'_c, by $k_3 f'_c$. Referring to Fig. 2.1(c), the average compressive stress can be stated as $k_1 k_3 f'_c$, where k_1 is the ratio of average stress to maximum stress. Then the compressive force C is the volume of the stress solid.

$$C = k_1 k_3 f'_c bc \tag{2.1}$$

where b = width of beam and
  c = distance from compression face to neutral axis.

For a ductile failure when the steel has reached the yield strain,

$$T = A_s f_y \tag{2.2}$$

For equilibrium, $T = C$, and equating Eqs. 2.1 and 2.2 and then solving for c,

$$c = \frac{A_s f_y}{k_1 k_3 f'_c b} \tag{2.3}$$

The nominal flexural strength M_n is given by the internal couple

$$M_n = T(d - k_2 c)$$

Substituting $A_s f_y = T$ and the value of c from Eq. 2.3,

$$M_n = A_s f_y \left[d - \frac{k_2}{k_1 k_3} \times \frac{A_s f_y}{f'_c b} \right] \tag{2.4}$$

where k_2 is a ratio in which the numerator is the distance from the compression face to the centroid of compression force, and the denominator is the distance from the compression face to neutral axis c.

It is not necessary to determine individual values for the arbitrary ratios k_1, k_2, and k_3. Experimental results give values of $k_2/k_1 k_3$ ranging from about 0.58 to 0.64 when f'_c ranges from 3 to 5 ksi [1].

Although Eq. 2.4 gives satisfactory results, it is cumbersome for design. In the 1930s, C.S. Whitney proposed a rectangular stress distribution, as shown in Fig. 2.1(d). It has been widely adopted due to its simplicity of application. Dimensions of the stress block have been empirically determined so that the magnitude and location of the compressive force C coincides with experimentally determined data. This results in the average stress = $0.85 f'_c$ and the depth of the compressive stress block, $a = \beta_1 c$ and $a/2 = k_2 c$. Whitney suggested and the ACI Code Committee has accepted values of $\beta_1 = 0.85$ for concrete $f'_c = 4$ ksi or less [2]. For strengths above

4 ksi, β_1 is reduced at the rate of 0.05 for each 1 ksi over 4 ksi, but β_1 need not be less than 0.65 (Code 10.2.7).

Using the rectangular stress block distribution,

$$C = 0.85f_c'ba \tag{2.5}$$

For equilibrium, $T = C$. Equating Eqs. 2.2 and 2.5 and solving for a,

$$a = \frac{A_sf_y}{0.85f_c'b} \tag{2.6}$$

The nominal flexural strength M_n is given by the internal couple

$$M_n = A_sf_y(d - a/2) \tag{2.7}$$

Substituting the value of a from Eq. 2.6,

$$M_n = A_sf_y\left[d - \frac{0.59A_sf_y}{f_c'b}\right] \tag{2.8}$$

Note the similarities between Eqs. 2.1 and 2.5, between Eqs. 2.3 and 2.6, and between Eqs. 2.4 and 2.8. The Code explicitly permits other assumed shapes of the compressive stress block provided they give results that agree with test results (Code 10.2.6). For the remainder of this text, however, the rectangular stress block will be used because it is widely accepted and it is simple [3, 4]. Reference 5 gives additional insights into the behavior of concrete in the compressive stress block, especially under long-term loading.

Minimum Reinforcement

Before cracking, a reinforced concrete flexural member acts essentially as a plain concrete member and the steel adds little to the moment resistance. After cracking, the reinforcing steel must resist the entire tensile force. If the bending moment that first cracks the concrete is larger than the moment capacity of the reinforced concrete section, an immediate, brittle failure could occur. To guard against this eventuality, in flexural members other than solid slabs, the ACI Code requires a minimum amount of positive steel (reinforcement resisting positive moment) such that

$$\rho_{min} = \frac{200}{f_y}$$

where f_y is in psi and ρ_{min} is the minimum value of ρ permitted by the ACI Code (Code 10.5). The symbol ρ (rho) is called the reinforcement ratio. In Fig. 2.1(a),

$$\rho = \frac{A_s}{bd} \tag{2.9}$$

For $f_y = 60$ ksi, $\rho_{min} = 0.33\%$. In computing ρ_{min}, the width of the tensile face of the member is used.

Alternatively, ρ_{min} can be disregarded if the area of steel provided at every section, positive and negative, is at least one-third greater than that required by analysis.

For solid slabs, the minimum positive reinforcement is that required for shrinkage and temperature reinforcement (see Section 3.2).

Figure A2.1 summarizes minimum reinforcement requirements for flexural members using $f_y = 60$ ksi. Although the above requirements of the ACI Code do not apply to negative reinforcement, it is conservative to apply these requirements to negative steel in cantilever beams, isolated footings, and other situations in which negative steel is the primary reinforcement in a member.

Balanced Beam

A *balanced beam* is one in which stress in tensile steel reaches its yield strength f_y just as the concrete reaches its assumed ultimate strain ϵ_{cu} of 0.003. If the steel reaches yield strength f_y before concrete reaches ϵ_{cu}, the beam is called *under-reinforced* and exhibits a ductile failure. As the moment increases and the steel continues to yield, the neutral axis moves toward the top of the beam and reduces the size of the compression block until a compression failure in the concrete occurs. In the process, very large deflections and cracking take place and give warning of impending failure.

If concrete reaches its ultimate strain ϵ_{cu} before steel reaches yield strength f_y, the beam is called *over-reinforced* and exhibits a brittle failure. Such failures occur with little warning. The ACI Code ensures a ductile failure by limiting the tensile reinforcement in flexural members to no more than 75% of the amount required for a balanced condition (Code 10.3.3). For members with compression reinforcement, the portion of ρ_b equalized by compression reinforcement need not be reduced by 75%.

The reinforcement ratio ρ_b for the balanced condition can be computed from equilibrium and compatibility conditions. From Fig. 2.1(*b*),

$$\frac{c_b}{d} = \frac{\epsilon_c}{\epsilon_c + \epsilon_y} = \frac{0.003}{0.003 + f_y/29{,}000} = \frac{87}{87 + f_y} \tag{2.10}$$

where f_y is in ksi and c_b is the value of c for balanced conditions.

The compression force for a balanced condition C_b is

$$C_b = 0.85 f'_c \beta_1 c_b b \tag{2.11}$$

The tensile force for a balanced condition T_b is

$$T_b = A_{sb} f_y = \rho_b b d f_y \tag{2.12}$$

Equating C_b to T_b gives

$$0.85 f'_c \beta_1 c_b b = \rho_b b d f_y$$

Inserting $87/(87 + f_y)$ for c_b/d and solving for ρ_b gives

$$\rho_b = \frac{0.85 \beta_1 f'_c}{f_y} \times \frac{87}{87 + f_y} \tag{2.13}$$

Table A2.1 gives values for $0.75\rho_b$ (the maximum steel ratio permitted by the Code for singly reinforced beams) and the depth of the compressive stress block at $0.75\rho_b$ for several common combinations of steel and concrete strengths.

2.3 T-BEAM ANALYSIS

Slabs and beams are normally cast together resulting in a "tee"-shaped beam or, simply, T-beam. The extra width at the top is called the flange, and the portion below the flange is called the web. When the compression face of the beam is on the flange side, it provides extra compression area to resist moments. Compressive stresses in the flange are not uniform, being highest directly over the web and declining with distance from the web. This is due to shear lag or longitudinal shearing deformations as the compressive force spreads from the web to the outermost fibers of the flange, as shown in Fig. 2.2.

The Code assumes a uniform compressive stress in the flange for simplicity but limits the width of the flange to assure satisfactory performance (Code 8.10). The rules governing flange width are given in Fig. 2.3. If the neutral axis at balanced conditions is within the flange, a T-beam can be treated the same as a rectangular beam for flexure, using the width of the flange as the width of the beam. Note, however, that the web width must be used for shear calculations (see Chapter 6).

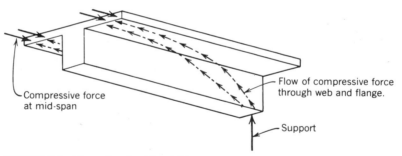

FIGURE 2.2   View of underside of T-beam showing spread of compressive force from support to outer edges of flange.

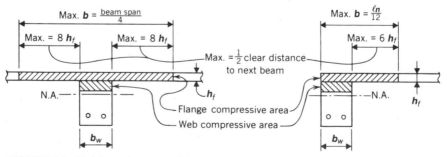

FIGURE 2.3   Reinforced concrete T-beam.

Location of the neutral axis in a rectangular beam at balanced conditions can be computed from Eq. 2.10. For $f_y = 40$, $c_b/d = 0.68$ and for $f_y = 60$, $c_b/d = 0.59$. Most T-beams have thinner flanges.

If the neutral axis is below the flange, the total compressive force can be divided into two components; one for the flange and one for the web. The two components can be computed separately and then added together for the total compressive force resistance. To meet ductility requirements, the tensile force $T = A_s f_y$ cannot exceed 75% of the total compressive force thus computed.

For simplicity, the flange can be disregarded if its area is small compared to that of the web. Alternatively, that portion of the web that is below the flange can be disregarded. Considering balanced conditions with tension reinforcement $A_s = \rho_b bd$ and the neutral axis at the bottom surface of the flange, the height of the compressive stress block a is $\beta_1 h_f$, where h_f is the thickness of the flange. Because ρ must not exceed $0.75\rho_b$ to meet ductility requirements, the height of the compressive stress block a must not exceed $0.75\beta_1 h_f$ for ductility.

Example 2.1   Given the beam in Fig. 2.4 with a 20-ft span, concrete $f_c' = 3$ ksi and steel $f_y = 60$ ksi, compute the maximum nominal moment strength M_n and allowable factored moment M_u.

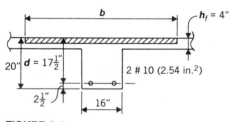

FIGURE 2.4

Solution   Maximum flange width is $8 \times 2 \times 4 + 16 = 80''$* but not more than $20 \times 12/4 = 60''$. Because beam spacing is not given, assume it is not critical. From Eq. 2.6, height of the compressive stress block a is

$$a = \frac{2.54 \times 60}{0.85 \times 3 \times 60} = 1.0''$$

Because $0.75\beta_1 h_f = 0.75 \times 0.85 \times 4 = 2.55'' > 1.0''$, requirements for ductility are met, and

$$\rho = \frac{2.54}{60 \times 17.5} = 0.24\%$$

---

*In the solutions to all examples throughout this text and in tables and figures, shorthand abbreviations are used for units involving feet ('), inches ("), and kips (k). Square feet and square inches are denoted by ᵒ' and ᵒ", respectively (see Appendix C). In a lifetime of engineering, the time saved by using these short forms is considerable. They are universally used by practicing engineers.

For purposes of minimum reinforcement, however,

$$\rho = \frac{2.54}{16 \times 17.5} = 0.91\% > \frac{200}{f_y} = 0.33\%^\dagger \qquad \underline{OK^\ddagger}$$

Using Eq. 2.7, the nominal moment strength is

$$M_n = \frac{2.54 \times 60(17.5 - 1.0/2)}{12} = 216'k$$

Because the capacity reduction factor $\phi = 0.9$ for flexure,

$$M_u = 0.9M_n = 194'k$$

For consistency of units in Solutions, moments are always considered to be given in units of foot-kips ($'k$). In practice, this requires the introduction of a factor of 12 in. per foot in computing steel areas, checking concrete strength, and in other computations, but it also eliminates the same factor in computing moments. Some engineers prefer to work with moments in units of inch-kips ($''k$).

Example 2.2 Suppose the beam in Fig. 2.5 has holes in the slab that limit the flange width to 20 in. and is made with concrete $f'_c = 3$ ksi and steel $f_y = 60$ ksi. Compute the maximum nominal moment strength M_n and allowable factored moment M_u.

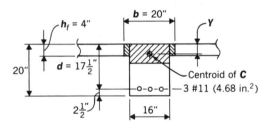

FIGURE 2.5

Solution   From Eq. 2.6, $a > \dfrac{4.68 \times 60}{0.85 \times 3 \times 20} = 5.5''.$

The neutral axis is below the flange, so the resistance of both the flange and web must be considered. Compute the compressive force capacity of the flange and the web separately using Eq. 2.5. From Table A2.1, $a/d = 0.38$ and, for balanced conditions, $a_b = 0.38/0.75 = 0.50$. Thus,

web $C = 0.85 \times 3 \times 16 \times (0.5 \times 17.5)$	$= 357$ k
flange $C = 0.85 \times 3 \times (20 - 16) \times 4$	$= \underline{41}$ k
total C	$= 398$ k

Using Eq. 2.2,   steel $T = 4.68 \times 60 = 281$ k

---

†Throughout this text, abbreviated statements of mathematical verities are used. In this example, a more complete but tedious statement would be: $\rho = 0.91\%$; $200/f_y = 0.33\%$; $0.91\% > 0.33\%$. Therefore, $\rho > 200/f_y$ and computed value of ρ is satisfactory.

‡OK = acceptable.

Checking for ductility,

$$T/C = \frac{281}{398} = 0.70 < 0.75 \qquad \underline{\text{OK}}$$

Note that for the web alone,

$$T/C = \frac{281}{357} = 0.79 > 0.75 \qquad \underline{\text{NA*}}$$

The height of the stress block can be computed accurately using a long procedure by subtracting the flange force from the steel force and using the remainder in Eq. 2.6 to obtain the required height. Thus,

required web $C = 281 - 41 = 240$ k

$$\text{and } a = \frac{240}{0.85 \times 3 \times 16} = 5.9''$$

To compute the nominal moment strength, first compute the location of the centroid of the compressive forces.

$$Y = \frac{(20 - 16)4 \times 2 + 16(5.9)^2 \times 0.5}{(20 - 16)4 \quad + 16 \times 5.9} = 2.81''$$

$$\text{Then, } M_n = \frac{4.68 \times 60(17.5 - 2.81)}{12} = 344'\text{k}$$

Alternatively, compute moment capacities of web and flange separately and then add them to obtain M_n. Thus,

web $M_n = 240(17.5 - 5.95/2)/12 = 290'$k
flange $M_n = \quad 41(17.5 - 2)/12 \qquad = \underline{\ 53}$
total $M_n \qquad\qquad\qquad\qquad\qquad = 343'$k
and $M_u = 0.9M_n \qquad\qquad\qquad = 309'$k

The difference in the answer in the third significant figure is due to rounding.

For a short procedure, simply assume that the centroid of the compressive force is at the centroid of the web compressive block. Then,

$$M_u = \frac{0.9 \times 281(17.5 - 5.95/2)}{12} = 306'\text{k}$$

This is about 1% less than the value given by long procedures. The short procedure is sufficiently accurate for most purposes.

2.4 DOUBLY REINFORCED BEAM ANALYSIS

Beams with both tension and compression steel are called "doubly reinforced beams." Normally, compression steel is not needed for strength but is added for

*NA = not acceptable.

control of deflections (see Chapter 8), for stirrup support bars, or in areas of negative moment to continue bottom reinforcement into the support. In such cases, the compression steel adds little to the flexural strength. To account for any contribution it may make requires laborious calculations.

When compression steel is required because a beam would otherwise be over-reinforced, that is, when $\rho > 0.75\rho_b$, the nominal moment capacity M_n can be calculated as the sum of two parts. The first part M_{n1} is the nominal moment capacity of a singly reinforced beam reinforced to 75% of balanced conditions, $0.75\rho_b$.

Referring to Fig. 2.6,

$$M_{n1} = 0.75A_{sb}f_y \left(d - \frac{a}{2}\right) \tag{2.14}$$

where A_{sb} = area of tension reinforcement required to produce balanced conditions.

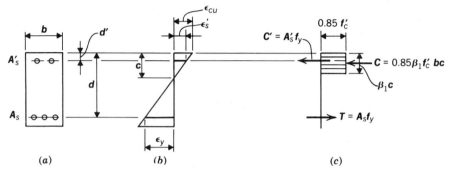

(a) (b) (c)

FIGURE 2.6   Doubly reinforced concrete beam.

Equation 2.14 is the same as Eq. 2.7, except for the substitution of $0.75A_{sb}$ for A_s. Equation 2.14 gives the maximum allowable moment on a singly reinforced beam.

The second part is the nominal moment capacity provided by a couple between the remaining tension steel and the compression steel.

$$M_{n2} = M_n - M_{n1} = (A_s - 0.75A_{sb})f_y(d - d') \tag{2.15}$$

but not more than

$$M_{n2} = A_s'(f_s' - 0.85f_c')(d - d') \tag{2.16}$$

where f_s' = stress in compression reinforcement.

Stress in compression steel is reduced by the stress in concrete because steel has replaced concrete.*

Stress in compression steel will reach yield if the beam is deep enough but not if the beam is too shallow. To determine the depth required for compression steel to reach yield, take the sum of horizontal forces in Fig. 2.6(c),

*Strictly speaking, Eqs. 2.15 and 2.16 are not quite correct. The concrete carries less of the compressive force in proportion to the area of concrete replaced by compression steel, and the compression steel carries more of the compressive force in the same proportion. The total M_n remains the same. The procedure outlined above is far simpler than a strict analysis of forces resisted by each material.

$$\boldsymbol{A_s f_y} = 0.85\boldsymbol{\beta_1 f'_c bc} + \boldsymbol{A'_s f_y}.$$

Rearranging,

$$\boldsymbol{A_s} - \boldsymbol{A'_s} = 0.85\boldsymbol{\beta_1}\frac{f'_c}{f_y}\boldsymbol{bc}$$

From the geometry of Fig. 2.6(b),

$$\boldsymbol{c} = \boldsymbol{d'}\,\frac{\epsilon_{cu}}{\epsilon_{cu} - \epsilon'_s} = \boldsymbol{d'}\,\frac{\epsilon_{cu}}{\epsilon_{cu} - \epsilon_y}$$

where ϵ_{cu} is the ultimate strain in concrete and assuming $\epsilon'_s = \epsilon_y$ as a limiting value.

Inserting this value of $\boldsymbol{c}$ into the previous equation and taking values of $\epsilon_{cu} = 0.003$, $\epsilon_y = f_y/E_s$, and $\boldsymbol{E_s} = 29{,}000$ ksi yields

$$\boldsymbol{A_s} - \boldsymbol{A'_s} = 0.85\boldsymbol{\beta_1}\frac{f'_c}{f_y}\boldsymbol{bd'}\,\frac{87}{87 - f_y} \tag{2.17}$$

As noted in Section 2.2 in the discussion on balanced beams, the ACI Code assures ductility by limiting tension reinforcement to 75% of that required for balanced conditions. This limitation need not apply to compression reinforcement because steel in compression will deform in a ductile manner, as well as steel in tension.

Assuming $\boldsymbol{A_s} - \boldsymbol{A'_s} = 0.75\boldsymbol{A_{sb}} = 0.75\boldsymbol{\rho_b bd}$, inserting the value of $\boldsymbol{\rho_b}$ from Eq. 2.13, and equating to Eq. 2.17 yields

$$0.75\boldsymbol{bd}0.85\boldsymbol{\beta_1}\frac{f'_c}{f_y} \times \frac{87}{87 + f_y} = 0.85\boldsymbol{\beta_1}\frac{f'_c}{f_y}\boldsymbol{bd'}\,\frac{87}{87 - f_y}$$

Solving for $\boldsymbol{d'/d}$

$$\frac{\boldsymbol{d'}}{\boldsymbol{d}} = 0.75 \times \frac{87 - f_y}{87 + f_y} \tag{2.18}$$

Table A2.2 gives values of $\boldsymbol{d'/d}$ at which compressive reinforcement will reach yield. These values must not be exceeded.

If $\boldsymbol{d'/d}$ is less than the ratio given by Eq. 2.18, compression steel will reach yield. Conversely, if $\boldsymbol{d'/d}$ is more, compression steel will not reach yield and the actual steel stress (which will be less than the yield stress) must be used in the analysis. Engineers should avoid flexural members in which compression steel will not reach yield, not only because of the calculation effort involved but also because the section will probably have the following characteristics:

1. A congestion of steel that makes placement of concrete difficult with a potential loss of concrete quality,
2. Large deflections,
3. High sensitivity to small steel placing errors possibly resulting in much higher stresses than calculated (see Examples 2.5 and 2.6),

4. High shear stress that is difficult to resist properly with reinforcement.

As a result of all these difficulties, the section will not be economical.

An interesting, useful, but simple conclusion can be drawn from Eq. 2.18. If both tension and compression steel have 2 in. cover and use #11 bars with $f_y = 60$, then the minimum total depth of a beam in which compression steel will reach yield is 22.3 in. Furthermore, for $f_y = 60$ ksi, the centroid of compression steel will be above the centroid of concrete at $0.75\rho_b$ so that the area of both tension and compression steel can be computed using the internal lever arm of $(d - a/2)$. The results will be slightly conservative. If the engineer uses other bar sizes, concrete cover, or steel strengths on a regular basis, a minimum beam depth can be computed easily for other conditions.

Compression steel bars in a beam can buckle outward by spalling the face shell of concrete. To prevent buckling failures, compression bars must be restrained with lateral ties throughout the distance where compression reinforcement is required. Size and spacing of ties must meet the requirements for lateral ties in columns, which are discussed in Chapter 9. Stirrups (see Chapter 6) meeting the size and spacing requirements for ties will provide the necessary lateral restraint if they restrain the compression bars.

Example 2.3   Determine the nominal negative moment capacity of the beam in Fig. 2.7 made with concrete $f'_c = 3$ ksi and steel $f_y = 60$ ksi.

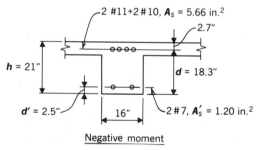

FIGURE 2.7

Solution   From Table A2.1, $0.75\rho_b = 1.6\%$ and $a/d = 0.38$.

$$0.75A_{sb} = 0.016 \times 16 \times 18.3 = 4.68^{\square ''}$$

Without compression reinforcement, the beam would be over-reinforced and would not be permitted by the ACI Code.

$$d'/d = 2.5/18.3 = 0.137 < 0.138 \text{ from Table A2.3.}$$

Therefore, compression steel will reach yield. Using Eq. 2.14, the moment capacity of $0.75A_{sb}$ is

$$M_{n1} = \frac{4.68 \times 60(18.3 - 0.38 \times 18.3/2)}{12} = 347'k$$

The excess of tension steel over $0.75A_{sb}$ is

$$(A_s - 0.75A_{sb}) = 5.66 - 4.68 = 0.98^{\square''}$$

The moment couple between compression steel and excess tension steel $(A_s - 0.75A_{sb})$ is limited by tension steel to

$$M_{n2} = \frac{0.98 \times 60(18.3 - 2.5)}{12} = 77'k$$

The moment couple is limited by compressive steel (Eq. 2.16) to

$$M_{n2} = \frac{1.20(60 - 2.55)(18.3 - 2.5)}{12} = 91'k$$

Using the lower value for M_{n2} and adding it to M_{n1} gives a total $M_n = 424'k$,

and factored moment $M_u = 0.9 \times 424 = 382'k$

2.5 DESIGN PROCEDURES

Determining Tension Steel Area (A_s)

The initial step in reaching the first objective of selecting the tension steel is to determine the cross-sectional area (or, simply, "the area") of steel required. By solving Eq. 2.8 for A_s and substituting $M_u = 0.9M_n$,

$$A_s = \frac{0.85f_c'bd}{f_y} - \sqrt{\left(\frac{0.85f_c'bd}{f_y}\right)^2 - \frac{1.7f_c'bM_u}{\phi f_y^2}} \tag{2.19}$$

Although Eq. 2.19 is satisfactory for a computer, it is prohibitively awkward for manual design. The area of tension steel A_s can be determined manually much more easily by an iterative procedure.

The procedure can be simplified considerably if A_s can be determined directly by a simple formula. After each iteration (including the first), the engineer can evaluate the advantages of proceeding with additional iterations.

The desired form of the equation is

$$A_s = \frac{M_u}{Kd} \tag{2.20}$$

where M_u = factored service moment,
and K is a factor that takes into account:
1. yield strength of steel, f_y
2. ratio of internal lever arm to effective depth j, and
3. strength of concrete.
To evaluate K, recall Eq. 2.7

$$M_n = A_sf_y\left(d - \frac{a}{2}\right) = A_sf_yd\left(1 - \frac{a}{2d}\right)$$

Substituting $M_u/\phi = M_n$ and solving for A_s gives

$$A_s = \frac{M_u}{(\phi f_y j)d} \tag{2.21}$$

where $j = 1 - \dfrac{a}{2d}$ $\hfill$ (2.22)

Values of $(\phi f_y j)/12$ are given in Table A2.3 for use with factored moments in ft-kips for $f'_c = 3, 4,$ and 6 ksi and for $f_y = 40$ and 60 ksi. Figure A2.2 gives the same information for $f'_c = 3$ and 4 ksi and $f_y = 60$ ksi.*

Checking Concrete Strength

The strength of concrete in the compression zone can be checked directly by computing the concrete moment capacity at balanced conditions and limiting the factored moment to 75% of capacity at balanced conditions (see the discussion in Sections 2.2 and 2.3). This procedure is laborious for design but is nevertheless advisable for T-beams with relatively small flanges and other nonrectangular sections. Fortunately, such conditions are encountered infrequently in practice, and when they are, the engineer can usually use conservative approximations that allow use of the shorter procedures for rectangular sections.[†]

Simple indirect methods can be used for rectangular beams. The easiest method in final design is to limit the tension steel area (excluding that portion used to balance compression steel) to 75% of the steel required for balanced conditions. For this method, Eq. 2.13 is useful and Table A2.1 gives values for $0.75\rho_b$ for common combinations of steel and concrete strengths.

Another method that is useful for preliminary design is to limit the moment resisted by the concrete section alone (excluding any compression steel) to the moment resisting capacity R_c of the concrete section. Thus,

$$M_u = R_c bd^2 \tag{2.23}$$

where M_u = factored moment and the limiting value of R_c depends on concrete strength f'_c and steel yield strength f_y.

To evaluate R_c, remember that the maximum moment on a beam is 75% of balanced conditions. Substituting $A_{sb} = 0.75\rho_b bd$ for A_s in Eq. 2.8 and rearranging,

---

*For the remainder of this text, the symbol $(\phi f_y j)$ without the units conversion factor of $\frac{1}{12}$ will be used for the factor $(\phi f_y j)/12$ to emphasize its physical meaning to readers. In practice, many engineers will abbreviate $(\phi f_y j)/12$ to K_{As}, K_s or, simply, K.

[†]For example, in a T-beam with a small flange, the flange can be conservatively ignored in computing the resisting moment capacity of the concrete.

$$\frac{M_n}{bd^2} = 0.75\rho_b f_y \left(1 - 0.59 \times 0.75\rho_b \frac{f_y}{f_c'}\right) \qquad (2.24)$$

but in Eq. 2. 13,

$$\rho_b = \frac{0.85\beta_1 f_c'}{f_y} \times \frac{87}{87 + f_y}$$

For 75% of balanced conditions,

$$0.75\rho_b = \frac{K f_c'}{f_y} \qquad (2.25)$$

where

$$K = 0.75 \times 0.85\beta_1 \frac{87}{87 + f_y} \qquad (2.26)$$

Substituting Eq. 2.25 into Eq. 2.24 and letting $M_u = \phi M_n$ yields

$$R_c = \frac{M_u}{bd^2} = \phi K f_c'(1 - 0.59K) \qquad (2.27)$$

Figure A2.3 and Table A2.4 give values of R_c for common combinations of steel and concrete strengths.

Determining Compression Steel Area (A_s')

If it has been determined that the concrete has inadequate strength to resist the moment and that addition of compression steel is the best remedy, the area of compression steel required can be determined in a step-by-step procedure.

Step 1   Will the compression steel reach yield? This can easily be determined by comparing d'/d to the values in Table A2.2.

Step 2   If compression steel does not reach yield, compute the stress in the compression steel and proceed to step 5. (As noted previously, beams that are too shallow for compression steel to reach yield are not recommended.) To determine stress in the compression steel, first compute the distance from the compression face to the neutral axis c and then compute the stress at the level of compression steel by the geometry of the strain diagram. (See Fig. 2.6(*b*), Eq. 2.10, and the discussion in Section 2.3, T-Beam Analysis).

Step 3   If the compression steel does reach yield, is $d'/d \le 0.5a/d$?

Step 4   If $d'/d \le 0.5a/d$, compute the tension steel area required using the full moment M_u and an internal couple arm of $(d - a/2d)$ for 75% of balanced conditions. Compute the compression steel area required as the difference between the required tension steel area and the tension steel required for balanced conditions.

$$A_s' = A_s - 0.75A_{sb} \qquad (2.28)$$

This step may result in a slight excess of steel. If greater accuracy is desired, use step 5 procedure instead.

Step 5   Compute the area of tension steel required by adding the area of steel required for 75% of balanced conditions A_{sb} to the area required for the couple with compression steel. That is.

$$A_s = 0.75A_{sb} + (A_s - 0.75A_{sb}) \tag{2.29}$$

where $0.75A_{sb} = 0.75\rho_b bd$, taking $0.75\rho_b$ from Table A2.1,

$$(A_s - 0.75A_{sb}) = \frac{M_u - M_{n1}}{\phi f_y(d - d')} \tag{2.30}$$

and M_{n1} = the nominal moment strength of balanced conditions from Eq. 2.14.

Equation 2.30 is an inversion of Eq. 2.15 in which the nominal moment strength has been converted to factored moment by inserting the strength reduction factor ϕ.

Then compute A_s' using the moment in excess of 75% of balanced conditions, the internal lever arm between tension and compression reinforcing bars, and the stress in the compression steel when concrete reaches its ultimate strain. Thus,

$$A_s' = \frac{M_u - M_{n1}}{\phi(f_s' - 0.85f_c')(d - d')} \tag{2.31}$$

The above steps have been shown in flowchart form in Fig. 2.8.

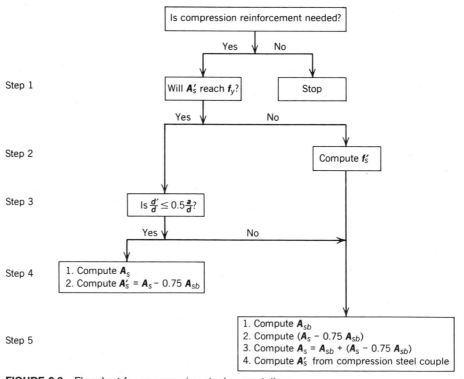

FIGURE 2.8   Flowchart for compression steel computations.

Example 2.4 Consider the beam in Fig. 2.9 spanning 25 ft, spaced 22 ft on centers, $f'_c = 4$, $f_y = 60$, and service positive moments at mid-span of

Dead load moment = 104′k
Live load moment = <u>86</u>
Total load moment = 190′k

Determine the area of tension steel required and check the capacity of the concrete in compression.

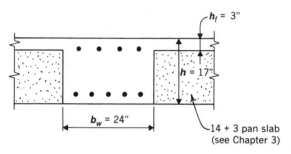

FIGURE 2.9

Solution Converting to factored moment,

$$^+M_u = 104 \times 1.4 + 86 \times 1.7 = 292'k*$$

From Table A2.3 or Fig. A2.2, select

$$(\phi f_y j) = 4.1 \text{ for } \rho \approx 0.8\%$$

$$d = 17 - 2.5 = 14.5''$$

$$^+A_s = \frac{292}{4.1 \times 14.5} = 4.91^{\square''}$$

As a T-beam, $b = 16 \times 3 + 24 = 72''$ <u>USE</u>

or $b = 22 \times 12 = 264''$

or $b = 25 \times 12/4 = 75''$

$$\rho = \frac{4.91}{14.5 \times 72} = 0.47\% < 0.8\% \text{ assumed} \quad \underline{OK}$$

$$a = \frac{4.91 \times 60}{0.85 \times 4 \times 72} = 1.20'' < 3 \times 0.85 \times 0.75 = 1.9'' \quad \underline{OK}$$

Minimum $A_s = 0.0033 \times 24 \times 14.5 = 1.16^{\square''} < 4.91^{\square''} \quad \underline{OK}$

*Throughout this text, as well as in engineering practice, a superscript plus sign (e.g., ^+M_u) before a symbol for moment or area of steel indicates positive moment or positive steel area. Likewise, a superscript minus sign ($^-$) indicates negative moment or negative steel area.

If greater accuracy is desired, try a second iteration. From Table A2.3, or Fig. A2.2, select $(\phi f_y j) \simeq 4.3$ for $\rho \simeq 0.50\%$

$$^+\!A_s = \frac{292}{4.3 \times 14.5} = 4.68^{□''}$$

$$\rho = \frac{4.68}{72 \times 14.5} = 0.45\% < 0.50\% \qquad \underline{OK}$$

Neglecting differences resulting from bar selections, the second iteration saved about $(4.91 - 4.74)3.4 \times 25$ or 15 lbs of steel,* which cost about \$3.75 per beam at 1985 prices. From Eq. 1.7,

$$T = \frac{\$3.75 \times 1 \times 0.5}{\$0.50 \times 3 \times 2} = 0.6 \text{ min}$$

Thus, the engineer can afford to spend no more than about one-half minute per beam on the second iteration to save 15 lbs of steel. With a little practice, the first iteration will be closer to the minimum steel required and thus reduce the incentive for a second iteration.

Example 2.5   Consider the same beam as in Ex. 2.4, with service negative moments at each end of

Dead load moment = 151'k
Live load moment = 125
Total load moment = 276'k

Determine the area of tensile steel required, check the capacity of the concrete in compression, and determine the area of compression steel required, if necessary.

Solution   Converting to factored moments,

$$^-\!M_u = 151 \times 1.4 + 125 \times 1.7 = 424'k$$

From Table A2.3 or Fig. A2.2, select

$$(\phi f_y j) = 3.65 \text{ for } \rho = 2.14\%$$

$$^-\!A_s = \frac{424}{3.65 \times 14.5} = 8.01^{□''}$$

$$\rho = \frac{8.01}{24 \times 14.5} = 2.30\% > 2.14\% = 0.75\rho_b$$

Therefore, compression steel is required

$$d'/d = \frac{2.5}{14.5} = 0.17 > 0.138 \text{ from Table A2.2,}$$

*The weight of a steel bar 1 in. sq. in size is 3.4 lbs/ft.

and compression steel will not reach yield.

From Table A2.1, $a/d = 0.38$, and

$$d'/d = 0.17 < a/2d = 0.5 \times 0.38 = 0.19$$

The $(d' - d)$ lever arm is greater than the $(1 - a/2d)$ lever arm so that no adjustment is needed for the tensile steel area A_s. In this case, the reduction in A_s to account for the greater lever arm in the compression steel couple would be so small that the cost of design effort would exceed the cost of steel saved.

To determine the area of compression steel required, consider the geometry of the strain diagram in Fig. 2.10.

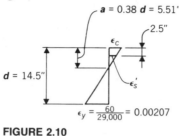

FIGURE 2.10

$$\epsilon_c = 0.38(\epsilon_c + 0.00207)$$

$$\epsilon_c = \frac{0.38 \times 0.00207}{1 - 0.38} = 0.00127$$

$$\epsilon_s' = \epsilon_c \frac{(5.51 - 2.5)}{5.51} = 0.000693$$

$$f_s' = \epsilon_s' E_s = 0.000693 \times 29,000 = 20.1 \text{ ksi}$$

$$A_{sb} = 0.0214 \times 24 \times 14.5 = 7.45^{\square\prime\prime}$$

$$M_{n1} = A_{sb}(\phi f_y j)d = 7.45(3.65)14.5 = 394'k$$

$$M_u - M_{n1} = 424 - 394 = 30'k$$

Using Eq. 2.31,

$$A_s' = \frac{30 \times 12}{0.9(20.1 - 0.85 \times 4)(14.5 - 2.5)} = 2.00^{\square\prime\prime}$$

Figure 2.11 summarizes the areas of steel required at each critical section for Exs. 2.4 and 2.5.

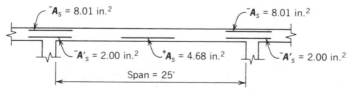

FIGURE 2.11

Example 2.6 Consider the same beam as in Ex. 2.5 but with an overall depth of 16 in. and $f'_c = 3$. Compute the steel area required top and bottom at the support.

Solution From Ex. 2.5, $^-M_u = 424'k$, and from Table A2.3, $(\phi f_y j) = 3.65$

$$d = 16 - 2.5 = 13.5''$$

$$^-A_s = \frac{424}{3.65 \times 13.5} = 8.60^{□''}$$

$$\rho = \frac{8.60}{24 \times 13.5} = 2.66\% > 1.60\% = 0.75\rho_b$$

Therefore, compression steel is required if the concrete outlines are to be maintained.

$$d'/d = \frac{2.5}{13.5} = 0.185 > 0.138 \text{ from Table A2.2,}$$

and compression steel will not reach yield.
From Table A2.1, $a/d = 0.38$

$$\text{and } d'/d = 0.185 < a/2d = 0.5 \times 0.38 = 0.19$$

The $(d - d')$ lever arm is about the same as the $(1 - a/2d)$ lever arm, so ^-A_s computed above is adequate.

To determine the area of compression steel required, consider the geometry of the strain diagram in Fig. 2.12.

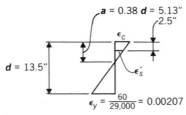

FIGURE 2.12

$$\epsilon_c = 0.38(\epsilon_c + 0.00207)$$

$$\epsilon_c = \frac{0.38 \times 0.00207}{1 - 0.38} = 0.00127$$

$$\epsilon'_s = \epsilon_c \frac{(5.13 - 2.5)}{5.13} = 0.000651$$

$$f'_s = \epsilon'_s E_s = 0.000651 \times 29,000 = 18.9 \text{ ksi}$$

$$A_{sb} = 0.016 \times 24 \times 13.5 = 5.18^{□''}$$

$$M_{n1} = A_{sb}(\phi f_y j)d = 5.18(3.65)13.5 = 255'k$$

$$M_u - M_{n1} = 424 - 255 = 169'k$$

$$A_s' = \frac{169 \times 12}{0.9(18.9 - 0.85 \times 3)(13.5 - 2.5)} = 12.53^{\square''}$$

From Examples 2.5 and 2.6, it is evident that when d'/d is close to the values in Table A2.2, small reductions in effective depth and concrete strength make large differences in the area of compression steel required. This is true whether the specified depth of the beam is less or the tension steel is accidentally displaced to reduce the effective depth and whether the specified strength is lower or actual strength of concrete is lower than specified.

2.6 BAR SELECTION

Skillful selection of longitudinal bars for beams is based on the following criteria:

1. The area of bars furnished must equal or exceed the area of steel required.

2. Some full-length straight bottom bars must extend into supports at least 6 in. The ACI Code requires that at least one-third the positive moment reinforcement in simple members and one-fourth the positive moment reinforcement of continuous members extend into both supports (Code 12.11.1). If the flexural member is part of a primary lateral load resisting system, such reinforcement must be anchored to develop yield strength f_y at the face of the support (Code 12.11.2) (see Chapter 5, Anchorage and Bar Development).

3. At least two full-length straight bottom bars extending to the supports must be furnished wherever stirrups are used and placed in the bend of the stirrup. In addition, at least two straight top bars should be used to hold stirrups in position and extend from supports to as far as stirrups are required. If stirrups have more than two vertical legs, one straight bar must be furnished in each bend between anchored ends (Code 12.13.3). Some stirrups require a straight bar in the hook at the end of stirrup bars (see Chapter 6). Although the Code does not require that these bars be used to resist moment, economy dictates that they be used for that purpose if possible.

4. Bars must be spaced close enough to control the size and spacing of flexural cracks. This requirement will be conservatively met if the spacing of bars does not exceed the spacing given in Table A2.5 for the largest bar (see the discussion below on flexural cracking).

5. Stress in the bars must be developed. The ACI Code assures this development by limiting the maximum size of bar that can be used (Code 12.11 and 12.12). This requirement will be met if the bar size does not exceed sizes given in Table A5.6 (see the discussion in Chapter 5).

6. Clear distance between bars shall not be less than the nominal bar diameter or 1 in., or less than $1\frac{1}{3}$ times the maximum size of coarse aggregate (see Section 1.8). More space should be provided where possible. This limits the number of bars that can be placed in one layer.

7. Bars must clear the steel in intersecting beams and columns; otherwise, adjustments may be made in the field during construction without the engineer's knowledge. As a result, steel may be placed in a less effective position to resist moments, shears, and forces. For example, the longitudinal top and bottom bars in intersecting beams cannot be placed in the same level. When column and longitudinal beam bars must pass each other, clearance should be based on assumed bar diameters at least $\frac{1}{4}$ in. larger than their nominal diameters to allow for bar deformations (e.g., $1\frac{3}{8}$ in. for a #9 bar). Bar setting will be facilitated if $\frac{1}{2}$- to $\frac{3}{4}$-in. clearance is provided instead of only $\frac{1}{4}$ in. Slab bars are less likely to interfere with beam bars because slab bars usually require only $\frac{3}{4}$ in. cover, whereas beam bars must have $1\frac{1}{2}$ in. cover.

8. If the depth of a beam web is more than 3 ft, additional longitudinal steel must be placed in the zone of flexural tension, as discussed in Section 6.5 (Code 10.6.7). Such reinforcement is not normally included in strength computations because of the additional tedious calculations.

9. As few bars as possible should be used for economy. Fewer and larger bars are more likely to be placed in one layer and, thus, be used most effectively. A worker can set a large bar in about as little time as it takes to set a smaller bar.

10. All straight bottom bars in one beam should be the same size for economy in fabrication, setting, and "shaking out" or sorting the bars on the job site. Likewise, all straight top bars and all truss bars in one beam should be one size.

11. Experienced engineers will normally not use bar combinations in one beam where the difference in size between the largest and smallest bar is more than three or four numbers (e.g., #7 to #10 or #11). Large differences in bar size are inefficient for crack control.

12. Truss bars (see Fig. 2.13) may be used to provide part of the top and bottom tensile reinforcement rather than all straight bars, at the engineer's option. Reasons in favor of using truss bars are (1) some steel will be saved because the total length of bars will be less, (2) truss bars assure anchorage of the bottom tensile steel (Code 12.10.1), (3) truss bars provide some shear reinforcement (Code 11.5.1.2), and (4) shape of the truss bars helps keep bars in the top of the beam where they belong. Reasons against the use of truss bars are (1) they cost more than straight bars per pound for fabrication, (2) the engineer may take more design time to select truss bars, (3) truss bars are very difficult to place where closed stirrups are used or where

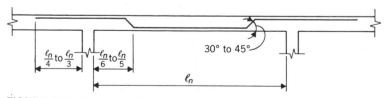

FIGURE 2.13   Truss bar with typical dimensions.

under floor electrical ducts are used and (4) tolerances in fabrication of bars and formwork may cause interferences and make placement difficult. When the span-to-depth ratio ℓ_n/d is less than about 8 or 10, it may not be possible to satisfy both top and bottom reinforcement requirements with truss bars. Because of these difficulties, truss bars are usually not used in beams today.

Crack Control

When high-strength steel with service load stresses in the range of 60% of the yield strength is used, wide cracks from flexural stresses may occur. Such cracks are undesirable when visible in the finished structure because laymen intuitively believe that large cracks are evidence of an unsafe structure. If the structure is in an environment in which water or other corrosive liquids can enter the cracks, the steel may corrode and damage the structure. Cracks cannot be eliminated, but closely spaced tension bars will cause more, but smaller, cracks to form. If sufficiently small, the cracks will not be noticeable and will inhibit the entry of liquids. (The design of tanks to hold liquids requires other considerations not covered here.)

To control flexural cracking, the ACI Code requires (Code 10.6.4) that flexural reinforcement in beams and slabs be closely spaced so that

$$f_s \sqrt[3]{d_c A} \qquad (2.32)$$

where z does not exceed 175 kips per inch for interior exposure and 145 kips per inch for exterior exposure (for structures subject to very aggressive exposure or designed to be watertight, special investigations and precautions are required),

f_s = steel stress at working loads but not more than $0.6 f_y$,

d_c = concrete cover to the center of the outermost bar,

A = effective tensile area of concrete surrounding the tension reinforcement and having the same centroid as that reinforcement divided by the number of bars N. When the reinforcement consists of different size bars, N equals the total area of reinforcement divided by the area of the largest bar used.

The effective tensile area A is the crosshatched area in Fig. 2.14. Assuming a 2 in. cover for a single layer of reinforcement with all bars the same size,

$$A = 2d_c S \qquad (2.33)$$

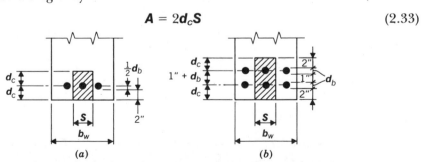

FIGURE 2.14 Effective tensile area of concrete surrounding tension reinforcement.

where S = the average spacing of bars = b_w/N

Substituting Eq. 2.33 into Eq. 2.32 and solving for the spacing S,

$$S = \frac{(z/0.6f_y)^3}{2d_c^2} \qquad (2.34)$$

For two layers of reinforcement with all bars the same size and equal numbers of bars in each layer,

$$A = (2d_c + d_b + 1)S \qquad (2.35)$$

Substituting Eq. 2.35 into Eq. 2.32 and solving for the spacing S,

$$S = \frac{(z/0.6f_y)^3}{d_c(2d_c + d_b + 1)} \cdot \qquad (2.36)$$

Values for Eqs. 2.34 and 2.36 have been tabulated in Table A2.5. Note that maximum spacing increases rapidly as service load stress decreases below 36 ksi (0.6×60 ksi). For example, the maximum spacing of #8 bars in one layer for interior exposure when service load stress is 30 ksi is $9.2(36/30)^3 = 15.9$ in.

If steel is placed in two layers and the number of bars in one layer is not the same as in the other layer, it is conservative to use the maximum tabulated bar spacing for two layers. This is normally an irrelevant consideration because bars will be at minimum spacing for clearance so that as many bars as possible will be in the outermost layer.

More than two layers of steel are rarely required. However, when three or more layers are required, the engineer should check the spacing using Eq. 2.32.

If more than one bar size is used in a layer, it is satisfactory to reduce the required average width from Table A2.5 for the smaller bars in proportion to their area compared to the largest bar. The maximum beam width then becomes the sum of the maximum spacings for the individual bars. Thus,

$$\text{Maximum } b_w = \Sigma S_{lb} + \Sigma S_{sb}$$

where S_{lb} = maximum spacing of the largest bar and

$$S_{sb} = \text{maximum spacing of the smaller bar} \times \frac{A_s \text{ smaller bar}}{A_s \text{ largest bar}}$$

Concrete cover over reinforcement of more than 2 in. may be considered sacrificial. Thus, the maximum concrete cover d_c that need be considered in Eqs. 2.32, 2.34, and 2.36 is 2 in. However, if large concrete cover (> 2 in.) is required, cracks might occur at the concrete surface that are larger than those occurring with 2 in. cover, regardless of strength and spacing of reinforcement. Engineers should inform the owner and others that visible cracks in concrete might occur if cover is more than 2 in.

Example 2.7   Select bottom bars with $f_y = 60$ ksi to furnish $A_s = 4.62^{□''}$ and fit in one layer in a 16-in.-wide beam with a clear span of 25 ft. Meet crack control requirements for exterior exposure.

Solution   From Table A5.6, for a 25′ span, #11 and smaller bars can be used. Try 2 #11 + 2 #8 bars (see Appendix E for area of bars).

$$A_s \text{prov.} = 3.12 + 1.58 = 4.70^{□''} > 4.62^{□''}* \qquad \text{OK}$$

Referring to Table A2.5 for maximum bar spacing,

$$\text{Max } b_w = 2 \times 4.5 + 2 \times 5.2 \times 0.79/1.56 = 9.0 + 5.3$$

$$= 14.3'' < 16'' \qquad \text{NA}$$

Try 3 #10 + 1 #8 bars.

$$A_s \text{prov.} = 3.81 + 0.79 = 4.60^{□''} > 4.62 \times 0.99 \qquad \text{OK}$$

$$\text{Max } b_w = 3 \times 4.7 + 5.2 \times 0.79/1.27$$
$$= 14.1 + 3.2 = 17.3'' > 16'' \qquad \text{OK}$$

Example 2.8   Select bars to fit in the beam shown in Fig. 2.9 to meet minimum steel area requirements shown in Fig. 2.11 using straight and truss bars. Assume interior exposure and f_y = 60 ksi.

Solution   From Table A5.6, stress in #11 bars can be developed.
Bottom steel ^+A_sreqd. = $4.68^{□''}$

By inspection of Table A2.5, 3 #10 bars minimum are required in one layer with an average 8 in. spacing.
Straight bars must have A_s = 4.68/3 = $1.56^{□''}$ each.
One possible combination is

$$2 \text{ #8 } \quad \text{bottom} \quad +2 \text{ #11} \quad \text{truss} \qquad ^+A_s \text{prov.} = 4.70^{□''} \qquad \text{OK}$$

$$b_w = 2 \times 7.8 + 2 \times 9.2 \frac{0.79}{1.56} = 24.8'' > 24'' \qquad \text{OK}$$

This seems to be the optimum selection because it uses the fewest number of bars, just meets the steel area required, and provides the most steel in truss bars to help resist negative moment.
Top steel ^-A_sreqd. = $8.01^{□''}$

Because no information is given on adjacent beams, assume that this beam is in the middle of a string of beams and that the same beam with identical steel is required at each end of this beam.
From bottom steel, 4 #11 truss bars are supplied.

$$^-A_s \text{ reqd.} = 8.01$$
$$\text{Truss } ^-A_s \text{prov.} = \underline{6.24^{□''}}$$
$$\text{Straight top } ^-A_s \text{reqd.} = 1.77^{□''}$$
Use 2 #9 straight top bars

$$\text{Total } ^-A_s \text{prov.} = 6.24 + 2.00 = 8.24^{□''} \qquad (3\% \text{ over}) \qquad \text{OK}$$

*Abbreviations used throughout this text and in engineering practice are $A_s$prov. = area of steel provided, and $A_s$reqd. = area of steel required.

Example 2.9   Select bars for a beam the same as that in Ex. 2.8 but with an exerior exposure.

Solution   <u>Bottom steel</u> ^+A_sreqd. = 4.68$^{□''}$

By inspection of Table A2.5, 5 #9 bars or smaller at 5.0 in. spacing or less are required.

$$\text{Straight bars must have } A_s = \frac{4.68}{5} = 0.94^{□''}$$

Following are some possible combinations:

1.   2 #9 bottom
     +3 #9 truss ^+A_sprov. = 5.00$^{□''}$ (7% over).
2.   2 #9 bottom
     +2 #9 truss
     +1 #8 truss ^+A_sprov. = 4.79$^{□''}$ (2% over).
3.   2 #8 bottom
     +4 #8 truss ^+A_sprov. = 4.74$^{□''}$ (1% over).

Option 1 uses fewer styles of truss bars than option 2 and fewer truss bars than option 3. Options 2 and 3 use less steel. The selection depends on top bar selection.

<u>Top steel</u> ^-A_sreqd. = 8.01$^{□''}$

From bottom steel,

1. 6 #9 truss bars furnished, ^-A_sprov. = 6.00$^{□''}$
2. 4 #9 + 2 #8 truss bars furnished, ^-A_sprov. = 5.58$^{□''}$
3. 8 #8 truss bars furnished, ^-A_sprov. = 6.32$^{□''}$

Some possible combinations of top bars:

1.   2 #9 top
     +6 #9 truss ^-A_sprov. = 8.00$^{□''}$ (0% over/under)
     average spacing = 3″
2.   2 #10 top
     +4 #9 truss
     +2 #8 truss ^-A_sprov. = 8.12$^{□''}$ (1% over)
     average spacing = 3″
3.   2 #9 top
     +8 #8 truss ^-A_sprov. = 8.32$^{□''}$ (4% over)
     average spacing = 2.40″

All three options are satisfactory. Options 1 and 2 use fewer bars in the top giving more generous space between bars. Options 1 and 3 use only one style of truss bar making fabrication and construction a little easier. Option 1 uses the least amount of steel. The two styles of truss bars in option 2 might be bent at different locations, giving shear resistance along a greater length of beam.

PROBLEMS

1. Using Whitney's method, what is the ultimate moment capacity of a beam 12 in. wide by 20 in. deep reinforced with two #10 bars in the bottom? Assume $f'_c = 3.5$ ksi, $f_y = 50$ ksi, and 2 in. cover to the reinforcement.

2. What is the balanced reinforcement ratio ρ_b for a beam using $f'_c = 5.3$ ksi and $f_y = 33$ ksi? What maximum reinforcement ratio is permitted by the Code in this beam?

3. What is the maximum width of a flange in a 16×16-in. spandrel beam spanning 22 ft center to center (c/c) of 12-in. columns, supporting a 5½-in. slab with a clear span of 14 ft?

4. Given a spandrel beam 12 in. wide by 27 in. deep spanning 29 ft, reinforced with 3 #11 bars, supporting a 4½-in. slab at least 10 ft wide, what is the maximum factored moment capacity? Assume $f'_c = 3.75$ ksi and $f_y = 50$ ksi.

5. What is the factored moment capacity of the spandrel beam given in Problem 4, neglecting the flange?

6. In a 36-in.-wide by 20½-in.-deep beam, how much flexural tension reinforcement is required if $M_u = 590$ ft-kip? Assume $f'_c = 3$ ksi and $f_y = 60$ ksi.

7. If the beam in Problem 6 has flanges on each side 3½ in. thick, how much flexural tension reinforcement is required? Assume beam span and spacing do not limit flange width.

8. What is the maximum moment permitted on the beams in Problems 6 and 7 without compression reinforcement?

9. If the beam in Problem 6 must resist $M_u = 750$ ft-kips, what area of reinforcement is required?

10. Select bars to meet the required steel area in Problems 6, 7, and 9 for both interior and exterior exposure.

Selected References

1. Eivind Hognestad, N.W. Hanson, and Douglas McHenry. "Concrete Stress Distribution in Ultimate Strength Design," *ACI Journal*, V.52, pp 455–479, December 1955.

2. Charles S. Whitney and Edward Cohen. "Guide for Ultimate Strength Design of Reinforced Concrete," *ACI Journal*, V.53, pp 455–490, November 1956.

3. ACI-ASCE Committee 327. "Ultimate Strength Design," *ACI Journal*, V.52, pp 505–524, January 1956.

4. Alan H. Mattock, Ladislav B. Kriz, and Eivind Hognestad. "Rectangular Concrete Stress Distribution in Ultimate Strength Design," *ACI Journal*, V.57, pp 875–928, February 1961.

5. Hubert Rüsch. "Researches Toward a General Flexural Theory for Structural Concrete," *ACI Journal*, V.57; pp 1–28, July 1960.

CHAPTER 3

ONE-WAY SLAB FLEXURE

3.1 OBJECTIVES

Slabs carry a large area of uniformly distributed load or many small concentrated loads, whereas beams carry heavy concentrated loads, line loads, and loads delivered to beams by slabs. Slabs rarely carry beams or column loads. Slabs are usually more lightly stressed than beams. Flexural design of slabs can be performed similarly to design of beams if these differences are recognized.

One-way slabs are slabs that are supported along only two edges or have a width-to-span ratio (aspect ratio) of at least two. Slabs supported along all edges, with a larger aspect ratio, carry almost all of the load by spanning in the short direction. Moments and shears in the long direction are relatively low compared with slab capacity provided by minimum reinforcement, except in the vicinity of stiff supports along the sides of the slab. Slabs with a smaller aspect ratio are treated as two-way slabs and are covered in Chapter 4. Slabs on grade are not considered here.

Design procedures for slabs are based on the assumption that the concrete outlines have been established in preliminary design (Chapter 14) and that moments acting on critical sections have been determined (Chapter 13). Following are the objectives of this chapter.

For solid slabs,

1. Select the size and spacing of longitudinal tension bars (bars in the direction of the span) required at critical sections. Normally it is not necessary for the design engineer to compute or show the actual number of bars required.

2. Select the size and spacing of temperature bars transverse to longitudinal bars.

3. Verify the adequacy of the slab in compression at critical sections. Normally this is not a problem as other requirements will control the depth of the slab.*

*If calculations indicate a problem in compression, the engineer has made a mistake in computations or in trying to make a slab too thin. Try again.

For joists (see Section 3.3 and Fig. 3.1),

1. Select the number and size of bars in individual joists.
2. Select the transverse distribution rib size and reinforcement.
3. Verify the adequacy of the joist in compression at critical sections, normally sections in negative moment at or near the support.
4. Specify the location of double and triple joists if and where necessary.

The length and bending of slab bars is discussed in Chapter 5, Anchorage and Bar Development. Deflection, discussed in Chapter 8, controls the thickness and design of many slabs. Shear rarely controls the design of solid slabs but frequently controls the design of joists (see Chapter 6, Shear and Torsion in Beams).

3.2 SOLID SLABS

Solid slabs normally have a uniform thickness. Exceptions are made for small depressions in the slab top surface (usually 1 to 2 in.) to accommodate terrazzo, ceramic tile, or other floor finishes. Other exceptions are slabs haunched near the supports and slabs sloped for drainage without a corresponding slope in the bottom surface. In any case, variations in slab thickness must be considered in the design.

FIGURE 3.1 Typical Pan Slab. *(a)* View of underside of a completed slab. *(b)* View of formwork before placing concrete with some reinforcement in place. (Photographs provided by The Ceco Corp.)

The design of slabs for flexure is very similar to that for beams, with the following minor exceptions:

1. Slabs normally carry much lighter loads than do beams, except for slabs used for foundation mats (Chapter 11) and for very heavily loaded industrial slabs.

2. Slabs provide wide areas of support for floor and roof surfaces, whereas beams support loads concentrated at a line or at a point.

3. Slabs not exposed to the weather or the ground need have only ¾ in. cover instead of the 1½ in. required for beams.

4. When ¾ in. of cover is used, the provisions for crack control (Code 10.6.4) will not govern the design as other bar spacing limitations apply.

5. Slabs are usually designed for a unit width, normally 1 ft. Thus, slabs that are the same in every respect, except their width or lateral extent, can use the same design.

6. Slabs are usually more slender than beams, that is, they have higher span-to-depth ratios. This slenderness leads to greater sensitivity to deflection.

7. Slabs have low shearing stresses and do not normally require shear reinforcement. The relatively small depth and thin cover in slabs make stirrups impractical, and truss bars probably would not reinforce a sufficient length of the slab to be effective. In those rare cases (heavily loaded slabs with short spans) where shear reinforcement is required, slabs should be designed as wide beams.

8. Compression reinforcement is not used in slabs because it is impractical to provide lateral support to prevent spalling of the face shell and buckling of the bars.

9. Minimum reinforcement in solid slabs is limited only by the amount required for temperature reinforcement. These requirements are discussed later in this section. Some conservative engineers increase positive reinforcement by one third when $^+\rho < 200/f_y$, where f_y is in psi (or 0.33% for $f_y = 60$ ksi), even though the ACI Code does not require it (see Fig. A2.1).

The first step in design of slabs for flexure is to select the cover over reinforcing steel. For interior exposure, ¾ in. is adequate unless greater cover for fire protection is required (Section 14.3). For exterior exposure, as in parking garages and slabs in contact with earth, cover of 1½ to 2 in. is required (see Section 1.8). Industrial slabs with heavy traffic may require additional cover to allow for erosion of the slab surface during the service life of the structure.

One-way solid slabs are almost always designed using a width of 1 ft. Loading given in pounds per square foot (psf) can be used without modification as the load per lineal foot (plf) of the 1-ft-wide slab strip. The area of steel required is calculated based on 1 ft of width. The area of steel furnished is calculated on the same basis without regard to the area of individual bars. For example, #4 bars at 8 in. c/c provide 0.30 sq. in. of steel per foot of width (area of 1 #4 = 0.20 sq. in., and 0.20 × 12/8 = 0.30 in.²).

Bar Selection

Selection of longitudinal bars for one-way slabs is based on the following criteria:

1. The area of bars furnished must equal or exceed the area of steel required.

2. At least one-third the area of positive moment reinforcement in simple members and one-fourth the area of positive moment reinforcement in continuous members must extend into the supports. Additional anchorage is required if the slab is part of the primary lateral load resisting system (Code 12.11).

3. Bars can be spaced not farther than three times the slab thickness nor 18 in. (Code 7.6.5). The purpose of these limits is to prevent excessive cracking and to prevent concentrated loads from breaking through the slab between reinforcing bars. When reinforcement with a specified yield strength f_y greater than 40 ksi is used with exterior exposure, cover of 1½ or 2 in. is required (see Section 1.8). When concrete cover on bars is more than ¾ in., the spacing is further limited by crack control provisions (Code 10.6) (see Table A2.5 and the discussion below on crack control).

4. Stress in the bars must be developed. The ACI Code assures this development by limiting the size of bar that can be used. This requirement will be met if the bar size does not exceed sizes given in Table A5.6 (see the discussion in Chapter 5).

5. Clear distance between bars shall not be less than the nominal bar diameter nor 1 in. nor less than 1⅓ times the maximum size of coarse aggregate (see Section 1.8). These requirements do not control bar spacing, except when a few bars are grouped on each side of an opening in the slab.*

6. It is more economical to specify fewer and larger bars in a slab rather than a greater number of smaller bars. Fewer bars are more quickly placed. Construction workers will not bend larger bars as easily when they walk on them before concrete is placed, and small bars cost more per pound.

7. Only one size bottom bar, one size top bar, and one size truss bar (if any) should be used. A mixture of sizes is more expensive.

8. Some engineers who avoid truss bars in beams will use them in slabs because they are easily selected and result in fewer pieces of steel to be placed; hence, they may reduce construction costs. For typical interior spans, if truss bars are made one size larger than the straight bottom bars, a good balance between top and bottom steel can result.

9. Whether or not truss bars are used, all bars should be spaced uniformly to the nearest inch to minimize placement errors. When truss bars are used, if extra top bars are used, they should be spaced at a multiple of the bottom bar spacing, not an independent spacing.

*Bars at minimum spacing indicate an error somewhere. Start over.

Crack Control

With a few modifications, the discussion on crack control in Chapter 2 applies to slabs.

1. Bars are never placed in two layers in one-way slabs as they sometimes are in beams.
2. Thinner cover in slabs means that bars may be spaced farther apart than in beams. For ¾ in. cover the allowable spacings for crack control all exceed the 18 in. maximum spacing permitted by the ACI Code. Using Eq. 2.34, maximum spacings have been tabulated in Table A2.5 for 1½ in. cover.
3. The ACI Code Commentary suggests, but does not require, that the value of **z** in Eq. 2.32 be decreased by the ratio of 1.20/1.35 to account for the smaller distance from tension steel to neutral axis in slabs. This has the effect of reducing the maximum allowable bar spacings to 70% of the values given by Eq. 2.34 and in Table A2.5.

Temperature Reinforcement

All concrete expands as the temperature rises and shrinks as the temperature drops. Concrete shrinks as it dries. A reinforced concrete structure will not expand or shrink uniformly because of the different dimensions of its component parts and different exposure conditions. Expansion and shrinkage is resisted by the steel embedded in the concrete, by columns, walls, foundation, and other elements of construction. As a result, when concrete shrinks, it will crack as the stress reaches the tensile strength of the concrete. To encourage small, well-spaced cracks, the ACI Code requires temperature reinforcement (sometimes called shrinkage reinforcement) perpendicular to the main reinforcement in solid slabs. The temperature reinforcement must be at least equal to 0.002 times gross concrete area for f_y = 40 or 50 ksi and 0.0018 times gross concrete area for f_y = 60 ksi (Code 7.12). Spacing must not exceed five times the slab thickness, nor 18 in. Temperature reinforcement is not subject to provisions of the Code on spacing for crack control.

Temperature bars should be located on top of the bottom bars or under the top bars to maintain the maximum effective depth of the main reinforcement. When both top and bottom main reinforcements are present, temperature steel may alternate between the two. Temperature bars should not be placed in the top if no other bars are present as it would be difficult to keep them in position.

Example 3.1 What temperature reinforcement is required in a 5-in. slab using either grade 40 or grade 60 steel?

Solution

Maximum spacing = 5 × 5″ = 25″ c/c, but not more than 18″ c/c.
For grade 40 bars,

$A_s$reqd. = 0.002 × 5 × 12 = 0.12□″/′
Use #3 bars at 11″ c/c, $A_s$prov. = 0.12□″/′

For grade 60 bars,
$$A_s \text{reqd.} = 0.0018 \times 5 \times 12 = 0.108^{\square\prime\prime}/\prime$$
Use #3 bars at 12" c/c, $A_s\text{prov.} = 0.11^{\square\prime\prime}/\prime$

A better selection would be fewer and larger bars—#4 @ 18" c/c—even though slightly more steel would be used.

Example 3.2 Consider a simple span 6-in. slab with interior exposure using $f'_c = 3$ ksi and $f_y = 60$ ksi, with mid-span ultimate moment, $M_u = 2.4'\text{k/ft}$. What reinforcement is required?

Solution   From Table A2.3 or Fig. A2.2,
$$\text{try } (\phi f_y j) = 4.3.$$
Assuming #4 bars and ¾ in. cover,

$$d = 6 - 0.75 - 0.5/2 = 5''$$

Using a 1′ wide strip,

$$A_s = \frac{2.4}{4.3 \times 5} = 0.11^{\square\prime\prime}/\prime$$

Although not required by the Code, conservative engineers would consider

Minimum flexural steel = $(200/60,000) \times 5'' \times 12'' = 0.20^{\square\prime\prime}/\prime$.

For minimum reinforcement, use one-third more steel than required for flexure, and

$$A_s \text{reqd.} = 0.11 \times 1.33 = 0.15^{\square\prime\prime}/\prime$$

Minimum temperature steel $A_s\text{min.}$

$$A_s\text{min} = 0.0018 \times 6 \times 12 = 0.13^{\square\prime\prime}/\prime < 0.15^{\square\prime\prime}/\prime \quad \underline{\text{OK}}$$

(In a design office, engineers would select temperature bars from a table.)
Using the fewest number of bars possible,
select #4 bars at 16" c/c,

$$A_s\text{prov.} = 0.20 \times 12/16 = 0.15^{\square\prime\prime}/\prime \quad \underline{\text{OK}}$$

Example 3.3 Given a 7-in. slab spanning 16 ft clear, with interior exposure, using $f'_c = 3$ ksi and $f_y = 60$ ksi, with ultimate moments at mid-span $^+M_u = 4.7'\text{k/}\prime$, and at the ends $^-M_u = 6.8'\text{k/}\prime$, what reinforcement is required? Assume that this slab adjoins slabs at each end with similar reinforcement requirements.

Solution   From Table A2.3 or Fig. A2.2,

$$\text{try } (\phi f_y j) = 4.25 \text{ at } \rho \simeq 0.5\%$$

Assuming #5 bars, $d = 7 - 0.75 - 0.625/2 = 5.9''$

$$^+A_s = \frac{4.7}{4.25 \times 5.9} = 0.19^{\square}{}''/{}'$$

$$^-A_s = \frac{6.8}{4.25 \times 5.9} = 0.27^{\square}{}''/{}'$$

$$\rho = \frac{0.27}{5.9 \times 12} = 0.38\% < 0.5\% \text{ assumed} \quad \underline{\text{OK}}$$

Neither minimum flexural reinforcement nor temperature reinforcement controls.

Use #4 bars at 32″ c/c bottom = 0.075$^{\square}{}''/{}'$
#5 bars at 32″ c/c truss = 0.116
Total ^+A_sprov. = 0.19$^{\square}{}''/{}'$ $\underline{\text{OK}}$

Add #3 bars at 32″ c/c top = 0.041
#5 bars at 16″ c/c truss = 0.232
Total ^-A_sprov. = 0.27$^{\square}{}''/{}'$ $\underline{\text{OK}}$

Note that on a second iteration, $(\phi f_y j) = 4.3$ could be used for ^-M and $(\phi f_y j) = 4.33$ for ^+M, but no practical change in bar selections and spacing would result.

Example 3.4   What reinforcement is required given the same conditions as those in Ex. 3.3 but with exterior exposure on the underside and interior exposure on the top (e.g., in a building with open space at ground level and occupied space above)?

Solution   Required cover = 1½″ on bottom and ¾″ on top.

and $^+d = 7 - 1.5 - 0.625/2 = 5.2''$
$^-d = 7 - 0.75 - 0.625/2 = 5.9''$

and $^+A_s = \dfrac{4.7}{4.25 \times 5.2} = 0.21^{\square}{}''/{}'$

$^-A_s = \dfrac{6.8}{4.25 \times 5.9} = 0.27^{\square}{}''/{}'$

From Table A2.5, the maximum spacing of #4 bottom bars is 10.7″.
Use #4 bars at 10½″ c/c alternate straight and truss

^+A_sprov. = 0.23$^{\square}{}''/{}'$ $\underline{\text{OK}}$

Add #3 bars at 31½″ c/c top, ^-A_sprov. = 0.04$^{\square}{}''/{}'$
plus #4 truss bars at 10½″ c/c = 0.23
Total ^-A_sprov. = 0.27$^{\square}{}''/{}'$ $\underline{\text{OK}}$

Transverse Reinforcement

One-way slabs normally have little or no moment transverse to the main span. However, when a stiff element such as a beam or a wall is parallel to span of the slab, some load will be drawn to the beam or wall and cause transverse moment in the slab. To guard against large cracks beside the parallel beam or wall, reinforcement perpendicular to the beam or wall must be provided in the top of the slab and designed to carry the slab load as a cantilever for the width of the slab, which is permitted to act as flange on the beam (see Fig. 3.2). Transverse reinforcement must be spaced not farther apart than five times the slab thickness, nor 18 in. (Code 8.10.5). Temperature steel will provide adequate reinforcement in the bottom of the slab in most cases.

Irregular Slabs

Some slabs do not have parallel or straight supports. In such cases, the span length varies across the width of the slab, and there will be varying steel area requirements. For economy, it is wise to group the bars in the same size and spacing for a reasonable distance before changing the pattern. Engineers may want to allow the contractor to order stock lengths of bars and cut them on the job to fit rather than sort bars of many lengths on the job site.

Stair Slabs

Structural behavior of a stair slab is rather complex, resembling that of a folded plate. Its three-dimensional nature makes a stair slab stronger than a comparable slab in one plane. Engineers commonly ignore this additional strength because stair spans are short and potential construction cost savings are small compared to the additional calculations necessary. Furthermore, design for three-dimensional behavior might introduce reinforcing patterns or construction procedures that are more expensive than those required by simplified design methods.

Stairs are usually designed as simple span one-way slabs spanning from landing to landing. The landings are supported on the two sides of the stairwell, the end, or on all three edges. Intermediate landings are sometimes supported by hanger rods. Landings at floor levels may be an integral part of the floor framing.

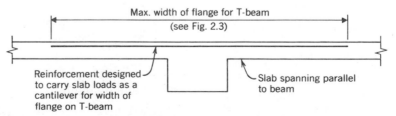

FIGURE 3.2 Transverse reinforcement in slabs spanning parallel to beams.

Example 3.5 In a multistoried building, what reinforcing steel is required in the stair slab shown in Fig. 3.3 supporting 100 psf live load? Assume $f'_c = 3$ ksi and $f_y = 60$ ksi.

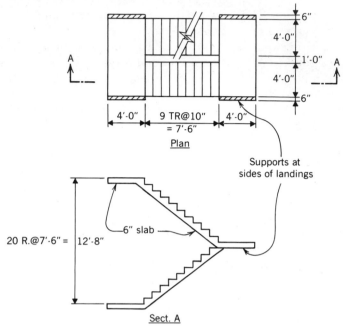

FIGURE 3.3

Solution Assume that the stair is in one plane and that the landing slabs act as wide slab-beams to span from side to side and carry their own load as slabs plus the load from the stair slabs (see Fig. 3.4 for load, shear, and moment diagrams).

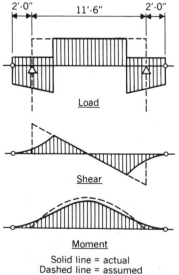

Solid line = actual
Dashed line = assumed

FIGURE 3.4

Common assumptions are made in the distribution of loads to simplify subsequent calculations. For the stair slab,

Live load: 100 psf × 1.7   = 170 psf

Dead load: 6″ slab = $75\left(\dfrac{12.56}{10}\right)1.4 = 132$

Tread = 47.5 × 1.4   = 66

Total w_u =   368 psf

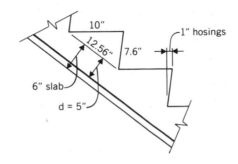

$$M_u = \frac{w_u \ell^2}{8} = \frac{0.368 \times 11.5^2}{8} = 6.1\text{'k/'}$$

Assume $(\phi f_y j) = 4.2$,

$$A_s = \frac{M_u}{(\phi f_y j)d} = \frac{6.1}{4.2 \times 5} = 0.29^{□″}/'$$

Use #5 at 12″ c/c. $A_s$prov. = 0.31$^{□″}/'$ <u>OK</u>

For landing slab,

Live load: 0.100 × 4 × 1.7 = 0.68 k/'

Dead load: slab = 0.075 × 4 × 1.4 = 0.42

Stair total load = $\dfrac{0.368 \times 7.5}{2}$ = 1.38

Total w_u = 2.48 k/'

$$M_u = \frac{w_u \ell^2}{8} = \frac{2.48(9.5)^2}{8} = 28.0\text{'k}$$

Assume $(\phi f_y j) = 4.2$,

$$A_s = \frac{M_u}{(\phi f_y j)d} = \frac{28.0}{4.2 \times 5} = 1.33^{□″}$$

Use 5 #5 @ 12″ c/c including one extra bar at the head of the stairs.

$A_s$prov. = 1.55$^{□″}$ <u>OK</u>

Stair slab temperature steel

$$= 0.0018 \times 12 \left(6 + \frac{7.6 \times 10}{2 \times 12.56} \right) = 0.19^{□″}/'$$

Use #4 @ 12″ c/c. $A_s$prov. = 0.20$^{□″}/'$ <u>OK</u>

Sometimes engineers specify 1 #3 or 1 #4 in the nosing of each tread, in which case temperature steel in the slab can be lighter.

Arrangement of steel is shown in Fig. 3.5. To prevent the stair slab steel from straightening and spalling the face shell of concrete at the bend of the stairs at the top, the stair slab bars must be extended into the compression zone of the slab and anchored rather than simply bending them around the corner (also see Chapter 10, Joints). Some engineers will place bent bars matching the slab steel in top of the slab at the head of the stairs (see Note A in Fig. 3.5) to prevent cracks if the slab acts as a folded plate and develops negative moment at this line.

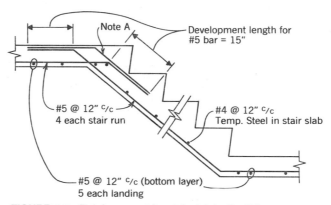

FIGURE 3.5   Reinforcement for stair slab in Ex. 3.5.

Main steel is in the bottom layer for greatest effectiveness; thus, the bottom layer is in one direction in the stair slab and in the perpendicular direction in the landing slab. Transverse steel in the landing need not be as heavy as shown, but the stair slab steel (#5 at 12″ c/c in this example) is usually extended across the landing for simplicity.

Sometimes an architect will want a stair that is freestanding, for example, suported only at the top and bottom landings (left side in Fig. 3.3) but not at the intermediate landing (right side in Fig. 3.3). Axial forces and in-plane bending moments are introduced all into slabs and transverse bending moments into the landing slabs. Torsional moments may also exist. Engineers can visualize structural behavior of free-standing stairs by building a simple paper model.

3.3 JOISTS

Because most solid slabs have low shear stresses and low flexural compression stresses, it is feasible to reduce the dead load by removing part of the concrete to form ribs, or small T-beams. The resulting member is called a "ribbed slab," a "pan slab" or, simply, a "joist." The ribs are most commonly formed by inverted U-shaped steel sheets called "pans." Sometimes fiber-reinforced plastic (FRP) pans are used for formwork. Pans are reusable and have tapered sides to make form removal easier.

The top slab over joists must have a thickness equal to at least 1/12 of the clear distance between ribs but not less than 2 in. over removable forms. Thicker top slabs may be necessary to meet fire resistance requirements (see Chapter 14) or to support concentrated loads. Ribs must be at least 4 in. wide and not deeper than 3½ times the minimum width. Clear spacing of ribs must not exceed 30 in. If these dimensional limitations are met, joists can be designed with shear stress 10% higher than permitted on beams (Code 8.11).

The most common forming system, metal pans, is furnished in depths of 6, 8, 10, 12, 14, 16, and 20 in. The deeper pans are used most often. The standard widths are 20 and 30 in. with special widths of 10 and 15 in. for forming extra width joists. The side taper is 1 in 12, but the ends are normally square. Special end pans are available that decrease the width from 30 to 25 in. in 3 ft or 20 to 16 in. in 3 ft. Joists can be any width of 4 in. or more, but widths of 5, 6, and 7 in. are most common. Pans need a minimum distance to the edge of beams to permit form removal. This is usually 1½ to 2 in. Additional clearance can be provided to form a larger flange on the beam or girder, if desired. (See Fig. 3.6 for a summary of standard joist dimen-

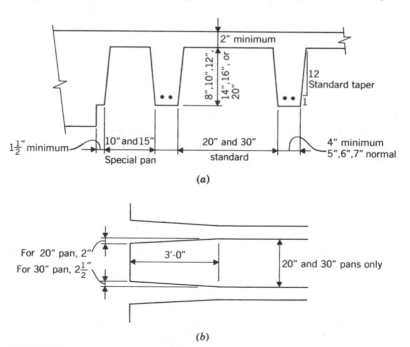

FIGURE 3.6 Standard pans for ribbed slabs.

sions and Table A3.1 for the cross section properties of the slab for various combinations of pan size and depths [1]. Thirty-inch pans in the deeper sizes are used most commonly today because they use less concrete and are more efficient on longer spans.

Coordination with other trades, as well as the structural design, will be simplified if the joist spacing is kept uniform throughout the building. For example, if a small opening to pass vertical piping misses the joists on one floor, the pipes will miss the joists on all floors if the joist spacing is held constant.

With some exceptions, design of joists for flexure is similar to that for T-beams.

1. Joists carry lighter loads than do T-beams.
2. Joists are used to carry floor and roof surfaces, whereas beams are rarely so used.
3. Joists not exposed to the weather or the ground need only ¾ in. cover instead of the 1½ in. required for beams.
4. Crack control provisions do not govern the number or spacing of bars in the bottom of joists (with one minor exception, see Table 3.1). For top bars and ¾ in. cover, 2 bars per joist will meet crack control requirements. For 1½ in. cover, see Table A2.5.
5. Joists are designed as individual units but are repeated as many times as possible despite minor variations in loading. Variations in loading are transferred between adjacent joists by a transverse distribution rib at mid-span or at the third points of the span. The weight of distribution ribs, when present, should be added to the weight of pan slabs. See Table A3.2 for average weights.
6. If shear stress in joists is critical, tapered end pans or wider joists may be used. Stirrups are not used.

TABLE 3.1 Maximum Size Bars for Clearance in Joists[a]

		Clearance (in.)[b]										
	¾[c]				1½[c]				2[d]			
Joist width (in.)	4	5	6	7	4	5	6	7	4	5	6	7
One bar (#)	11	11	11	11	8	11	11	11	NP	8	5	NP
Two bars[e] (#)	6	9	11	11	NP	4	8	10	NP	NP	4	8

[a] Bars tabulated are much heavier than would be required for flexure, except under unusual circumstances.

[b] NP = not permitted.

[c] For interior exposure only.

[d] For exterior exposure. Crack control provisions control for one bar in 6-in.- and 7-in.-wide joists.

[e] Clear distance between bars = one bar diameter but not less than 1 in.

7. When compressive stresses in negative moment at the supports are critical, tapered end pans to form a haunch may be used. It is impractical to provide lateral support in joists for compression bars; hence, bars in compressive zones are not considered effective in resisting compression.

8. Joists are thinner sections with lower volume-to-surface ratios than most beams. This leads to more shrinkage, creep, and cracking and higher deflection than beams comparably loaded with the same span-to-depth ratios (see Chapter 8, Deflection). However, joists are governed by the same span-to-depth ratios as beams (see Table A8.3).

9. Minimum reinforcement in joists is determined in the same manner as for T-beams (see Fig. A2.1). The requirements for solid slabs do not apply to joists.

Welded wire fabric (WWF) is normally placed in the top slab for temperature reinforcement. If the minimum slab thickness will not support a concentrated load in the middle of the slab, a pan one size shallower can be used to increase the slab thickness by 2 in. Many engineers use a 3-in. top slab as the minimum thickness needed to support concentrated loads from partitions.

In recent years, a modified pan slab system called "skip-joists" has been used. Every other joist of a normal pan slab is omitted or skipped so that center-to-center joist spacing becomes 4, 5, or 6 ft. Skip-joists are used when the top slab must be 4½ in. thick or more as it will span between joists at the wide spacing with minimum reinforcement and carry normal loading. Only deeper pans are used for skip-joists, from 14 to 24 in. deep. Before designing a skip-joist system, engineers should consult formwork suppliers for available sizes of pan forms.

Skip-joists must be designed as beams; thus, 1½ in. cover over reinforcement is required, and shear stress may not be increased 10% as it is for pan slab joists. Shear reinforcement in the form of stirrups or truss bars is more likely to be required than in pan slab joists.

Example 3.6 Using 30-in.-wide by 12-in.-deep pans, with a 3-in. top slab, determine the steel required in the top slab if a 4-in. wall weighing 400 plf is located in the center of the 3-in. slab as in Fig. 3.7. Live load equals 100 psf. Use $f'_c = 3$, $f_y = 60$ ksi and assume a simple span because WWF is unlikely to be in top of the slab at supports.

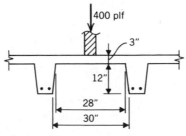

FIGURE 3.7

Solution Compute moments in slab:

$$\text{Dead load, slab } \boldsymbol{M_u} = 0.038 \times \frac{1.4(2.33)^2}{8} = 0.036'k/'$$

$$\text{Dead load, wall } \boldsymbol{M_u} = 0.40 \times \frac{1.4(2.33)}{4} = 0.327$$

$$\text{Live load } \quad \boldsymbol{M_u} = 0.10 \times \frac{1.7(2.33)^2}{8} = \underline{0.115}$$

$$\text{Total } \boldsymbol{M_u} \qquad\qquad = 0.478'k/'$$

$\boldsymbol{d} = 3'' - 1'' = 2''$

Assume $(\boldsymbol{\phi f_y j}) = 4.4$

$$\boldsymbol{A_s}\text{reqd.} = \frac{0.478}{4.4 \times 2} = 0.054^{\square''}/'$$

Minimum steel $= 0.0018 \times 3 \times 12 = 0.065^{\square''}/'$

Minimum steel controls because it is more than $\boldsymbol{A_s}$reqd.

Provide WWF, W4 wires at 6" c/c. (WWF of this style is likely to be in warehouse stock; see Appendix E for wire properties.)

$\boldsymbol{A_s}$prov. $= 0.08^{\square''}/'$

As with solid slabs, the first step in design of joists is to select the cover over the reinforcing steel. For interior exposure, ¾ in. is used, and 1½ to 2 in. is used for exterior exposure. Loads per square foot are multiplied by the joist spacing to obtain the loads per lineal foot of joists. Flexural compression in positive moment is never critical unless holes remove most or all of the top slab. Flexural compression is sometimes critical in negative moment. Additional compressive resistance is provided most easily by tapered end pans. Compressive stresses should then be checked at the face of the support and at the interior end of the tapered pan.

Concentrated point loads or line loads parallel to the joists can be supported by double or triple joists. A narrower pan is used, and one or two extra sets of joist bars are placed in the wider joist. This is more economical than designing special bars different from the joist bars to carry heavier loads. Sometimes double joists and headers are required to frame around openings in the floor slab. Figure 3.8 shows a typical pan slab floor with a large opening.

For floors, no distribution rib is used in spans less than 20 ft. One distribution rib is usually used at mid-span for spans from 20 to 30 ft, and two ribs at the third points for spans over 30 ft. No distribution rib is needed in roof slabs unless concentrated loads or occupancy live loads are expected. Ribs are the same depth as the joists and usually the same width, although they can be made any desired width of 4 in. or more. Their purpose is to distribute concentrated line or point loads so that joists can be reinforced to resist average loading and made identical over large areas.

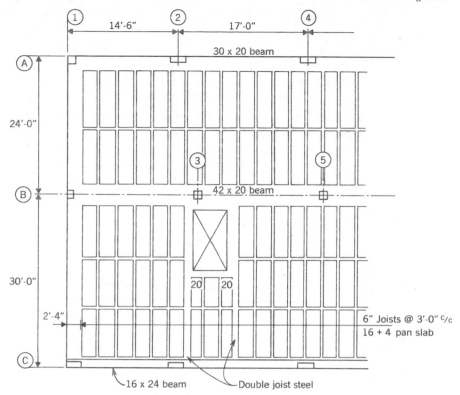

FIGURE 3.8 Layout of typical pan slab floor.

It is not economical to vary the reinforcing from joist to joist except in special circumstances (e.g., when variations in loading cannot be accommodated by cross ribs or by appropriately placed double or triple joists). In most cases, minimum reinforcement of one bar in the top and one bar in the bottom of the rib will be adequate, using bars the same size as the smallest bar in the joists or at least a #4 bar. Engineers should design the cross rib when loads are highly variable or conditions are unusual [2].

Bar Selection

Selection of bars in joists is based on the following criteria:

1. The area of steel furnished must equal or exceed the area of steel required.
2. At least one-third the area of positive moment reinforcement in simple members and one-fourth the area of positive reinforcement in continuous members must extend into the supports. Additional anchorage is required if the joists are part of the primary lateral load carrying system (Code 12.11).
3. Stress in the bars must be developed. The ACI Code assures this development by limiting the size of bar that can be used. This requirement will be met

if bar size does not exceed sizes given in Table A5.6 (see the discussion in Chapter 5).

4. Clear distance between bars must not be less than the nominal bar diameter nor 1 in., nor less than 1⅓ times the maximum size of coarse aggregate. These requirements and the required clear distance to the face of the concrete will be met if the size of bars is limited to the sizes in Table 3.1.

5. Truss bars are occasionally used in joists because they result in fewer pieces of steel to be placed and, hence, may reduce construction costs. Truss bars are more easily selected for joists than for beams. For typical interior spans, if truss bars are made one size larger than the straight bottom bars, a good balance between top and bottom bars usually results. Crack control provisions will be met if one top bar is provided in the middle of each pan.

Example 3.7   Given a 12 + 4 pan slab (12-in.-deep pans with a 4-in. slab over the pans) with 5-in. joists at 35 in. c/c, in an interior span with joist spans requiring similar reinforcement on each end, and interior exposure, select reinforcing bars and check compression stresses. Assume ^-M = 67.7'k and ^+M = 46.6'k and material strengths are f'_c = 3 ksi and f_y = 60 ksi.

Solution d = 16 − 0.75 − 0.75/2 = 14.9″

At mid-span, try $(\phi f_y j)$ = 4.4 at ρ = 0.2%

$$^+A_s = \frac{46.6}{4.4 \times 14.9} = 0.71^{□″}$$

Use 1 #5 bott.

   + 1 #6 truss, $A_s$prov. = 0.75$^{□″}$ OK

Check for one-third ^+A_s into support:

$$\frac{0.31}{0.71} = 0.44 > 0.33 \quad \text{OK}$$

Check steel percentage for selection of $(\phi f_y j)$:

$$\rho = \frac{0.71}{14.9 \times 35} = 0.14\% < 0.2\% \quad \text{OK}$$

At ends, try $(\phi f_y j)$ = 3.65 at ρ = 1.6%

$$^-A_s = \frac{67.7}{3.65 \times 14.9} = 1.24^{□″}$$

Check for concrete compression capacity:

$$\rho = \frac{1.24 \times 100}{5 \times 14.9} = 1.67\% > 1.6\% \text{ from Table A2.1} \quad \text{NA}$$

or $\quad R_c \times 10^3 = \dfrac{M}{bd^2} = \dfrac{67.7 \times 12}{5 \times 14.9^2} = 732$

$$732 > 702 \text{ from Fig. A2.3} \quad \underline{\text{NA}}$$

Therefore, use tapered end pans, $b = 5 + 5 = 10''$

In a second iteration, try $(\phi f_y j) = 4.1$ at $\rho = 0.8\%$

$$^-A_s = \dfrac{67.7}{4.1 \times 14.9} = 1.11^{\square''}$$

Use 1 #5 top bar

$\quad\quad + 2$ #6 truss bars, ^-A_sfurn. $= 1.19^{\square''}$

$\quad\quad\quad\quad\quad$ and $\rho = 0.74\% < 0.8\% \quad \underline{\text{OK}}$

PROBLEMS

1. Select reinforcement for a simple span slab 8 in. thick with $M_u = 11.3'$k/ft. and interior exposure. Use $f'_c = 3$ ksi and $f_y = 50$ ksi.
2. Design a two-span 7-in. slab with equal spans, $^+M_u = 5.9'$k/ft. and $^-M_u = 9.6'$k/ft, exterior exposure, using all straight bars. Select another reinforcement pattern using truss bars. Use $f'_c = 3$ ksi and $f_y = 60$ ksi.
3. Select reinforcement for the stair shown in Fig. 3.3 if support is provided only at the far ends of the landings (clear span $= 15'$-$6''$). Use $f'_c = 3.75$ ksi and $f_y = 40$ ksi.
4. Select reinforcement using truss bars for a three-span $16 + 4\frac{1}{2}$ pan slab with 6-in. joists at 36 in. c/c, shown in Fig. 3.9, with interior exposure. Also select reinforcement using straight bars only. Use $f'_c = 3$ ksi and $f_y = 60$ ksi.

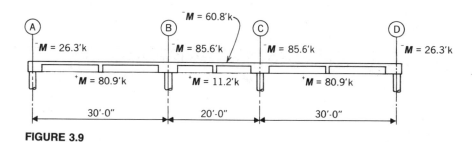

FIGURE 3.9

References

1. *CRSI Handbook*, Concrete Reinforcing Steel Institute, Chicago, 1982.
2. Paul F. Rice. "Effective Width of One-Way Monolithic Joist Construction as a Two-Way System," *Structural Bulletin*, No.8, Concrete Reinforcing Steel Institute, March 1983.

CHAPTER 4

TWO-WAY SLAB FLEXURE

4.1 OBJECTIVES

Assume that concrete outlines, including slab thickness and any drop panels (Fig. 4.2), column capitals (Fig. 4.3), and beams, have been established in preliminary design as described in Chapter 14. Methods for distributing moments in two-way slab systems is peculiar to this form of construction and no other. Therefore, moment distribution is discussed in this chapter as one step in reaching the following objectives:

1. Decide the placing sequence of reinforcing bars.
2. Select the size and number (or size and spacing) and distribution of reinforcing bars in each panel in each direction. Preferably, the reinforcing pattern selected should permit easy placement of the bars. For two-way joist systems, select the number and size of bars in each joist.
3. Select distribution of reinforcing bars across the width of the panel, especially near the column.

The length and bending of slab bars is covered in Chapter 5, Anchorage and Bar Development. Deflection, covered in Chapter 8, controls the thickness of many slabs. Adequate shear strength is crucial to the structural integrity of two-way slab systems, especially flat plates. Shear and torsion are covered in Chapter 6.

4.2 DESCRIPTION

Two-way slabs are those with significant bending in two directions. This occurs when the ratio of the long side to the short side (aspect ratio) is between one and two. Slabs with larger aspect ratios are treated as one-way slabs and are discussed in Chapter 3.

In two-way slab systems, a panel is a portion of the slab bounded by column, beam, or wall centerlines on all sides (Code 13.2.3).

Historically, two-way slabs have been divided for structural design purposes into two distinct types with markedly different behavior. One type is slabs supported by stiff, unyielding beams or walls on all four sides. These have been called "two-way slabs." The other type is slabs without beams. These have been called "flat slabs" or "flat plates," depending on whether or not there was a thickening of the slab around the column. For ease of reference, this nomenclature will be continued in this text as it is still widely used. Sozen and Siess have written an excellent summary of the historical development of two-way slab systems [1].

The distinction between two-way slabs and flat slabs is completely valid only for the two extremes, that is, for the cases cited above. When flexible beams are introduced, perhaps not on all four sides of the slab, the structural behavior lies between the limiting extremes. As the size and flexibility of the beams vary from very small and highly flexible to very large and very stiff, the structural behavior varies smoothly from that of a flat slab to that of a two-way slab.

In the first half of this century, flat slabs were widely used for industrial buildings to carry heavy or very heavy loads over what was then considered long spans. Conservative stresses were used for concrete and reinforcement. To accommodate the heavy bending and shear stresses around columns, the slab was thickened and the column enlarged at the top under the slab. The slab thickening is called a drop panel, and the column enlargement is called a column capital. The current ACI Code contains rules for proportioning drop panels (Code 13.4.7) and capitals (Code 13.1.2). (See Fig. 4.1 for an example of a flat slab, and Fig. 4.2 for a summary of drop panel requirements.)

A column capital may be square, rectangular, or round. The shape of the capital usually matches the shape of the column for simplicity of formwork. The ACI Code has no limits on the size of a column capital, except that no portion of the capital or bracket that lies outside the largest right circular cone, right pyramid, or tapered wedge that fits within the concrete outlines may be considered for structural purposes. The angle between column axis and slope of the cone, pyramid, or wedge must not exceed 45° (see Fig. 4.3).

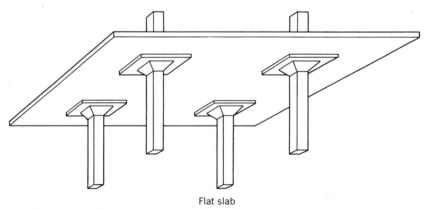

Flat slab

FIGURE 4.1 Perspective view of the underside of a flat slab with drop panels, column capitals, and square columns.

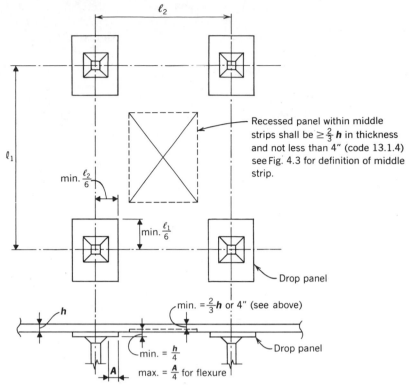

Recessed panel within middle strips shall be $\geq \frac{2}{3}h$ in thickness and not less than 4" (code 13.1.4) see Fig. 4.3 for definition of middle strip.

min. $\frac{\ell_2}{6}$

min. $\frac{\ell_1}{6}$

Drop panel

min. $= \frac{2}{3}h$ or 4" (see above)

h

min. $= \frac{h}{4}$

Drop panel

A max. $= \frac{A}{4}$ for flexure

FIGURE 4.2   Summary of drop panel requirements.

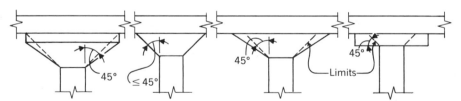

FIGURE 4.3   Limits on column capitals for structural purposes (see Code 13.1.2).

To simplify the placing of reinforcement, it has long been the practice to divide a two-way slab system into column strips and middle strips. A column strip is a design strip with a width on each side of a column centerline equal to one quarter of the transverse span length or the span length in the direction that moments are being determined, whichever is less. A middle strip is a design strip bounded by two column strips (Code 13.2) (see Fig. 4.4).

Since World War II, many buildings have been built for light loading, especially for residential occupancy. Drop panels and column capitals are no longer needed. When they are eliminated, the resulting structure is called a flat plate to distinguish it from a flat slab (see Fig. 4.5 for an example of a flat plate structure).

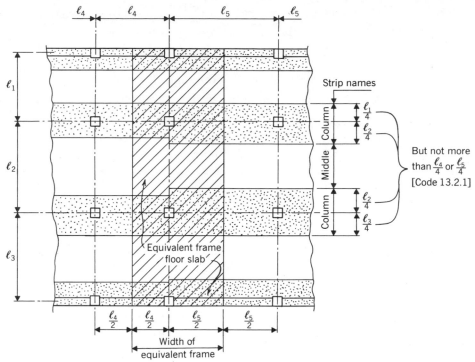

FIGURE 4.4 Two-way slab definitions.

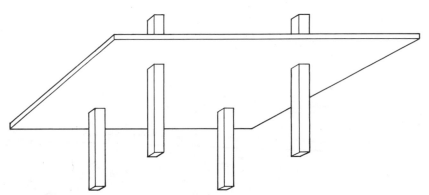

FIGURE 4.5 Perspective view of the underside of a flat plate without drop panels or column capitals.

Although flat slabs are far less common today than earlier in this century, there are still occasions for their use, despite their higher cost of formwork. Following are some of the uses of drop panels:

1. To increase the shear strength of the slab with a minimum of interference with work of the mechanical trades installed between the ceiling and floor above.

2. To increase the negative moment capacity of the slab.

3. To stiffen the slab and reduce its deflection.

Following are uses of column capitals:

1. To increase the shear strength of the slab.

2. To reduce the moment in the slab by reducing the clear span.

Historically, drop panels and column capitals have been used together on all columns and slabs in a flat slab structure. However, this need not be so today. Either drop panels or column capitals can be used alone. Either or both can be used at columns in a structure where needed to meet required stress conditions but need not be used at other columns.

When spans are long, and especially if loads are heavy, thick slabs are needed to resist moments and shears and to minimize deflection. To reduce the weight of the structure and eliminate some unnecessary concrete, square pans or domes are used to produce a coffered soffit in the slab. The result is a two-way joist system, popularly called a waffle slab. Some architects believe that this makes an attractive ceiling and specify its use for that reason alone. (See Fig. 4.6 for a typical waffle slab.)

Waffle slabs must use a drop panel or a solid slab acting as a drop panel around the columns because the joists alone are inadequate to resist all shear and bending stresses at the face of columns. Size and shape of drop panels depend on the ingenuity of the engineer in satisfying shear and bending stresses at all points. In addition, beams can be formed by eliminating a row or two of domes on the column centerlines. This is especially appropriate if the column spacing is not evenly divisible by the dome spacing. A beam as wide as the fractional spacing can maintain an attractive ceiling and keep a joist or a dome centered on each column.

Another option for long spans is the flat-banded slab shown in Fig. 4.7. Beams,

FIGURE 4.6 Typical waffle slab. *(a)* View of underside of a completed slab.

FIGURE 4.6 (*continued*) (*b*) View of formwork before placing concrete, with some reinforcement in place. (Photographs provided by The Ceco Corp.)

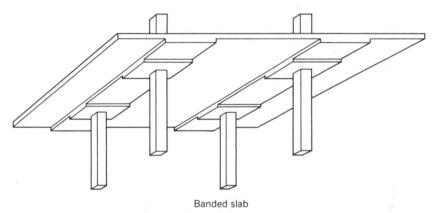

Banded slab

FIGURE 4.7 Perspective view of the underside of a banded slab with bands in one direction only, plus drop panels.

or bands, are provided one and one-half to two times as deep as the slab and as wide as necessary, up to the width of the column strip. Concrete and, hence, dead weight, is minimized. Minimum depth is provided while meeting deflection limitations. A drop panel can be used if necessary to further reduce deflection and increase moment capacity. However, use of this system is limited to projects where the high cost of formwork can be amortized over many uses.

4.3 STRUCTURAL BEHAVIOR

The structural engineer's most important responsibility is to design a structure that does not fail under any required loading condition. The central problem in two-way slab systems to assure safety is the connection between slab and column, which is a small area or portion of the structure. All loads supported by the slab converge on the column. Moments are usually the largest over the column, and shear stresses also reach a peak there. Because the connection between slab and column is largely a problem in shear and torsion, it is covered in Chapter 7. Engineers are well advised to give this small but important area their primary attention.

Some research engineers who have tested flat slabs and flat plates to destruction have commented informally that such structures seldom fail in flexure and that they usually fail in shear almost regardless of the amount of flexural reinforcement. The distribution of flexural reinforcement seems to have little effect on the ultimate flexural capacity. Other research engineers indicate that near maximum capacity the distribution of moments is controlled largely by the distribution of steel adopted by the designer [2]. The problem, then, in distributing flexural reinforcement is to assure proper behavior of the slab at service loads. That is, deflection should be limited, and cracks in slabs, beams, and columns should not be obtrusive. This is accomplished by distributing reinforcement approximately in proportion to moments that occur in the slab system at service loads. ACI Code procedures are intended to determine the distribution of moments that would occur in a linearly elastic structure of the same dimensions.*

Two-way slabs are highly redundant in the sense that concentrated loads are carried by a larger portion of the slab than the strips immediately under the load. Also, in a multipaneled structure, one panel loaded beyond capacity cannot fail without receiving some support from adjacent panels. Partly for this reason, the ACI Code before 1971 required that flat slabs be designed for only 72% to 80% of static moment ($w\ell^2/8$). Currently, the ACI Code requires that slabs resist 100% of static moment, based on the clear span.

Some two-way slab systems are not redundant in the sense that a shear failure around one column may allow the surrounding portion of the slab, or even the entire slab system, to drop unless provision is made to support the slab after shear failure. Without such a provision, when shear failure occurs at one column, load on nearby columns will be higher and may lead to shear failures at these columns also. When this happens, a progressive collapse of the entire slab system could occur with catastrophic results. A progressive collapse can progress both horizontally and vertically through a structure. This subject is discussed further in Section 4.8.

Flat plates supported by columns at the four corners deflect in double curvature, as shown in Fig. 4.8(a). Curvature and moment are greatest along the column centerlines. Along a line at mid-panel, curvature is less and moments are smaller. By contrast, a two-way slab with unyielding supports along the full length of all four sides deflects in a dish-shaped double curvature, as shown in Fig. 4.8(b). Curvature and moment are greatest at mid-span and gradually reduce to zero (or to the curvature of the stiff beam) at the edges. Note that this is the opposite of a flat plate on columns. For clarity of presentation, the slabs in Fig. 4.8 have been shown as

---

*Because concrete is not linearly elastic, a high degree of refinement in distribution of moments in two-way slab systems without interior beams seems unnecessary. It is far more important to provide total moment resistance capacity, somewhere in positive and negative moment regions, at least equal to the static moment.

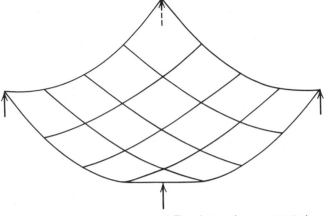

(*a*) Flat plate, column supported;

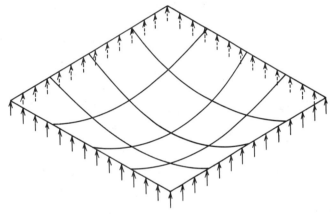

(*b*) two-way slab, edge supported.

FIGURE 4.8   Simple span, two-way slab system deflections.

simple spans without negative moment along the support lines. With multiple spans, negative moment and reverse curvature would occur at the interior supports. Slabs connected to rigid supports at the exterior edge would experience negative moment also.

The longitudinal distribution of moment in a flat plate is shown in Fig. 4.9(*a*). Along the column centerline, the moment curve has the familiar shape of a moment curve in a beam. At the support (Note 1), the size of the rounded peak depends on the width of the support. With a large column capital, the width can be quite large. At mid-span, the moment curve is much shallower with no distinct peak at the support line (Note 2).

The lateral distribution of peak moments at column centerline and at mid-span are shown in Fig. 4.9(*b*). Also shown are the distribution of moments assumed by the ACI Code to make layout of reinforcement practical. Lateral distribution of negative moment in the vicinity of the column is also affected by the width of the column

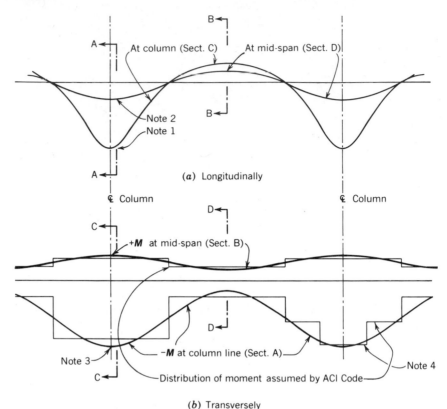

FIGURE 4.9 Approximate distribution of moment in one direction in slabs supported on columns at corners (See text for notes 1, 2, 3, and 4.)

(Note 3). If large moments are transferred between column and slab, further concentration of steel at the column is required (Note 4).

The longitudinal distribution of moment in a two-way slab on unyielding supports on all four sides is shown in Fig. 4.10(a). The maximum moments are on a line at mid-span. As with flat plates, the shape of the peak moment curve at the supports is influenced by the width of the supports (Note 5). On a line near the support, the moments approach zero.

The lateral distribution of peak moments at column centerline and at mid-span, as well as the distribution of moments assumed by the ACI Code, are shown in Fig. 4.10(b). The distribution curves for maximum moments near the support have been simplified in that there is actually significant moment at 45° to the principal axes near the corners of the slab (Note 6). The Code requires that slabs be reinforced for these moments (Code 13.4.6).

In flat plates, the slab must carry 100% of the load in each direction. In two-way slabs, the slab carries less than 50% of the load in each direction and the beams carry the remainder to the columns. As a result, mid-span moments in middle strips in both systems are similar, but moments in column strips for flat slabs are much higher than moments in column strips in two-way slabs.

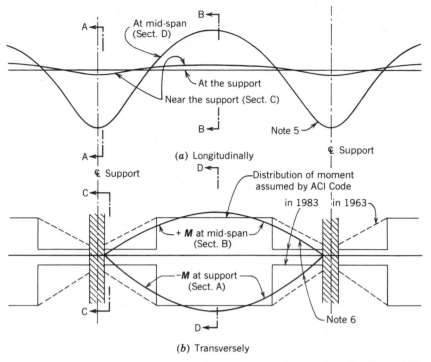

FIGURE 4.10 Approximate distribution of moment in one direction in slabs with unyielding supports on four sides. (See text for notes 5 and 6.)

Distribution of moments in slab systems with flexible beams falls between the two extremes just described above and depends on the flexibility of slabs, beams, and columns in the system. The ACI Code treats both two-way slabs and flat slabs/flat plates in the same procedure and provides for a smooth transition between the two extremes.

In concept, the design of two-way slab systems involves dividing the entire structure into a series of frames in each direction by making vertical cuts midway between columns. The slabs, including any beams present, form the horizontal members, and the columns form the vertical members of each rigid frame. These rigid frames are called "equivalent frames." Equivalent frames can be analyzed by one of several methods described in more detail in the following sections.

Normally, the beam and column members of a rigid frame are about the same width, and the joint between them is assumed to be rigid. That is, the angle between beam and column is assumed to remain constant under load. By contrast, in an equivalent frame, the slab is normally much wider than the column, and the assumption of joint rigidity is not valid except for that portion of the slab framing directly into the column. Other portions of the slab must transmit moment to the column by torsion in the slab or beam perpendicular to the slab span for which moments are being considered. Such a perpendicular slab or beam must also frame directly into the column. Thus, the greater the distance of a slab strip from a column and the greater the torsional flexibility of the transverse member, the greater the rotation at the end of the slab strip and the lower the moment transferred between the slab and column. Consideration of torsional rotation of perpendicular members at a joint introduces an additional consideration in frame analysis of two-way slab systems not normally present in rigid frame analysis.

4.4. DESIGN PROCEDURES

The basic design procedure of a two-way slab system encompasses five steps.

1. Determine moments at critical sections in each direction, normally the negative moments at supports and positive moment near mid-span.
2. Distribute moments transversely at critical sections to column and middle strips, and, if beams are used in the column strip, distribute column strip moments between slab and beam.
3. Determine the area of steel required in the slab at critical sections for column and middle strips.
4. Select reinforcing bars for the slab and concentrate bars near the column, if necessary.
5. Design beams, if any, using procedures discussed in Chapters 2, 5, and 6.

The four procedures described in the following sections are concerned primarily with steps 1 and 2. The first two steps for Procedure 1 are shown in Fig. 4.11 as a flowchart. Steps 1 and 2 are similar for other procedures.

In step 2, engineers may consider distributing the entire negative moment to the column strip in order to simplify placement of top bars. If the negative moment that would normally be allocated to the middle strip would cause flexural stresses that are less than the flexural strength of plain concrete, if the structure is in a noncorrosive environment, and if the appearance of cracks is not objectionable, there is little risk in omitting top steel in the middle strip. Tests indicate that performance of the

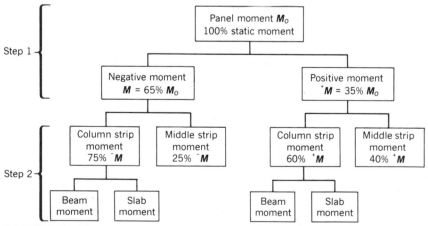

FIGURE 4.11 Flowchart for distribution of moments in a two-way slab system, using Procedure 1 showing, as an example, the numerical distribution in an interior span flat plate without beams.

structure is satisfactory [3]. Although the specific requirements of the ACI Code for design of two-way slabs do not permit omission of middle strip top bars, it could be approved as a special system, the adequacy of which has been shown by successful use, analysis, and test (Code 1.4).

Before continuing with step 3, the engineer must decide the placement order for the reinforcing bars, as this determines the effective depth of the reinforcement. It is common practice to place bars in the direction of the long span in the outer layer (bottom layer in the bottom of the slab and top layer in the top of the slab) to gain the greatest effective depth because the long span has the largest moment. If the panel is nearly square, the average effective depth is usually used. Reinforcement in one direction will be slightly overstressed, and reinforcement in the other direction will be slightly understressed, but capacity of the slab will remain the same. If a building has rectangular panels with different orientations, the engineer must decide which orientation to favor, as the placement order should be the same throughout the structure.

After the effective depth is determined, the area of steel can be computed in the same manner as for one-way slabs (see Chapter 3), including the requirements for minimum reinforcement for either solid slabs or pan slabs.

Bar selection in step 4 is based on the following criteria:

1. The area of bars furnished must equal or exceed the area of steel required.

2. Bars can be spaced no farther than two times the slab thickness (Code 13.4.2) (not three times the slab thickness as in one-way slabs), nor 18 in. (Code 7.6.5). Crack control provisions do not apply to two-way slabs systems, as the requirements just mentioned are more stringent.

3. Stress in the bars must be developed. This requirement will be met if the bar size does not exceed sizes given in Table A5.6.

4. As few bars at as wide a spacing as possible should be used for economy and for the same reasons as cited in Chapter 3.

5. For bar-placing simplicity, arrange bottom bars at a uniform spacing throughout the slab with extra bars in column strips, if necessary.

6. Top bars should be spaced far enough apart to avoid interference with column verticals and to permit easy placement of concrete between the bars even though the minimum spacing is the same as for beams and one-way slabs (Section 1.8).

7. Truss bars are used infrequently today as they require a precise and intricate placing procedure with as many as 12 separate steps to allow placing bars in the proper layer without threading some bars beneath previously placed bars. Additional steps are required for placing bar support chairs in the proper order. Even then it is possible for some bar interferences to occur.

8. Bottom bars, #4 in size or larger, are advisable to resist deformation and displacement by foot traffic during construction. For top bars, #5 size and larger are desirable because small bars several inches above formwork are easily bent by foot traffic. Alternatively, for light steel requirements, consider using welded wire fabric (WWF).

If WWF is considered, the engineer should consult manufacturers before selecting the wire size and spacing. Mat size influences the economy of production runs and shipping. Weight of mats should be considered for job handling. If a panel size is too large to be covered with one WWF mat, mats must be lapped. The placement order and method of lapping may determine the effective depth of the steel. WWF mats can be supplemented with bars in column strips to meet required reinforcement.

When unbalanced moment must be transferred between slab and column, a portion will be transferred by flexure and the remainder must be transferred by torsion in the slab surrounding the column. Moment transferred by torsion causes an eccentric shear in the slab around the column (see Section 7.3 on torsional shear in slabs). The fraction of the unbalanced moment to be carried by flexure γ_f is given by an empirical equation (Code 13.3.3),

$$\gamma_f = \frac{1}{1 + \dfrac{2}{3}\sqrt{\dfrac{c_1 + d}{c_2 + d}}} \tag{4.1}$$

where c_1 = size of rectangular or equivalent rectangular column, capital, or bracket measured in the direction of the span for which moments are being determined

c_2 = size of rectangular or equivalent rectangular column, capital, or bracket measured transverse to the direction of the span for which moments are being determined

and d = distance from extreme compression fiber to centroid of longitudinal tension reinforcement.

Flexural reinforcement to resist unbalanced moment must be concentrated over the column between lines that are one and one-half slab or drop panel thickness (1.5h) outside opposite faces of the column or capital, as in Fig. 4.12. Equation 4.1 is evaluated in Table A4.1 for common ranges of values. The aspect ratio of the critical section $(c_1 + d) / (c_2 + d)$ is limited to three for reasons discussed in Section 7.3.

Research indicates that negative moment reinforcement is most effective when

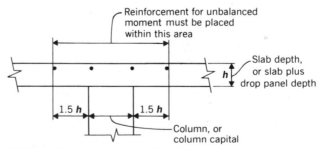

FIGURE 4.12   Concentration of reinforcement over column to resist unbalanced moment by flexure.

it is located at or near the column as just described. Even when concentration of reinforcement near the column is not required by Code, some engineers prefer to space top bars at the maximum spacing across the width of the column strip and concentrate the remaining required bars at or near the column.

Columns or supporting elements must also resist the unbalanced slab moments, whether moments are transferred by flexure or by torsion, or whether unbalanced loads arise from gravity loads or lateral loads.

In addition to column strip moments allocated to beams in accordance with one of the following design procedures, beams must be proportioned to resist moments caused by concentrated and linear loads applied directly on the beams. This would include the weight of the beam itself, as well as walls and other construction superimposed upon it. Such loads are not considered in the design of the slab system.

Three methods for designing a two-way slab system are allowed by the ACI Code. If conditions are the same for all three methods, they should yield similar results.

The first, Direct Design Method (DDM), makes simplifying assumptions for the sake of reducing design effort. To qualify for design by DDM, a slab must meet certain limitations for which the simplifying assumptions are valid. Requirements for DDM are covered in ACI Code Section 13.6. DDM is discussed, and two procedures based on DDM are presented in Section 4.5 of this text.

The second, Equivalent Frame Method (EFM), is more rigorous than DDM and usually takes more design effort. A slab not meeting the limitations for DDM must be designed by EFM or by the third method. Requirements for EFM are covered in ACI Code Section 13.7. EFM is discussed, and a procedure based on EFM is presented in Section 4.6 of this text.

The third method is any procedure satisfying conditions of equilibrium and geometrical compatibility if the required strength at every section and all serviceability conditions are met (Code 13.3.1). This permits the engineer to design a slab through the combined use of classic solutions based on a linearly elastic continuum, numerical solutions based on discrete elements, or yield line analysis. The third method is especially helpful for slabs with random column spacing that cannot be organized into equivalent frames. The third method also permits approximate solutions given in earlier editions of the ACI Code, one of which is presented in Section 4.7 of this text.

Regardless of the method or procedure used for design, the ACI Code explicitly permits openings up to a certain size in slabs without beams, without special analysis for flexure (Code 13.5.2). The size of openings in intersecting middle strips is unlimited. The size of openings in intersecting column strips is limited to one eighth of the width of the column strip in either span. Openings in intersecting column and middle strips must not interrupt more than one quarter of the reinforcement in either direction. Thus, the maximum size of opening is one quarter of the column or middle strip width, plus two bar spacings, less 2 in. for cover on the bars. Limits on such openings are summarized in Fig. 4.13.

If reinforcement is interrupted by an opening, it must be replaced, one half on each side of the opening. The total amount of reinforcement required for the panel without the opening must be maintained. Larger openings and openings in slabs

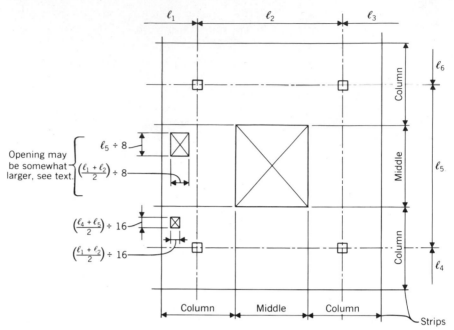

FIGURE 4.13 Limits on openings in slabs without beams and without special analysis (Code 13.5.2).

with beams may be provided but require a special analysis (Code 13.5.1). Requirements for shear and torsion discussed in Chapter 7 must be met regardless of size and location of openings in slab.

Table 4.1 is a guide for selection of two-way slab system design procedures presented in this text. For comparison purposes, Table 4.2 gives a summary of the first two design steps for Procedure 1 and the first three design steps for Procedure 2. The procedures are discussed in detail in Section 4.5.

TABLE 4.1 Guide for Selection of Two-Way Slab System Design Procedures

Procedure No.	Description[a]	Beams[b] Ext.	Int.	Best Application
1	DDM	O	O	Slabs with interior beams and approximately uniform panels
2	DDM-short	O	NP	Slabs without interior beams Preliminary design
3	EFM	O	O	Slabs not meeting limits for DDM Systems resisting lateral loads
4	ACI 318-63/2	R	R	Slabs with stiff beams on all four sides

[a] DDM = Direct Design Method
   EFM = Equivalent Frame Method

[b] O = Optional
   NP = Not permitted Ext. = Exterior exposure
   R = Required Int. = Interior exposure

TABLE 4.2   Summary of DDM Procedures[a]

Step 1	Step 2	Step 3
Basic Design Procedure		
1. Compute $+M$ and $-M$ at critical sections.	2. Distribute moment to column and middle strips and to beams.	
Procedure 1—DDM		
1.1 Compute M_o.	2.1 Compute EI of slab and beam and ratio α_1. Compute β_t.	
1.2 Compute parameters and distribute M_o to $+M_u$ and $-M_u$.	2.2 Distribute moment to column and middle strips.	
1.3 If $\beta_a < 2.0$, compute parameters and increase $+M$.	2.3 Distribute column strip moment between beam and slab.	
Procedure 2 DDM—Short[b]		
1.1 Compute $+M_u$/ft in middle strip.	2.1 Compute $+M_u$/ft in column strip.	3.1 Compute $-M_u$ for column and middle strips.
1.2 If $\beta_a < 2.0$, increase $+M_u$.	2.2 If $\beta_a < 2.0$, increase $+M_u$.	3.2 Compute $-A_s$.
1.3 Compute $+A_s$.	2.3 Compute $+A_s$.	3.3 Select bars.
1.4 Select grid bars.	2.4 Select added bars.	3.4 Concentrate bars at column.

[a] First two steps of Procedure 1 and three steps of Procedure 2.

[b] Procedure 2 combines the first four steps of the basic procedure for each critical section in each strip.

4.5 DIRECT DESIGN METHOD (DDM)

A two-way slab system that meets certain limitations summarized in Fig. 4.14 may be designed using DDM. With these limitations, simplifying assumptions can be made that reduce the amount of design time without increasing construction cost. The DDM uses empirical procedures and distribution factors derived for a range of conditions meeting these limitations and may not be appropriate for slab systems outside the limitations. Engineers may use DDM even if the structure does not fit the limitations if it can be shown by analysis that the particular limitation does not apply. For example, if the load is applied uniformly on all spans as in a water reservoir, live load is not limited.

Two procedures are presented below that follow the DDM. Within the limitations of DDM, Procedure 1 is the most flexible and uses ACI Code requirements without modification. It is most appropriate for slab systems with interior beams or drop panels. Procedure 2 makes simplifying assumptions in order to reduce design time for slab systems without interior beams or drop panels.

Procedure 2 proceeds directly from a single calculation of moment to the computation of reinforcement area required and the selection of bars. Visualizing practical, systematic patterns of reinforcement is easiest with Procedure 2 but may

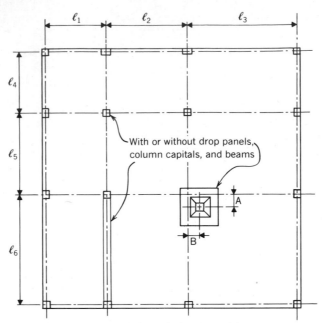

1. Three continuous spans in each direction, minimum.
2. Rectangular panels with aspect ratio ≤ 2.0. e.g., $\frac{\ell_6}{\ell_1} \le 2.0$.
3. Span lengths differ by $\frac{1}{3}$ or less of longer span. e.g., $\frac{\ell_5}{\ell_6} \ge 0.67$.
4. Column offset a maximum at 10%. e.g., $A \le 0.1\, \ell_6$.

5A. Gravity loads only, uniformly distributed, not lateral loads.
5B. Live load $\le 3 \times$ dead load.

6. Beam stiffness: $0.2 \le \frac{\alpha_1 \ell_2^2}{\alpha_2 \ell_1^2} \le 5.0$.
7. No moment redistribution as permitted by ACI Code Section 8.4.

FIGURE 4.14 Flat slab limitations for use of Direct Design Method (Code 13.6.1).

become unnecessarily intricate or conservative if there are many special conditions such as offset columns, varying bay sizes, and openings in slabs. As one can see from the examples given later, Procedure 2 uses much less design time than Procedure 1. Furthermore, the reduced volume of computations gives less opportunity for arithmetical errors or errors in logic.

Procedure 1 (DDM)

The five steps of the basic design procedure outlined in Section 4.4 can be further defined for Procedure 1 as follows:

Step 1.1   Compute total factored static moment M_o for all spans in each direction.

$$M_o = \frac{w_u \ell_2 \ell_n^2}{8}$$

(4.2)

If $\ell_n < 0.65\ell_1$, a span $\ell_n = 0.65\ell_1$ should be used.

where w_u = total factored load per unit area

ℓ_n = length of clear span in direction that moments are being determined (measured face to face of supports)

ℓ_1 = length of the span in the direction that moments are being determined (measured center to center of supports)

ℓ_2 = length of the span transverse to ℓ_1 (measured center to center of supports).

If a column or supporting member does not have a rectangular cross section, it is treated as a square support with equal area for purposes of determining moments.

Step 1.2   Distribute the total static moment M_o to negative factored moment ^-M_u and positive factored moment ^+M_u using Table A4.2 (Code 13.6.3).

In Table A4.2, column 1 is for the case of a slab simply resting on top of a wall but not tied to it or restrained by being embedded in the wall. Column 5 represents a slab fully restrained by a concrete wall or other construction at least as rigid as an interior support. Column 2 applies to two-way slab systems with beams on all four sides. Columns 3 and 4 apply to flat slabs and flat plates without interior beams and without or with edge (spandrel) beams, respectively.

Negative and positive factored moments may be modified or redistributed at this step, but it is more useful to consider redistribution later in order to simplify reinforcement patterns or minimize quantity of steel. Redistribution procedures are discussed in Step 4.

Step 1.3   To provide for pattern loading of live load, the positive moment may have to be increased if columns are not stiff enough to provide adequate restraint to the slab. If $\beta_a < 2.0$, where β_a is the ratio of dead load per unit area to live load per unit area (in each case without load factors), compute α_c for column-slab joints and compare the value to values of α_{min} in Table A4.3 (Code 13.6.10). The factor α_c equals the ratio of flexural stiffness of columns K_c above and below the slab to combined flexural stiffness of the slabs K_s and beams K_b at a joint taken in the direction of the span for which moments are being determined. Flexural stiffness K is the moment per unit of rotation.

$$\alpha_c = \frac{\Sigma K_c}{\Sigma(K_s + K_b)} \tag{4.3}$$

The ratio α_{min} is the minimum α_c for which positive moments need not be increased. If α_c is less than α_{min}, compute the multiplier δ_s from

$$\delta_s = 1 + \frac{2 - \beta_a}{4 + \beta_a}\left(1 - \frac{\alpha_c}{\alpha_{min}}\right) \tag{4.4}$$

Alternatively, the columns may be conservatively assumed to have zero stiffness and the multiplier δ_s taken from Table A4.4.

Increase positive moment ^+M by multiplier δ_s.

For many buildings with residential occupancy, the ratio β_a will be greater than 2.0, and positive moments need not be increased. When $\beta_a < 2.0$, a short procedure is to increase the positive moment using Eq. 4.4 or using values in Table A4.4 by assuming $\alpha_c/\alpha_{min} = 0$. In many cases, the increase in construction cost will be less than the savings in computational labor.

Step 2.1   If beams parallel to the slab span are used, compute the flexural stiffness ***EI*** of both beams and slabs. To reduce the computation labor, Fig. A4.1 for spandrel beams and Fig. A4.2 for interior beams have been prepared. These figures give the ratio of moment of inertia for a T-beam to the moment of inertia of a rectangular beam β_{tb} in terms of two parameters: the ratio of flange thickness to beam depth h_f/h and the ratio of flange thickness to beam web width h_f/b_w. This is a special case of T-beams, because the width of the flange of beams in two-way slabs is limited to that portion of the slab on each side of the beam extending a distance equal to the projection of the beam above or below the slab, whichever is greater but not greater than four times the slab thickness (Code 13.2.4). Normally, accurate selection of factor β_{tb} is not required because it has little, or sometimes no, effect on the final selection of reinforcement.

Then compute α_1, the ratio of flexural stiffness of beam section to flexural stiffness of the slab bounded laterally by centerlines of adjacent panels (if any) on each side of the beam. Both the beam and the slab are spanning in the direction for which moments are being determined.

$$\alpha_1 = \frac{EI/\ell \text{ for beam}}{EI/\ell \text{ for slab}}$$

Also compute the ratio β_t of torsional stiffness of edge beam section to flexural stiffness of a width of slab equal to span length of the beam, center to center of supports. The slab is spanning in the direction for which moments are being determined, and the edge beam is spanning in a perpendicular direction.

$$\beta_t = \frac{C}{2I_s} \tag{4.5}$$

where I_s = gross moment of inertia of the slab using a width equal to the span length of the edge beam center to center of supports

and C = a cross-sectional constant to define torsional properties.

$$C = \Sigma \left(1 - \frac{0.63x}{y} \right) \frac{x^3 y}{3} \tag{4.6}$$

where x = the shorter overall dimension of the rectangular part of the cross section

and y = the longer overall dimension of the rectangular part of the cross section.

(See Fig. 4.15 for definitions of dimensions used in computing C.)

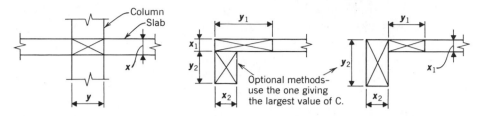

FIGURE 4.15   Dimensions of T-beams and slabs used in computing constant **C** to define torsional properties.

If a panel has beams between supporting columns on all sides, the relative stiffness of beams in two perpendicular directions must not be less than 0.2 or greater than 5.0. The relative stiffness is given by

$$\frac{\alpha_1 \ell_2^2}{\alpha_2 \ell_1^2}$$

where subscripts 1 indicate the span in one direction and subscripts 2 indicate the span in the other direction. If the beams do not meet this limitation, a more precise analysis must be made and the design costs may exceed any savings in construction costs realized. The purpose of this limitation is to ensure that the two-way slab system will perform approximately as anticipated when the empirical distribution factors of Table A4.5 were developed.

Step 2.2   Distribute total negative factored moment and total positive factored moment computed in step 1 to the column strip using Table A4.5 (Code 13.6.4). Distribute that portion of negative and positive factored moments not resisted by the column strip to the half middle strips on each side. Each middle strip must resist the sum of moments assigned to its two half middle strips. When little or no moment is assigned to a half middle strip because it is adjacent to and parallel to an edge with stiff supports (a wall or deep beam), the middle strip must be proportioned for twice the moment in the other half of the middle strip.

Step 2.3   If a beam parallel to the slab span is present, distribute 85% of the moments in the column strip to the beam if $(\alpha_1 \ell_2/\ell_1) \geq 1.0$. For values of $(\alpha_1 \ell_2/\ell_1)$ between 1.0 and 0, the proportion of moments distributed to beams must vary linearly between 85% and 0. As noted in Section 4.4, beams must also resist all moments caused by concentrated or linear loads applied directly to the beam.

At exterior edges, columns and supporting elements above and below the slab must be designed to resist the moment in the slab at the edge.

At an interior support, columns above and below a slab must resist the moment given by Eq. 4.7 in direct proportion to their stiffness, unless a general analysis is made.

$$M = 0.07[(w_d + 0.5w_\ell)\ell_2\ell_n^2 - w_d'\ell_2(\ell_n')^2] \tag{4.7}$$

where w_d', ℓ_2, and ℓ_n refer to the shorter span, and w_d, w_ℓ, and M are factored unit dead and live loads and factored moments, respectively.

Equation 4.7 allows for unbalanced moment caused by 50% of the factored live load on the longer span with no live load on the shorter span (Code 13.6.9). In most cases for nearly equal slab spans, the moment computed by Eq. 4.7 will not control and may be conservatively disregarded (see Chapter 9, Columns).

Step 3   Determine the steel area required (see the basic design procedure in Section 4.4).

Step 4   Select reinforcement as outlined in the discussion on the basic design procedure in Section 4.4. In this step, some redistribution of moment may be desirable to facilitate practical reinforcing bar patterns. Negative and positive factored moments may be modified by up to 10%, provided the total static moment for a panel in the direction considered is not less than that required by Eq. 4.2 (Code 13.6.7). It is best to redistribute moments from areas of high moments to areas of low moments (normally, negative moment to positive moment sections) because high moments lead to cracking, lower moment of inertia, and some stress relief. Increasing positive moment is especially appropriate in end spans where negative moments are resisted in part by torsion in the slab and spandrel beam at the exterior support.

Step 5   Apportion shear to beams as follows. Beams with $(\alpha_1 \ell_2 / \ell_1)$ equal to or greater than 1.0 shall be proportioned to resist shear caused by factored loads on tributary areas bounded by 45° lines drawn from the corners of the panels and the centerlines of the adjacent panels parallel to the long sides. Beams with $(\alpha_1 \ell_2 / \ell_1)$ less than 1.0 may be proportioned to resist shear obtained by linear interpolation, assuming beams carry no load at $\alpha = 0$ (see Fig. 4.16).

To these shears and the moments determined in Step 2.3 above, add the shears and moments from concentrated and linear loads superimposed directly on the beams and design the beams using the procedures in Chapters 2, 5, and 6.

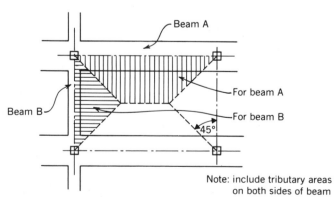

FIGURE 4.16   Tributary area for shear.

Procedure 2 (DDM—Short)

Procedure 2 separates column and middle strips, positive and negative moment regions, and completes the first four steps of the basic design procedure individually on each of the four critical section strips before proceeding to the next critical section strip. The order is generally as follows:

1. Positive moment, middle strip
2. Positive moment, column strip
3. Negative moment, column strip
4. Negative moment, middle strip

The steps for procedure 2 are further defined as follows:

Step 1   The purpose of this step is to select a grid of bars uniformly spaced in the bottom of the slab, to resist middle strip positive moment. The grid should extend across the full width and length of the slab. If any bars are interrupted by an opening, they must be replaced by other bars, one-half on each side of the opening.

Step 1.1   Compute the total factored static moment M_{oo} for all spans in each direction, using Eq. 4.8. Next, compute the positive factored moment per foot of width in the middle strip, using factors from Table A4.6. Moments are determined at each critical section (positive and negative) for each strip (column and middle) by multiplying the static moment M_{oo} by coefficients from Table A4.6. Moment M_{oo} is the static moment for a unit width of slab, usually 1 ft wide.

$$M_{oo} = \frac{w_u \ell_n^2}{8} \qquad (4.8)$$

The coefficients in Table A4.6 have been determined by combining the coefficients in Tables A4.2 and A4.5 for the case of no interior beams. The computations are summarized in Table 4.3. A ratio of two is normal for panel width to column strip width. Although this ratio could be larger than two in some cases, two is conservative if a column strip width equal to half the panel width is assumed in computing the area of steel required and if the additional column strip steel computed in Step 2 is concentrated within the actual width of the column strip.

Step 1.2   If the ratio of unfactored dead load per unit area to unfactored live load per unit area β_a is less than 2.0, increase the positive moment by multiplier δ_s from Eq. 4.4 or from Table A4.4, as in Step 1.3 of Procedure 1.

Step 1.3   Compute the area of reinforcement required and increase it if necessary to meet requirements for minimum reinforcement. (For solid slabs, see the discussion on minimum reinforcement in Section 3.2. For waffle slabs, see Section 2.2.)

TABLE 4.3 Computation for Table A4.6

		Values in Tables		
Spans		A4.2	A4.5	A4.6

Interior Spans

Column strip,	^+M =	$2 \times 0.35 \times 0.60 = 0.42$
	Int. ^-M =	$2 \times 0.65 \times 0.75 = 0.98$
Middle strip,	^+M =	$2 \times 0.35 \times 0.40 = 0.28$
	Int. ^-M =	$2 \times 0.65 \times 0.25 = 0.33$

End Span, Restrained

Column strip,	^+M	
Without beams =		$2 \times 0.52 \times 0.60 = 0.62$
With beams =		$2 \times 0.50 \times 0.60 = 0.60$
Int. ^-M =		$2 \times 0.70 \times 0.75 = 1.05$
Ext. ^-M		
Without beams =		$2 \times 0.26 \times 1.00 = 0.52$
With beams =		$2 \times 0.30 \times 0.75 = 0.45$
Middle strip,	^+M	
Without beams =		$2 \times 0.52 \times 0.40 = 0.42$
With beams =		$2 \times 0.50 \times 0.40 = 0.40$
Int. ^-M =		$2 \times 0.70 \times 0.25 = 0.35$
Ext. ^-M		
Without beams =		$2 \times 0.26 \times 0 \quad = 0$
With beams =		$2 \times 0.30 \times 0.25 = 0.15$

End Span, Unrestrained

Column strip,	^+M =	$2 \times 0.65 \times 0.60 = 0.76$
	Int. ^-M =	$2 \times 0.75 \times 0.75 = 1.13$
	Ext. ^-M =	$2 \times 0 \quad \times 1.00 = 0$
Middle strip,	^+M =	$2 \times 0.63 \times 0.40 = 0.50$
	Int. ^-M =	$2 \times 0.75 \times 0.25 = 0.38$
	Ext. ^-M =	$2 \times 0 \quad \times 0.25 = 0$

Step 1.4 Select a grid of bottom bars of uniform size and spacing for all panels requiring the same (or similar) areas of steel.

Step 2 The purpose of this step is to select additional bottom bars to be placed in the column strip. If any bars are interrupted by an opening, they must be replaced by other bars, one-half on each side of the opening.

Step 2.1 Compute the positive factored moment per foot in the column strip using M_{oo} and factors from Table A4.6.

Step 2.2 If the ratio of unfactored dead load per unit area to unfactored live load per unit area β_a is less than 2.0, increase the positive moment by multiplier δ_s from Eq. 4.4 or from Table A4.4, as in Step 1.3 of Procedure 1.

Step 2.3   Compute the area of reinforcement required in the column strip and then the area required in addition to that provided by the grid bars selected in Step 1.4. Using the width of the column strip equal to one-half the panel width, compute the total additional area of steel required in the column strip.

Step 2.4   Select the additional bars required in the column strip. Place the additional bars centered on the column, midway between grid bars, at the same spacing as the grid bars.

Step 3   The purpose of this step is to select top bars in column strips. If any bars are interrupted by an opening, they must be replaced by other bars, one-half on each side of the opening.

Step 3.1   Compute the negative factored moment in the column strip using M_{oo} and factors from Table A4.6. Use the width of the column strip equal to one-half the panel width.

Step 3.2   Compute the area of reinforcement required for the column strip.

Step 3.3   Select the number and size of top bars for the column strip.

Step 3.4   Concentrate the spacing of bars near the column if necessary to resist unbalanced moment. (See Eq. 4.7, Section 4.4, and Eq. 4.1.)

Step 4   The purpose of this step is to select top bars in middle strips. If any bars are interrupted by an opening, they must be replaced by other bars, one-half on each side of the opening. Using M_{oo}, factors for the middle strip from Table A4.6, and the actual width of the middle strip, compute the required area of reinforcement and select the bars in the same manner as in Steps 3.1, 3.2, and 3.3.

Step 5   Design spandrel beams using Step 5 of Procedure 1.

Example 4.1   Consider a flat plate structure, shown partially in Fig. 4.17, and select the amount and distribution of reinforcing steel using Procedure 1. Assume a live load of 40 psf and a superimposed dead load of 30 psf due to partitions and floor and ceiling finishes. Use concrete $f'_c = 3$ ksi and steel $f_y = 60$ ksi for beams, slabs, and columns.

It is common practice to give numbers or letters to column lines, reading from top to bottom and from left to right. Individual columns are designated by the letter and number of intersecting column lines. This practice is followed here.

Solution   Figures 4.19 and 4.23 show calculations as they might be prepared in a design office. Following are the side calculations to determine values given in Fig. 4.18 and to explain each step.

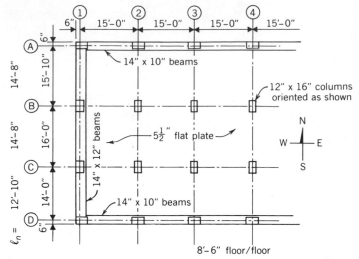

FIGURE 4.17

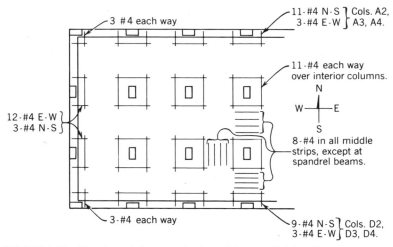

FIGURE 4.18   Showing reinforcement selected for top of slab in Ex. 4.1.

Line No.	Item	(D)		(C)		(B)		(A)
				Column Line 2				
1.	w_u		207 psf		207 psf		207 psf	
2.	ℓ_n		12'-10"		14'-8"		14'-8"	
3.	M_o		63.9'k		83.5'k		83.5'k	
4.	Distribution	0.30	0.50	0.70 | 0.65	0.35	0.65 | 0.70	0.50	0.30
5.	^-M_u	⁻19.2'k		⁻54.3'k		⁻58.4'k		⁻25.0'k
6.	^+M_u		+32.0'k		+29.2'k		+41.7'k	
7.	Distribution to column strip	0.95	0.60	0.75 | 0.75	0.60	0.75 | 0.75	0.60	0.95
	Column strip							
8.	^-M_u	⁻18.2'k		⁻40.7'k		⁻43.8'k		⁻23.8'k
9.	^+M_u		+19.2'k		+17.5'k		+25.0'k	
	Middle strip							
10.	^-M_u	⁻1.0'k		⁻13.6'k		⁻14.6'k		⁻1.2'k
11.	^+M_u		⁺12.8'k		+11.7'k		+16.7'k	
	A_s Required, column strip							
12.	Top	1.02□‴		2.28□‴		2.45□‴		1.33□‴
13.	Bottom		1.08□‴		0.98□‴		1.40□‴	
	A_s Required, middle strip							
14.	Top	0.06□‴		0.76□‴		0.82□‴		0.07□‴
15.	Bottom		0.72□‴		0.65□‴		0.94□‴	
	Bar selection							
16.	Bottom		#4@11"c/c		#4@11"c/c		#4@11"c/c	
	Top							
17.	Column strip	(6) 10 #4		11 #4		11 #4		(7) 11 #4
18.	Middle strip	(1 #4)		8 #4		8 #4		(1 #4)
	Column strip bars concentrated at column							
19.		4@10"c/c						5@7"c/c
				Column Line 1				
20.	Distribution to beam	0.40	0.33	0.33	0.33	0.38	0.33	0.40
21.	Beam ^-M_u	⁻7.8'k		⁻17.8'k		⁻19.1'k		⁻10.1'k
22.	^+M_u		+10.5'k		+9.6'k		+13.6'k	
23.	Slab ^-M_u	⁻1.4'k		⁻3.1'k		⁻3.4'k		⁻1.8'k
24.	A_s	0.08□‴		0.17□‴		0.19□‴		0.10□‴
	Bar selection							
25.		3 #4		3 #4		3 #4		3 #4

FIGURE 4.19 Example 4.1 calculations, North-South direction.

Line   *Explanation*
No.

Step 1.1

1. Live load: 40 psf $\times$ 1.7 = 68 psf
   Dead load: slab   69 psf
              other $\underline{30}$
                     99 psf $\times$ 1.4 = $\underline{139}$
   Total factored load, w_u = 207 psf

2. Clear spans from Fig. 4.17.

3. Panels AB and BC, $M_o = 0.207 \times 15 \left(\dfrac{14' - 8''}{8}\right)^2 = 83.5'$k

   Panel CD, $M_o = 0.207 \times 15 \left(\dfrac{12' - 10''}{8}\right)^2 = 63.9'$k

Step 1.2

4. Distribution coefficients are taken from Table A4.2.

5. Negative moments at critical sections equal M_o times distribution factors on line 4. For example, $^-63.9 \times 0.30 = {}^-19.2'$k

6. Positive moments at mid-span equal M_o times moment coefficients on line 4. For example,
   $83.5 \times 0.35 = {}^+29.2'$k

Step 1.3 $\beta_a = \dfrac{129 \text{ psf}}{40 \text{ psf}} = 3.2 > 2.0$; therefore, no increase in ^+M is necessary.

Step 2.1

7. $\alpha_1 = 0$, because there are no beams in direction of moments.

   $I_s = 15 \times 12 \dfrac{(5.5)^3}{12} = 2450$

   Referring to Fig. 4.20 and using Eq. 4.6,*

   $C = \left(1 - 0.63 \times \dfrac{10}{14}\right) \dfrac{10^3 \times 14}{3} + \left(1 - 0.63 \times \dfrac{4.5}{5.5}\right) \dfrac{4.5^3 \times 5.5}{3}$
   $= 2567 + 81 = 2650$

---

*A major source of computational error in design of two-way slab systems arises from the confusion of beam, column, and slab dimensions in two directions. However, the consequences of error from this source usually involve, at most, the placement of a bar or two per panel. Of far more importance is (1) providing adequate slab shear capacity, (2) the placement of some steel wherever needed (even though it is not quite the right amount), and (3) the preparation of crystal clear drawings to direct construction workers. If workers are confused by the drawings and misinterpret them, the consequences could be far more serious than the mere misplacement of a few bars.

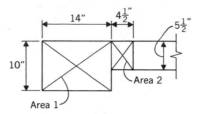

FIGURE 4.20

Using Eq. 4.5,

$$\beta_t = \frac{C}{2I_s} = \frac{2650}{2 \times 2450} = 0.54$$

The portions of moments distributed to column strips are selected from Table A4.5, interpolating where necessary.

<u>Step 2.2</u>

8, 9. Column strip moments are computed by multiplying moments at critical sections on lines 5 and 6 by the coefficients on line 7.

10, 11. The remainder of moments at critical sections on lines 5 and 6 not allocated to the column strip are allocated to the middle strip.

<u>Step 3</u>

12−15. Because the North-South span is almost equal to the East-West span, use the average effective depth, $d = 4\frac{1}{4}''$, as shown in Fig. 4.21.

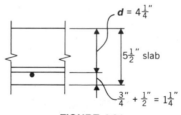

FIGURE 4.21

Use $(\phi f_y j) = 4.2$ at an assumed $\rho = 0.65\%$.
Compute the area of steel required at each section in each strip using Eq. 2.21.

For example, at line D, $A_s = \dfrac{18.2}{4.2 \times 4.25} = 1.02^{\square''}$

Check: Max $\rho = \dfrac{2.45}{90 \times 4.25} = 0.64\% < 0.65\%$ assumed in selecting $(\phi f_y j)$;

therefore <u>OK</u>

From Section 3.2, for a 7'−6"-wide strip,

Min A_s = 0.0018 × 90 × 4.25 = 0.89$^{□''}$

or $\dfrac{0.89^{□''}}{7.5}$ = 0.12$^{□''}$/'

Step 4

16. Minimum reinforcement requirements control for all areas of the slab for bottom reinforcement. The maximum spacing of bars is 2 × 5.5 = 11". Use minimum size bars for construction traffic or #4 at 11" c/c. This provides A_s = 0.2 × 12/11 = 0.22$^{□''}$/', or A_s = 1.64$^{□''}$/7'−6" strip.

17. In selecting top bars, reduce the negative reinforcement area requirements by 10% as permitted by a redistribution of moments from negative moment to positive moment sections. Note that the positive reinforcement area provided is more than 10% in excess of that required. For example, at line B, $A_s$reqd. = 2.45 × 0.9 = 2.20$^{□''}$. Use 11 #4 bars at 8" c/c providing A_s = 2.20$^{□''}$. Eight #5 bars at 11" c/c could be used, but the weight of steel would be much greater than with the #4 bars.

18. Requirements for minimum reinforcement in slabs do not apply to reinforcement for negative moment. To resist moment, the middle strip requires one top bar at the spandrel beam. However, if reinforcement is to be provided it must meet the maximum spacing requirements. Thus, 8 #4 bars in a 7'−6" strip is required, assuming a column strip bar is near the dividing line between column and middle strips. In this example, the practical solution would be to eliminate top bars in the middle strip at the spandrel beam and provide reinforcement in the column strip to meet the required area for both column and middle strip. Seven #4 bars at line A and 6 #4 bars at line D will meet this requirement.

At interior column lines B and C, the engineer may consider eliminating top bars in the middle strip and placing the steel required in the column strip. Maximum middle strip ^-M_u = 14.6'k for a 7 ft.−6 in. (= 90") wide strip. By considering the ratio of service loads to factored loads (see explanation of line 1), the maximum service moment ^-M,

$$^-M = {}^-M_u \frac{139}{207} = 9.8'k$$

Section modulus S of the slab is

$$S = 90 × \frac{5.5^2}{6} = 454 \text{ in.}^3$$

Flexural stresses in plain concrete f_b,

$$f_b = M/S = \frac{9800 × 12}{454} = 260 \text{ psi}$$

The modulus of rupture of plain concrete can be assumed to be that given for deflection considerations (see Section 8.4).

$$7.5\sqrt{f'_c} = 7.5\sqrt{3000} = 410 \text{ psi}$$

Because service load stresses are only 63% (260/410) of plain concrete modulus of rupture, top bars in middle strips could be omitted and the reinforcement placed in column strips.

Thus, required steel area at line B, $A_s = 2.45 + 0.82 = 3.27^{\square''}$,

and at line C, $A_s = 2.28 + 0.76 = 3.04^{\square''}$.

Top bars in the column strip would then be 15 #4 (or 10 #5) bars at line B and 14 #4 (or 9 #5) bars at line C. Note that documentation and approval of a building official is required for this deviation from the specific requirements of the ACI Code.

19. At column lines A and D,

    $(c_1 + d)/(c_2 + d) = (12 + 5.5)/(16 + 5.5) = 0.81$. From Table A4.1, or Eq. 4.1, $\gamma_f = 0.63$. That is, 63% of top bars (or, strictly speaking, 63% of required reinforcement area) at column lines A and D must be placed, centered on column, within a bandwidth

    $= c_2 + 2(1.5h) = 16 + 3 \times 5.5 = 32.5''$

    at line A, $0.63 \times 7 = 5$ bars at maximum spacing 7" c/c

    at line D, $0.63 \times 6 = 4$ bars at maximum spacing 10" c/c.

    The remaining bars should be spaced evenly in the remaining column strip space. The maximum spacing is still limited to two times the slab thickness, or 18 in., so that three bars on each side are required. The total number of column strip bars must be increased to 11 at column line A and 10 at column line D. The number and spacing of bars can be easily visualized with a quick sketch to scale as in Fig. 4.21.

    Placing bars at two spacings in one strip (e.g., 10 in. and 7 in. in Fig. 4.22) may be confusing to construction workers. Furthermore, it is better to concentrate reinforcement at the face of the column. Therefore, some engineers prefer to call for bars at one spacing and add extra bars near the column centerlion. For example, on line A, Code requirements will be met by specifying 9 bars at 11" c/c plus two additional bars near the centerline of the column. On line D, nine bars could be spaced at 11" c/c with the one remaining bar near the centerline of the column.

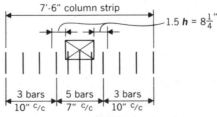

FIGURE 4.22

Because no significant moment is being transferred at interior columns B and C, top bars may be spaced uniformly across the column strip or at about 9″ c/c. By inspection, the moment being transferred at column C ($<$ $0.65 \times 83.5 - 0.70 \times 63.9 = 9.5'k$) is so small that uniform bar spacing will meet the requirements for concentration of bars to resist unbalanced moment.

20. On column line 1,

<u>Step 2.1</u>   The spandrel beam parameters are

$$\frac{h_f}{b_w} = \frac{5.5}{14} = 0.4 \qquad \text{and} \qquad \frac{h_f}{h} = \frac{5.5}{12} = 0.45$$

From Fig. A4.1, $\beta_{tb} = 1.2$.

$$\text{Beam } I_b = 1.2 \times 14 \times \frac{12^3}{12} = 2420$$

$$\text{Slab } I_s = (90 - 8 - 6.5) \times \frac{(5.5)^3}{12} = 1050$$

$$\alpha_1 = \frac{I_b}{I_s} = \frac{2420}{1050} = 2.3 > 1.0$$

$$\frac{\ell_2}{\ell_1} = \frac{15}{16} = 0.94$$

$$\frac{\alpha_1 \ell_2}{\ell_1} = 0.94 \times 2.3 = 2.2$$

From Table A4.5, the proportions of moments distributed to the column strip, by interpolation, are

> Interior ^-M_u = 77% times moment at critical section
> Exterior ^-M_u = 95%
> ^+M_u = 77%

Because $\alpha_1 \ell_2/\ell_1 > 1.0$, moment distributed to beams is 85% of column strip moment. Beam moments are most easily computed by ratio with critical section moments already computed for column line 2. Thus, the critical section moments on column line 1 can be computed by multiplying the critical section moments on column line 2 by the ratio of the strip width on line 1 to the strip width on line 2, or 0.50 in this example. The beam moments are equal to the critical section moments times the proportions computed just above times 85%. For example, on column line D, the distribution coefficient = $0.50 \times 0.95 \times 0.85 = 0.40$.

21, 22. Beam moments are computed from distribution coefficients on line 20 times moments on lines 5 and 6.

23. Because 85% of the column strip negative moment is carried by the beam, and the remainder, or 15%, is carried by the slab, the slab column strip negative moment can be most easily computed as 15%/85% = 0.176 of the beam moment. For example, on line D, slab $^-M_u = 8.0 \times 0.176 = ^-1.4'$k. As previously demonstrated, in this example the bottom reinforcement in the slab is more than adequate and need not be computed again.

24. The area of reinforcement is computed the same as for lines 12 and 13.

25. Bar selections are made the same as for line 17. To meet maximum spacing requirements, at least three bars are required.

Following are the side calculations to determine values given in Fig. 4.23 and to explain steps that are different from the explanation for Fig. 4.19.

Line   *Explanation*
No.

3A. $M_o = \dfrac{0.207 \times (15'-11'') \times 14^2}{8} = 80.7'$k

<u>Step 2.1</u>

7A. $I_s = \dfrac{191(5.5)^3}{12} = 2650$ in.^{4}

The beam on line 1 is similar to the beam in Fig. 4.20, using Eq. 4.6,

$$C = \left(1 - \frac{0.63 \times 12}{14}\right)\frac{12^3 \times 14}{3} + \left(1 - \frac{0.63 \times 5.5}{6.5}\right)\frac{5.5^3 \times 6.5}{3}$$

$$= 3709 + 168 = 3880 \text{ in.}^4$$

Using Eq. 4.5,

$$\beta_t = \frac{3880}{2 \times 2650} = 0.73$$

With no interior beams, $\alpha_1 = 0$.
The proportions of moments distributed to column strips are selected from Table A4.5, interpolating where necessary. Thus,

$$^+M_u = 60\% \text{ times moment at critical section}$$
$$\text{Interior } ^-M_u = 75\%$$
$$\text{Exterior } ^-M_u = 93\%$$

19A. Concentrate top bars at the column in the same manner as for the North-South strips.

$$63\% \times 7 = 5 \text{ bars at } 7'' \text{ c/c}$$

Line No.	Item	Column Line B					
		(1)		(2)			(3)
1A.	w_u		207 psf			207 psf	
2A.	ℓ_n		14'-0"			14'-0"	
3A.	M_o		80.7'k			80.7'k	
4A.	Distribution	0.30	0.50	0.70	0.65	0.35	0.65
5A.	^-M_u	−24.2'k		−56.5'k			−52.5'k
6A.	^+M_u		+40.4'k			+28.2'k	
7A.	Distribution to column strip	0.93	0.60	0.75		0.60	0.75
	Column strip						
8A.	^-M_u	−22.5'k		−42.4'k			−39.4'k
9A.	^+M_u		+24.2'k			+16.9'k	
	Middle strip						
10A.	^-M_u	−1.7'k		−14.1'k			−13.1'k
11A.	^+M_u		+16.2'k			+11.3'k	
	A_s Required, column strip						
12A.	Top	1.26$^{□''}$		2.37$^{□''}$			2.21$^{□''}$
13A.	Bottom		1.36$^{□''}$			0.95$^{□''}$	
	A_s Required, middle strip						
14A.	Top	0.09$^{□''}$		0.79$^{□''}$			0.73$^{□''}$
15A.	Bottom		0.91$^{□''}$			0.63$^{□''}$	
	Bar selection						
16A.	Bottom		#4@11"c/c			#4@11"c/c	
	Top						
17A.	Column strip	(7)11#4		11#4			10#4
18A.	Middle strip	(1#4)		8#4			8#4
	Column strip bars concentrated at column						
19A.		5@7"c/c					

Line No.	Item	Column Line A					
20A.	Distribution to beam	0.38	0.31	0.31		0.31	0.31
21A.	Beam ^-M_u	−9.2'k		−17.5'k			−16.3'k
22A.	^+M_u		+12.5'k			+8.7'k	
23A.	Slab ^-M_u	−1.6'k		−3.0'k			−2.9'k
24A.	A_s Required	0.09$^{□''}$		0.17$^{□''}$			0.16$^{□''}$
	Bar selection						
25A.		3#4		3#4			3#4

FIGURE 4.23 Example 4.1 calculations, East-West direction.

As in the North-South strips, the maximum bar spacing must be maintained beyond the concentration of bars at the column. Thus, the number of bars must be increased to 11 at column line 1.

20A. On column line A,

$$\frac{h_f}{b_w} = \frac{5.5}{14} = 0.4 \text{ and } \frac{h_f}{h} = \frac{5.5}{10} = 0.55$$

From Fig. A4.1, $\beta_{tb} = 1.5$

$$\text{Beam } I_b = 1.15 \times \frac{14 \times 10^3}{12} = 1340 \text{ in.}^4$$

$$\text{Slab } I_s = \frac{(95 - 8 - 4.5)(5.5)^3}{12} = 1150 \text{ in.}^4$$

$$\alpha_1 = I_b/I_s = \frac{1340}{1150} = 1.17 > 1.0$$

$$\frac{\ell_2}{\ell_1} = \frac{15.83}{15} = 1.06$$

$$\frac{\alpha_1 \ell_2}{\ell_1} = 1.17 \times 1.06 = 1.23$$

For the beams on line 1 (see explanation for line 7A), $\beta_t = 0.73$

From Table A4.5, the proportions of moments distributed to the column strip, by interpolation, are:

$$^+M_u = 73\% \text{ times moment at critical section}$$
$$\text{Interior } ^-M_u = 73\%$$
$$\text{Exterior } ^-M_u = 90\%$$

Determine the distribution coefficients in the same manner as for the explanation for line 20 in Fig. 4.19. For example, the distribution coefficient on line D = $0.5 \times 0.90 \times 0.85 = 0.38$

21A. For example, the beam moment on line 1 is

$$M_u = 0.38 \times 24.2 = 9.9'k$$

Final selection of slab top bars is shown in Fig. 4.18. Note that some additional steel has been provided over interior columns and on line 1 in order to simplify bar placing for economy and to reduce the chance for error in placing bars.

Example 4.2 Given the same structure, materials, and conditions as in Ex. 4.1, select the amount and distribution of reinforcement using Procedure 2.

Solution   See Figs. 4.24 and 4.25 and the side calculations below. Where no explanation is given, it is the same as for Ex. 4.1.

Line No.	Explanation

3B. $\boldsymbol{M}_{oo} = \dfrac{w_u \ell_n^2}{8}$

For example, in bay D–C,

$$\boldsymbol{M}_{oo} = \frac{0.207 \times 12.83^2}{8} = 4.3'k/'$$

4B. Moment distribution coefficients are taken from Table A4.7.

5B. For example, in bay D–C, $^+\boldsymbol{M}_u = 0.40 \times 4.3 = {}^+1.7'k/$,

6B. For example, in bay D–C, $\boldsymbol{A}_s = \dfrac{1.7}{(4.2 \times 4.25)} = 0.10^{\square''}/'$

Line No.	Item	Column Line 2			
		(D)	(C)	(B)	(A)
1B.	w_u	207 psf	207 psf	207 psf	
2B.	ℓ_n	12'-10"	14'-8"	14'-8"	
3B.	M_o/ft	4.3'k/'	5.6'k/'	5.6'k/'	
4B.	Distribution to middle strip	0.40	0.28	0.40	
5B.	^+M_u	$^+1.7'$k/'	$^+1.6'$k/'	$^+2.2'$k/'	
6B.	A_s/'	$0.10^{\square''}/'$	$0.09^{\square''}/'$	$0.13^{\square''}/'$	
7B.	Selection of grid bars in bottom of slab	#4@11"c/c - - - $A_s$prov. = $0.22^{\square''}/'$			
8B.	Distribution to column strip	0.60	0.42	0.60	
9B.	^+M_u	$^+2.6'$k/'	$^+2.3'$k/'	$^+3.4'$k/'	
10B.	A_s/'	$0.15^{\square''}/'$	$0.13^{\square''}/'$	$0.19^{\square''}/'$	
11B.	Extra A_s/'	0	0	0	
12B.	Selection of extra bars, bottom of column strip	0	0	0	
	Top steel, column strip				
13B.	Distribution	0|60	1.05|0.98	0.98|1.05	0|60
14B.	^-M_u	$^-20$|6'k	$^-43$|9'k	$^-47$|0'k	$^-26$|9'k
15B.	A_s Required	1|15$^{\square''}$	2|46$^{\square''}$	2|63$^{\square''}$	1|50$^{\square''}$
16B.	Bar selection, top, column strip	(5)|9 #4	11|#4	12|#4	(7)11|#4
17B.	Column strip bars concentrated at column	3@10"c/c			5@7"c/c
	Top steel, middle strip				
18B.	Distribution	0|15	0.35|0.33	0.33|0.35	0|15
19B.	^-M_u	$^-4$|5'k	$^-12$|9'k	$^-13$|7'k	$^-5$|9'k
20B.	A_s Required	0|25$^{\square''}$	0|72$^{\square''}$	0|77$^{\square''}$	0|33$^{\square''}$
21B.	Bar selection, top, middle strip	(2|#4)	8|#4	8|#4	(2|#4)

FIGURE 4.24   Example 4.2 calculations, North-South direction.

Line No.	Item	Column Line B		
		(1)	(2)	(3)
1C.	w_u		207 psf	207 psf
2C.	ℓ_n		14'-0"	14'-0"
3C.	$M_o/'$		5.1'k/'	5.1'k/'
7C.	Selection of grid bars in bottom of slab		#4@11"c/c - - - - - - - - - - - - - $A_s$prov. = $0.22^{\square''}/'$	
	Top steel, column strip			
13C.	Distribution	0.60	1.05 / 0.98	0.98
14C.	^-M_u	-24.3'k	-42.8'k	-40.0'k
15C.	A_s Required	$1.36^{\square''}$	$2.40^{\square''}$	$2.24^{\square''}$
16C.	Bar selection, top, column strip	(6) 11#4	11 #4	10 #4
17C.	Column strip bars concentrated at column	5@7"c/c		
	Top steel, middle strip			
18C.	Distribution	0.15	0.35	0.33
19C.	^-M_u	-6.1'k	-14.3'k	CNR*
20C.	A_s Required	$0.34^{\square''}$	$0.80^{\square''}$	CNR
21C.	Bar selection, top, middle strip	(2 #4)	8 #4	8 #4

*CNR = calculation not required.

FIGURE 4.25 Example 4.2 calculations, East-West direction.

11B. The extra reinforcement area required is the area on line 10B less the area of reinforcement provided by bars selected on line 7B. In this example, middle strip reinforcement is based on minimum requirements and is adequate throughout the slab. The material enclosed in the outlined box is shown for illustration purposes. In this example, an experienced engineer would proceed directly to lines 8B, 9B, and 10B in panel B−A, find that minimum steel is adequate for the most heavily loaded strip, and not perform computations enclosed in the box. If more than the minimum reinforcement is required in the most heavily loaded strip, the engineer can judge how many additional computations will be required.

13B. Distribution coefficients are taken from Table A4.6.

14B. Negative moment in the column strip equals M_{oo}/ft from line 3B, times strip width, times distribution coefficient in line 13B. On column line 2, the column strip width is 7'−6", but half the panel width is 8'−0". For example, at line D,

$$^-M_u = 4.3 \times 8.0 \times 0.60 = 20.6'\text{k}$$

15B. For example, at line D,

$$A_s = \frac{20.6}{4.2 \times 4.25} = 1.15^{\square''}$$

17B. For example, at column line D,

$$0.63 \times 6 = 3 \text{ bars at maximum spacing } 10'' \text{ c/c}$$

As in the previous example, maximum bar spacing requires three bars on each side of the concentration of bars at the column. Thus, a total of 10 bars is required at column line D and 11 bars at column line A.

19B. Negative moment in the middle strip is M_{oo}/ft from line 3B, times strip width, times distribution coefficient in line 18C. On column line 2, the middle strip width is $15'$ less $8' = 7'$. For example, at column line C, $^-M = 5.6 \times 7.0 \times 0.33 = {}^-12.9'$k

7C. By inspection of calculations for the North-South direction, the same minimum reinforcement will be adequate. Additional computations are not required.

14C. Column strip width $= 8'-0''$.
For example, on line 1, $^-M_u = 5.1 \times 0.60 \times 8 = 24.3'$k

4.6 EQUIVALENT FRAME METHOD (EFM)

The EFM is more general than DDM and therefore applicable to a wider range of situations. In its details EFM is similar to DDM but differs in the following ways (Code 13.7):

1. There are no limitations on dimensions or loading as there are for DDM (see Fig. 4.14). Thus, EFM can be used for a wider range of structures and loading conditions.

2. Moments are distributed to critical sections by elastic analysis rather than by factors. Critical arrangement of live load must be considered if it exceeds three quarters of the dead load, whereas pattern loading need not be considered in DDM. However, if a structure meets all the limitations for use of DDM but is analyzed by EFM, the total static moment need not exceed the static moment M_o required by DDM.

3. In making the elastic analysis, certain simplifications permitted for DDM are not permitted for EFM. Variations in moment of inertia along the axis of members must be considered, as well as the contribution of metal column capitals. Simplifications in computing the torsional stiffness of transverse members and at corner columns are not permitted.

4. Moment redistribution is permitted by general requirements (Section 13.5) but not by the simpler procedure of DDM.

5. EFM can be used for lateral load analysis, whereas DDM cannot.

6. EFM is appropriate for computer analysis, and DDM is appropriate for manual analysis.

Procedure 3 (EFM)

The five steps of the basic design procedure outlined in Section 4.4 can be further defined for Procedure 3 as follows:

Step 1.1   Assemble the data and compute the parameters necessary for an elastic analysis. The ACI Code Commentary describes the procedure for making these computations and tabulates various constants for commonly occurring structural members. In general, computation of the necessary parameters will proceed in the following order:

1. Flexural stiffness of the slab (K_s).
2. Flexural stiffness of the beam in the direction of the span for which moments are being determined (K_b).
3. Flexural stiffness of the column (K_c).
4. Torsional stiffness of torsional members, such as transverse beams and slabs (K_t).
5. Flexural stiffness of the equivalent column (K_{ec}).
6. Fixed end moments.
7. Carryover factors.

Step 1.2   Make an elastic moment distribution analysis. Reduce negative moments to the moment at the face of the support. If the structure meets the limitation for DDM, reduce the positive and negative moments proportionally so that the absolute sum does not exceed the value obtained from Eq. 4.2.

Step 2.1   Distribute positive and negative moments to column and middle strips in the same manner as for DDM using Table A4.5. For a panel with beams between supports on all sides, the relative stiffness of beams in perpendicular directions must be approximately equal within broad limits. (See the discussion of Step 2.1 for DDM in Section 4.5.)

Step 2.2   If beams are present, distribute moments in column strips to beams and column strip slabs using the procedures for DDM. (See the discussion of Step 2.3 for DDM in Section 4.5.)

Step 3   See basic design procedure (Section 4.4).

Step 4   See basic design procedure. In this step, some redistribution of moment may be desirable to facilitate practical reinforcing bar patterns.

Step 5   Apportion shear to beams using the procedures for DDM. (See the discussion of Step 5 for DDM in Section 4.5.) To these shears and the moments determined in Step 2.2 above, add the shears and moments from concentrated and linear loads superimposed directly on the beams and design the beams using procedures in Chapters 2, 5, and 6.

4.7 OTHER METHODS

With the rapid lowering of computer cost to benefit ratios, finite element and other theoretical approaches to analyzing the distribution of moments become more practical. The ACI Code permits any such method provided that strength and serviceability requirements are met (Code 13.3.1). Theoretical solutions using finite element, yield-line analysis, or other classic solutions are especially useful for slabs with random column spacing, unusual column shapes, and for other special cases not within the limitations of DDM or EFM. A special case of staggered columns has been studied and presented [4].

The ACI Code also permits shorter and simpler procedures. Short procedures are especially useful where the scope of construction does not justify a more rigorous analysis. One such condition is two-way slabs on rigid beams on all four sides. These slabs may be designed using one of the three procedures in the Appendix of the ACI 318-63 Building Code adopted in 1963 [5]. The second of the three methods is reproduced below with the permission of the American Concrete Institute.

The moments in two-way slabs supported on four sides is less than the static moment ($w\ell^2/8$) due to the contribution of torsion in the slab. Winter and Nilson [6] give an excellent explanation of this phenomenon.

To qualify for use of ACI 318-63 Method 2, beams supporting the two-way slab should be sufficiently stiff to mobilize the torsional stiffness of the slab. As a guideline, beams for which $\alpha_1\ell_2/\ell_1 \geq 1.0$ should be adequate, as this limit is used in the current ACI Code (Code 13.6.4). Assuming the same concrete is used in both slab and beam, $\alpha_1 = I_b/I_s$. Using the most critical condition where $\ell_2/\ell_1 = 0.5$, the required minimum moment of inertia for the beam is

$$I_b \geq 2I_s \tag{4.9}$$

Figure A4.3 gives the required moment of inertia for concrete beams, including the effects of the slab acting as a beam flange (See Figs. A4.1 and A4.2). If I_b is less than twice the moment of inertia of the slab I_s but more than $0.5I_s$, the beam might still be satisfactory because of the conservative assumption made in developing Eq. 4.9. A quick computation will determine whether $\alpha_1\ell_2/\ell_1 > 1.0$.

If the slab is supported on steel beams, it is conservative to assume that the modular ratio $n = E_s/E_c = 10$ and requires only one tenth of the moment of inertia for steel beams as required by Eq. 4.9 for concrete beams. Even though the elastic E_c can be higher than $0.1E_s$, creep will quickly reduce the effective E_c, whereas E_s will remain constant. Typical steel beams are shown in Fig. A4.3 to aid the engineer in a quick analysis.

ACI 318-63—Method 2

Notation

    C = moment coefficient for two-way slabs, as given in Table 4.4 (below)

m = ratio of short span to long span for two-way slabs

S = length of short span for two-way slabs. The span shall be considered as the center-to-center distance between supports or the clear span plus twice the thickness of slab, whichever value is the smaller

w = total uniform load per square foot

Limitations—These recommendations are intended to apply to slabs (solid or ribbed), isolated or continuous, supported on all four sides by walls or beams, in either case built monolithically with the slabs.

A two-way slab shall be considered as consisting of strips in each direction as follows:

A middle strip one-half panel in width, symmetrical about panel centerline, and

TABLE 4.4 Method 2—Moment Coefficients

Moments	Short span						Long Span, all Values of m
	Values of m						
	1.0	0.9	0.8	0.7	0.6	0.5 and less	
Case 1. Interior panels							
Negative moment at—							
Continuous edge	0.033	0.040	0.048	0.055	0.063	0.083	0.033
Discontinuous edge	—	—	—	—	—	—	—
Positive moment at mid-span	0.025	0.030	0.036	0.041	0.047	0.062	0.025
Case 2. One edge discontinuous							
Negative moment at—							
Continuous edge	0.041	0.048	0.055	0.062	0.069	0.085	0.041
Discontinuous edge	0.021	0.024	0.027	0.031	0.035	0.042	0.021
Positive moment at mid-span	0.031	0.036	0.041	0.047	0.052	0.064	0.031
Case 3. Two edges discontinuous							
Negative moment at—							
Continuous edge	0.049	0.057	0.064	0.071	0.078	0.090	0.049
Discontinuous edge	0.025	0.028	0.032	0.036	0.039	0.045	0.025
Positive moment at mid-span	0.037	0.043	0.048	0.054	0.059	0.068	0.037
Case 4. Three edges discontinuous							
Negative moment at—							
Continuous edge	0.058	0.066	0.074	0.082	0.090	0.098	0.058
Discontinuous edge	0.029	0.033	0.037	0.041	0.045	0.049	0.029
Positive moment at mid-span	0.044	0.050	0.056	0.062	0.068	0.074	0.044
Case 5. Four edges discontinuous							
Negative moment at—							
Continuous edge	—	—	—	—	—	—	—
Discontinuous edge	0.033	0.038	0.043	0.047	0.053	0.055	0.033
Positive moment at mid-span	0.050	0.057	0.064	0.072	0.080	0.083	0.050

extending through the panel in the direction in which moments are considered; a column strip one-half panel in width, occupying the two quarter-panel areas outside the middle strip.

Where the ratio of short to long span is less than 0.5, the middle strip in the short direction shall be considered as having a width equal to the difference between the long and short span, the remaining area representing the two column strips.

The critical sections for moment calculations are referred to as principal design sections and are located as follows: for negative moment, along the edges of the panel at the faces of the supporting beams; for positive moment, along the centerlines of the panels.

Bending moments—The bending moments for the middle strips shall be computed from the following formula:

$$M = CwS^2$$

The average moments per foot of width in the column strip shall be two thirds of the corresponding moments in the middle strip. In determining the spacing of the reinforcement in the column strip, the moment may be assumed to vary from a maximum at the edge of the middle strip to a minimum at the edge of the panel.

Where the negative moment on one side of a support is less than 80% of that on the other side, two thirds of the difference shall be distributed in proportion to the relative stiffnesses of the slabs.

Shear—The shear stresses in the slab may be computed on the assumption that the load is distributed to the supports in accordance with the next paragraph.

Supporting beams—The loads on the supporting beams for a two-way rectangular panel may be assumed as the load within the tributary areas of the panel bounded by the intersection of 45° lines from the corners with the median line of the panel parallel to the long side.

The bending moments may be determined approximately by using an equivalent uniform load per lineal foot of beam for each panel supported as follows:

For the short span: $\dfrac{wS}{3}$

For the long span: $\dfrac{wS}{3}\dfrac{(3 - m^2)}{2}$

4.8 SPECIAL PROBLEMS

Progressive Collapse

The ACI Code has no specific provisions relating to progressive collapse, although the subject has received some attention from researchers [7]. The ACI Code deals with the subject only indirectly, if at all. Flat plate structures are especially vulnerable, and several such structures in recent years have experienced catastrophic failures with large loss of life and property. Therefore, engineers may wish to provide redundant systems to guard against such failures, especially if this can be done with little or no increase in cost.*

In a flat plate structure, progressive collapse is usually initiated by shear failure around a column. One way to prevent a shear failure at one column from causing failure at nearby columns and propagating to the entire floor and perhaps to floors below is to support the floor at the column *after* shear failure even though serviceability requirements (deflection and cracking) are not met. This can be done rather simply by placing bottom slab bars through the column. After shear failure, these bars can support the slab in direct tension if they are of sufficient size and if they are anchored into the column and into the slab. For convenience, these bars will be called "hanger bars."

Figure 4.26 shows the mechanism after shear failure at an interior column. Note that top bars are ineffective as they tend to split off the top face shell of concrete and lose their anchorage in the slab. To mobilize the reserve strength of hanger bars, the slab must drop an inch or two. The slab is clearly unserviceable at that point, but a widespread failure may have been averted.

Reference 7 recommends, based on laboratory tests, the shear capacity ϕV_n of hanger bars. $V_u = 0.5\ A_v f_y$. By adding the capacity reduction factor ϕ,

$$\phi V_n = 0.5\ \phi A_s f_y \qquad (4.10)$$

Hanger bars may be provided by extending bottom bars from adjacent spans and lap splicing them over the column with a minimum lap length of ℓ_d as in Fig. 4.27a, provided the support width (column or column capital) is at least equal to ℓ_d. Alternatively, separate hanger bars may be provided and lap spliced with bottom slab bars outside the support with a minimum lap splice length equal to $2\ \ell_d$ as in Fig.

*Redundant systems have also been known to cure severe cases of insomnia.

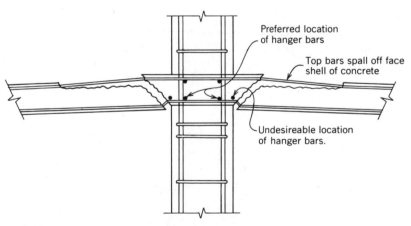

Preferred location of hanger bars

Top bars spall off face shell of concrete

Undesireable location of hanger bars.

FIGURE 4.26   Slab-column joint after shear failure.

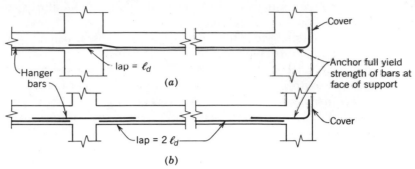

FIGURE 4.27   Anchorage of hanger bars.

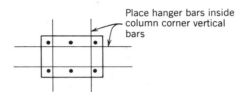

FIGURE 4.28   Placement of hanger bars.

4.27b. At discontinuous edges, hanger bars must develop the full yield stress at the inside face of the support. Hanger bars should be provided in both directions and at all faces of the support receiving slab loads. Detailing of bars should avoid congestion of bars over the support. In any case, it is best to place hanger bars between column verticals, as in Fig. 4.28.

As a minimum, hanger bars should be designed for dead loads and long-term or permanent live loads. The engineer may also wish to design for the weight of the slab above plus impact, to prevent a failure from progressing vertically down a multistory structure.

Example 4.3   Consider the same structure as in Example 4.1; design the hanger bars at interior column B2. For Case I, consider that 25% of the live load is permanent, and dead load is 10% higher than expected. For Case II, consider the same conditions for the floor in question plus the floor above. Assume that the floor above has failed and drops to the floor in question with a 50% impact factor.

Solution

Case I.   Live load: 40 psf × 0.25 =   10 psf
　　　　　Dead load: 99 psf × 1.1 = $\underline{109}$

　　　　　　　　Total load = 119 psf

　　　　Load on hangers: 15 × 16 × 0.119 = 28.6 k

Case II.   Load on hangers:　　　28.6 × 2.5 = 71.5 k

The capacity of a #4 hanger bar = $0.5 \times 0.9 \times 0.20 \times 60 = 5.4$ k. If the bar extends through the column and is effective on both sides, its capacity is 10.8 k.

Thus, $\dfrac{28.6}{10.8}$ = 3 bars required for Case I, and

$\dfrac{71.5}{10.8}$ = 6.6 bars required for Case II.

These bars should be placed in both directions. For example, for Case II, an engineer might specify four bars East-West (across the longest side of the column) and two bars North-South. Because only one slab bottom bar in each direction, at 11″ c/c spacing, will pass through the column, extra bottom bars passing through the column would be required: three bars East-West and one bar North-South. The extra bars should extend past the column on each side a distance equal to twice the bar development length (see Chapter 5).

An additional easy method for partial protection against progressive collapse is to make all bottom reinforcement continuous so that the slab can hang by catenary action if the support of one column is withdrawn. The method requires that splices in the bottom bars develop the tensile strength of the bars required by catenary action and that bars at edges be anchored to develop the full tensile strength. The edges must have sufficient strength as buttresses to resist the horizontal reaction of the catenary. Because spandrel beams, columns, and other supports at free edges may be weak in lateral strength, the method is most effective for interior columns and panels. [8]

Waffle Slabs

Although waffle slabs are designed by the same procedures as flat slabs, there are some differences in details; these will be discussed below.

Drop panels are produced by omitting the pans that create voids between joists. Depth of the drop panel is the same as depth of joists. Drop panel width and length are limited not only by structural and Code requirements but also by the pans themselves.* The engineer's creative imagination can be exercised to shape drop panels (see Fig. 4.29 for some examples).

Thickness and reinforcement for the top slab over the pans is subject to the same limitations as for pan slabs (see Section 3.3)

In selecting bottom reinforcement, width of the joists will usually not permit more than two bars per joist, but one bar could be used. As with flat slabs, truss bars are not advisable due to the intricate placing sequences required. Higher shear stresses in joists than in solid slabs make it more advisable to extend bottom bars full length of the span.

---

*A fraction of a pan is not available except at the cost of special fabrication, something most engineers and owners will not want to endure.

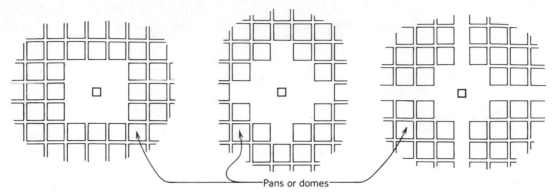

FIGURE 4.29 Example of waffle slab drop panels.

Top bars can be located over the joists and halfway between joists or spread at a spacing independent of joist spacing, as in a solid slab. In the column strip, top bars should be closely spaced, especially in the vicinity of the column.

The greater depth of concrete in waffle slabs around the column will permit the use of beam stirrups for shear reinforcement, if necessary.

Slab Corner Reinforcement

Slabs on stiff beams ($\alpha > 1.0$) between columns or on unyielding supports along the edges will develop moments at exterior corners at a 45° angle to the edges. These moments tend to open up cracks, as shown in Fig. 4.30(a). The ACI Code requires that reinforcement be provided as shown in Fig. 4.30(b). The special reinforcement in both top and bottom of the slab shall be sufficient to resist a moment equal to the maximum positive moment (per foot of width) in the slab (Code 13.4.6). Alternatively, the bars can be placed parallel to the sides of the slab. Such reinforcement is also good practice for one-way slabs. When negative moment reinforcement is placed over supporting beams and walls, it is not necessary to add additional corner reinforcement in the top.

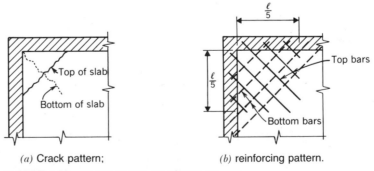

(a) Crack pattern;   (b) reinforcing pattern.

FIGURE 4.30 Corner of slab on stiff beams.

PROBLEMS

1. For the flat plate without beams shown in Fig. 4.31, select the size and spacing of bottom bars in each direction to meet middle strip moment requirements. Assume $f'_c = 3$ ksi and $f_y = 60$ ksi and use design Procedure 2. Live load = 60 psf. Other dead load = 40 psf inside the exterior walls that are at the outside face of the exterior columns (lines 1, 4, and A). Exterior walls weight 500 plf. Assume interior exposure for slabs inside the walls and exterior exposure for slabs cantilevered outside the walls.
2. Select additional bottom bars, if required, on line C for the structure in Problem 1.
3. Select additional bottom bars, if required, on line A for the structure in Problem 1.
4. Select column strip top bars at column A3 for the structure in Problem 1.
5. For what moment in the East-West direction must the connection between slab and column A3 be designed in problem 1?

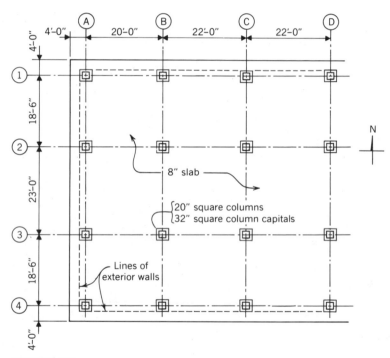

FIGURE 4.31

Selected References

1. Mete A. Sozen and Chester P. Siess. "Investigation of Multiple-Panel Reinforced Concrete Floor Slabs, Design Methods—Their Evolution and Comparison," *ACI Journal, Proc.* V.60, pp 999–1028, August 1963.
2. A.S. Hall and B.V. Rangan. "Moments in Edge Panels of Flat Plate Floors," *ASCE Journal of Structural Engineering*, V.109, No.11, pp 2638–2650, November 1983. See also B.

Vijaya Rangan and A.S. Hall. "Moment Redistribution in Flat Plate Floors," *ACI Journal*, V.81,No. 6: pp 601–608, November-December 1984.

3. Alex E. Cardenas and Paul H. Karr. "Field Test of a Flat Plate Structure," *ACI Journal*, V.68, pp 50–59, January 1971.

4. Maurice P. Van Buren. "Staggered Columns in Flat Plates," *ASCE Journal of the Structural Division*, V.97,No. ST6, pp 1791–1797, June 1971.

5. ACI Committee 318. *Building Code Requirements for Reinforced Concrete (ACI 318-63)*, American Concrete Institute, Detroit, Michigan 1963.

6. George Winter and Arthur S. Nilson. *Design of Concrete Structures*. McGraw-Hill, New York City, 1979.

7. Neil M. Hawkins and Denis Mitchell. "Progressive Collapse of Flat Plate Structures," *ACI Journal*, V.76, pp 775–808, July 1979.

8. Dennis Mitchell and William D. Cook. "Preventing Progressive Collapse of Slab Structures," *ASCE Journal of Structural Engineering*, V.110, No. 7, pp 1513–1532, July 1984.

CHAPTER 5

ANCHORAGE AND BAR DEVELOPMENT

5.1 OBJECTIVES

The force in steel reinforcement must be transferred to concrete by anchorage or bearing of bar deformations on concrete. Force in the steel at a critical section must be developed in some manner between that section and *each* end of the bar.

Assuring proper bar development and anchorage can be divided into several common conditions encountered in concrete design.

1. For flexural reinforcement, determine the points at which bars can be cut off (or, simply, "cutoff points") and where the bends in truss bars and other bent bars should occur.
2. When bars cannot be shipped in one piece or must be cut and spliced for other construction reasons, determine the length of lap splices and details of mechanical splices necessary to ensure splices of adequate strength.
3. Select the anchorage necessary at the ends of other bars.

5.2 INTRODUCTION

When a smooth bar is used to reinforce concrete, the bond between concrete and steel depends on chemical adhesion and friction. If these fail, the bar can slip and pull out of the concrete. Because chemical adhesion and friction are unreliable, the ACI Code does not permit the use of smooth bars. If the bar has deformations or lugs on its surface, the lugs will interlock with the concrete and provide a much greater resistance to pullout. However, adhesion or bond strength and lug-bearing strength are not additive.

As the lugs bear on the concrete, the bearing forces act at an angle with the bar (see Fig. 5.1). The diagonal bearing forces have a component perpendicular to the bar (a splitting force) that tends to split the concrete and thus reduce the capacity of the system to transfer force from steel to concrete. The amount of reduction depends on the amount of concrete cover, transverse reinforcement, tensile strength of concrete, and spacing of the bars. In extreme cases with soft aggregate, the lugs can pull out of the concrete, leaving a hole the size of the bar measured outside of the lugs. Reinforcing bars must meet specifications prepared by the American Society for Testing and Materials (ASTM) that have requirements for size and spacing of deformations (see Section 1.7).

The bond stress on loaded bars is not uniform for several reasons. For example, at the loaded end of the bar the steel stress is at a maximum, and at the free end, the steel stress is zero. Thus, the strain or movement along the bar also varies from a maximum to zero, tending to cause a similar variation in bond stress. Embedding a bar twice as deep in concrete does not double the pullout resistance. For this reason, the concept of bond stress has been abandoned as a design concept in the current ACI Code. Instead, a bar must be embedded into concrete a sufficient distance to anchor it, or it must be anchored in some other way to develop the design stress at critical sections.

Conceptually, the simplest test to determine the length of embedment needed to develop the full strength of a bar is a pullout test. Referring to Fig. 5.2, the tensile force, $T = A_s f_y$.

The tensile force is resisted by lugs bearing between concrete and steel so that

$$T = u\pi d_b \ell_d$$

where u = the unit equivalent bond stress on the contact surface between concrete and steel

d_b = diameter of bar

ℓ_d = development length, or length of embedment of a reinforcing bar required to develop the design strength of the bar at a critical section.

Combining and simplifying,

$$\ell_d = \frac{d_b f_y}{4u} \tag{5.1}$$

A pullout test does not accurately represent conditions in a real structure. Other tests have been devised to better simulate actual conditions, but the concept of "development length" ℓ_d remains. ACI Code requirements for ℓ_d have been deter-

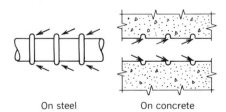

FIGURE 5.1 Forces between deformed bars and concrete tending to split the concrete.

On steel   On concrete

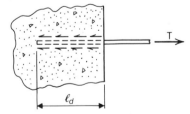

FIGURE 5.2 Bar anchorage.

mined empirically to fit test data and recognize the more important variables. These variables include the following:

1. Steel yield strength f_y. Higher strength steel requires a longer ℓ_d because the potential tensile force is larger.

2. Stress in steel. When only a part of the steel strength is used, ℓ_d can be reduced proportionally.

3. Bar size. The ℓ_d tends to be proportional to bar area and not bar diameter.

4. Bar location. Horizontal bars cast with more than 12 in. of concrete below them have lower bond strength because the concrete shrinks and settles, leaving a void or area of weakened lug-bearing strength on the underside of the bar.

5. Bar spacing. Close spacing increases the possibility of spalling off the face shell or splitting the concrete, thus reducing or eliminating anchorage of the bar. Closely spaced bars have less concrete between bars to resist the splitting forces. (See Fig. 5.3 for examples of possible splitting due to anchorage stresses.)

6. Cover on bars. Small cover increases the possibility of spalling the face shell.

7. Enclosure or confinement by lateral ties. Ties can prevent splitting and spalling of the concrete and, thus, develop higher anchorage strength of the bars. If the face shell does spall, ties can still hold bars to the concrete core and thus maintain at least part of the anchorage resistance. Other lateral reinforcement, such as temperature steel, also contributes to bar anchorage.

8. Confinement of concrete around a bar by a compressive reaction. For example, the reaction transverse to the bars at the end of a slab or beam resting on top of its support reduces the possibility of splitting the concrete. Thus, the anchorage strength of the bars is increased.

9. Type of aggregate. Lightweight aggregate concrete frequently has a lower tensile strength than normal weight aggregate concrete. The anchorage strength of bars will be lower if tensile strength of the concrete is lower because spalling and splitting of concrete is more likely.

10. Placement tolerances. By establishing minimum development lengths, the ACI Code implicitly recognizes that field tolerances in placing bars may reduce the intended anchorage length.

Further background on bar development and splice behavior is found in Refs. 1 and 2.

A failure in bond or anchorage would be a brittle failure with potentially serious results. Therefore, the ACI Code is more conservative for anchorage than for

FIGURE 5.3 Examples of spalling of face shell under high anchorage stresses.

flexural stresses in reinforcement. The 1983 ACI Code requires a development length about 20% longer than did the 1963 ACI Code, which used the bond stress concept.

Computation procedures for development length ℓ_d have been established by the 1983 ACI Code to provide the best fit with test data (Code 12.2 and 12.3). The development length ℓ_d is computed as the product of the basic development length ℓ_{db} and applicable factors to account for some of the variables discussed above.

For bars in tension, the basic development length ℓ_{db} must be

#11 bar and smaller	$40A_b f_y/\sqrt{f'_c}$
but not less than	$0.4d_b f_y$
#14 bar	$85f_y/\sqrt{f'_c}$
#18 bar	$110\,f_y/\sqrt{f'_c}$
Deformed wire	$30d_b f_y/\sqrt{f'_c}$

where d_c = nominal diameter of bar or wire, in inches and f'_c is in psi.

Following are applicable factors to be applied to ℓ_{db} for bars in tension to account for variables:

1. Top bars (horizontal bars so placed that more than 12 in. of fresh concrete is cast in the member below the bars)—   1.4.

2. Yield strength greater than 60 ksi $(2-60/f_y)$.

3. All-lightweight concrete   1.33.
   Sand-lightweight concrete*   1.18.

4. Bar spacing at least 6 in. c/c and at least 3 in. clear to the face of the member, measured in direction of the spacing—   0.8.

5. Reinforcement in flexural members in excess of that required by analysis (when development of f_y is not specifically required)— $(A_s\text{reqd.})/(A_s\text{prov.})$.

6. Bars enclosed by spirals at least ¼ in. in diameter and not more than 4 in. pitch—   0.75.

7. For deformed welded wire fabric, the yield stress f_y may be reduced by 20 ksi if there is at least one cross wire within the development length but not less than 2 in. from the critical section.

For bars in compression, the basic development length ℓ_{db} must be

All bar sizes	$20d_b f_y/\sqrt{f'_c}$
but not less than	$0.3d_b f_y$
where f_y is in psi.	

---

*"All-lightweight" concrete is concrete using lightweight aggregate without natural sand and with an air-dry unit weight not exceeding 115 pcf. "Sand-lightweight" concrete is concrete using lightweight coarse aggregate and natural sand. Linear interpolation between 1.18 and 1.33 may be used when partial sand replacement is used. If the lightweight concrete is tested for its average splitting tensile strength f_{ct}, ℓ_{db} may be modified by a factor equal to $6.7\sqrt{f'_c}/f_{ct}$ but not less than 1.0, where both f'_c and f_{ct} are expressed in psi.

Applicable factors to be applied to ℓ_{db} for bars in compression to account for variable are factors 5 and 6 in the list above. Factor 5 applies to all types of members. No increase or decrease for variables in factors 1 through 4 are necessary or permitted for bars in compression.

Because the basic development length ℓ_{db} is used so often in calculations for splice lengths (Section 5.3) and bar cutoff points (Section 5.4), it is tabulated in Table A5.1 for steel strength of $f_y = 60$ and concrete strengths of $f'_c = 3, 4, 5,$ and 6 ksi. Because ℓ_{db} is used in other computations, it is given in Table A5.1 to the nearest 0.1 in., but engineers should round off the length given on construction drawings to the nearest inch. Even though smaller values are given in Table A5.1, minimum development length ℓ_d is 8 in. for bars in compression and 12 in. for bars in tension, except for development of web reinforcement (Code 12.2.5 and 12.3.1). In computation of lap splices, a basic development length ℓ_{db} less than 12 in. may be used, but the lap splice must not be less than 12 in. for bars in tension or 8 in. for bars in compression. Development lengths do not require a strength reduction factor ϕ (Code 9.3.3), because a ϕ factor has already been applied in determining the size and area of reinforcing bars.

5.3 SPLICES

Some reinforcing bars must be spliced in almost all structures. For example, vertical bars in columns in multistoried buildings do not generally extend from foundation to roof nor are they embedded in the foundation. When the required length of a bar is more than the stock length of steel or when the bar is too long to be shipped conveniently, it must be placed in two or more pieces and spliced.

Bars can be spliced by butting and welding using American Welding Society procedures (Code 12.14.3.2). This is generally the most expensive and least reliable method due to the difficulty of field control and unfavorable chemistry of steel used in reinforcing bars.

Bars can also be spliced by a variety of mechanical connections (Code 12.14.3). Most of these are proprietary and consist of a sleeve to align the bars and hold them in position. For some compression mechanical connections, alignment is the only function of the sleeve. Compression forces are transferred by direct bearing of one bar on the other. For this reason, the bars in end-bearing splices must have ends cut square within $1\frac{1}{2}°$ (Code 12.16.6).

For tension connections and some compression connections, the sleeve transfers the tension or compression force from one bar to the other. The connection of sleeve to bar is made by threading, swaging, or filling the annular space between the bar and the sleeve with a molten metal. Mechanical connections must carry 125% of specified yield strength f_y of the bar. When bundled bars are spliced using sleeves, allowance must be made for the thickness of the sleeve. Engineers should consult manufacturers for test data, strength, availability, cost, and procedures for installation of mechanical connections.

The most common and, frequently, the most economical splice is simply to lap the two bars for a length adequate to transfer the tension or compression force. Lap splices are usually in contact, but in flexural members the bars can be separated by as much as 6 in. or one-fifth the lap length, whichever is less, as shown in Fig. 5.4, to meet practical construction conditions (Code 12.14.2.3).

Lap-splicing #14 and #18 bars to each other is not permitted by the ACI Code, because bar forces are so large they may split the concrete and destroy the effectiveness of the lap splice. Lap splices of #11 or smaller bars are permitted because the bar forces are smaller. It is permissible to splice #11 and smaller bars to #14 and #18 bars. In this case, the lap splice length required is that required for the smaller bar in the splice.

Sometimes two, three, or four bars are placed in contact in a compact bundle to act as a unit. In such cases, bundled bars are lap spliced individually, but the bundle may be anchored as a unit. Three-bar bundles require a 20% increase in development length, and four-bar bundles require a 33% increase in development length (Code 12.4) due to the difficulty of surrounding the individual bars with concrete. Lap lengths for two-bar bundles need not be increased. Bundles with more than four bars are not permitted because at least one bar in the bundle will not be surrounded by concrete.

Tests indicate that longer lap splices are required when the lapped bars have a higher stress and when more bars are spliced at the same location. The ACI Code requires a class A, B, or C splice in accordance with Table 5.1.

$$\text{Class A splice} = 1.0\ell_d$$
$$\text{Class B splice} = 1.3\ell_d$$
$$\text{Class C splice} = 1.7\ell_d$$

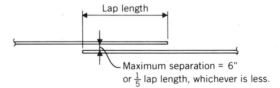

Maximum separation = 6″
or $\frac{1}{5}$ lap length, whichever is less.

FIGURE 5.4   Limits on spaced lap splice.

TABLE 5.1   Tension Lap Splices

$A_s$prov.[a]	Maximum percent of A_s spliced within required lap length[b]		
$A_s$reqd.	50	75	100
Equal to or greater than 2	Class A	Class A	Class B
Less than 2	Class B	Class C	Class C

[a] Ratio of area of reinforcement provided to area of reinforcement required by analysis at splice location.

[b] Class A splice = 1.0 ℓ_d
Class B splice = 1.3 ℓ_d
Class C splice = 1.7 ℓ_d

where ℓ_d is the tension development length of the spliced bars. Splices must always be at least 12 in. long (Code 12.15).

Regardless of the method of splicing, splices should be avoided at points of maximum stress in the bars and should be staggered wherever possible if more than one bar is spliced. The ACI Code requires that multiple welded or mechanical tension splices be staggered at least 24 in. If the area of reinforcement is more than twice that required by analysis, the tension strength at every section must be at least twice the calculated tension force but not less than the area of steel provided times 20 ksi (20 $A_s$prov.). The tension strength of unspliced reinforcement is equal to the yield strength f_y times the ratio of actual development length to required development length ℓ_d. A large number of highly stressed, closely spaced splices with minimum concrete cover might not be conservative and should be avoided.

ACI Code provisions for bar development and splices of reinforcement are general, but the following discussion is related to specific common situations.

Splices in Column Vertical Reinforcement

Compression forces are the primary forces to be resisted by most columns, but under some loading conditions, one or more faces of the column could be under tension. Regardless of computations, the ACI Code presumes that each face of each column could be in tension due to miscalculation, construction errors, temporary conditions during construction, or unexpectedly high seismic, wind, or other lateral loads. Thus, the Code requires that the tensile strength of bars in each face of a column must equal at least 25% of the tension strength of all bars in that face and at least double the computed tension force (Code 12.17). (Table A5.2 summarizes Code provisions.)

To simplify the selection of minimum required column splices, refer to Table A5.3. There are other splice possibilities not listed in Table A5.3. For example, lap splices longer than that required for compression but shorter than that required for tension could be used. In columns with many bars, less than 50% or more than 50% could be spliced at any one section. Mechanical connections with a tensile capacity less than $1.25f_y$ could be used. However, Table A5.3 will cover a large majority of all practical conditions.

In the selection of column splices, consideration should be given to congestion of steel in the splice region that might restrict the placement of concrete. Columns must have a clear space between vertical bars not less than $1.5d_b$ nor 1½ in. (Code 7.6.3). The clear space limitation applies between one contact lap splice and adjacent splices or bars. Experience shows that columns with an area of vertical steel more than about 4% of the gross area of the column will not meet this limitation easily, depending on bar and column sizes. Therefore, columns with heavy vertical reinforcement require end-bearing or mechanical connections. Where column vertical bars are less than 4% of column area, the engineer might still choose end-bearing or mechanical connections if they are less expensive or to reduce congestion of steel at beam-column joints.

Tests indicate that less length is needed for a compression lap splice than for a tension lap splice. The Code requires a compression lap splice length of $0.5f_y d_b$ ($30d_b$ for $f_y = 60$ ksi). When f_y is greater than 60 ksi, the required compression lap splice length is $(0.9f_y - 24)d_b$. When f'_c is less than 3 ksi, lap lengths must be increased by one-third. In any case, lap lengths must be at least equal to the compression development length (see Table A5.1), but this does not normally control lap lengths. (See Table A5.4 for required compression lap splice lengths under common conditions.)

In columns, bar sizes are frequently reduced in size from one story to the next higher story. In such cases, the compression lap length of the smaller bar (Table A5.4) or the compression development length of the larger bar (Table A5.1), whichever is longer, is required for the splice. The former criterion usually controls unless the bars differ in size by more than two numbers (e.g., #6 bar spliced to a #9 bar) or compression lap lengths are reduced in spiral columns as discussed below. Dowels between footing and column must not exceed the diameter of longitudinal column bars by more than one size (Code 15.8.2.3).

In spiral columns, the compression lap length may be reduced 25% (Code 12.16.4). In tied columns, the compression lap length may be reduced 17% if ties are sufficiently heavy (Code 12.16.3). An engineer may not want to take advantage of this provision because the amount of reduction in tied columns is only 6 in. for #9 and #10 bars, and 7 in. for #11 bars. The required area of ties is 0.0015 hs, where h is the largest overall thickness of the column and s is the tie spacing. Required tie spacings are shown in Table 5.2.

Unless all columns can use the shorter splice length and unless there is a large number (say 100 or more) of heavy bars (say #9, #10 or #11 bars), engineers may choose to disregard these provisions for shorter compression lap lengths.* Doing so will save design time and cost and reduce or eliminate the risk that bars with short splices are used where bars with long splices are required. Furthermore, if ties spaced closer than normal are used in the lap area, construction workers may inadvertently neglect to place the extra ties required to justify shorter lap lengths.

*The cost savings in 100 bars will be about the price of a good steak dinner for two, less the cost of the eningeer's own time spent in making the saving.

TABLE 5.2   Ties Required in Columns when Compression Lap Is Reduced 17%

Maximum Column Size (in.)	Single Tie Spacing	
	#3 ties (in.)	#4 ties[a] (in.)
12	12 c/c	NA
14	10 c/c	NA
16	9 c/c	16 c/c
18	8 c/c	14 c/c
20	7 c/c	13 c/c

[a] NA = not applicable.

For these reasons, engineers normally use shorter compression laps and extra ties in columns only where all columns in a structure can use the shorter splices or where column vertical bars have been accidentally fabricated a few inches short. If compression lap splices are reduced 17%, the column splices tabulated for typical conditions in Table A5.3 may not be conservative.

Sometimes in columns (and in other compression members as well) it is desirable to use a combination of splices that is less than 100% effective in transferring the entire tension capacity of the bars. In such cases, the Code requires that splices be staggered at least 24 in. and develop twice the tension force computed at every section and at least 20 ksi times the total area of reinforcement provided. Within the stagger distance, unspliced bars must develop the required stress by lapping (Code 12.15.4). The splice situation in Fig. 5.5 considers two groups of bars with splices staggered a distance ℓ_s. The two groups of bars are not necessarily spliced by the same means. The effectiveness R_t of the two groups together at section A-A can be taken as

$$R_t = R_1 E_1 + R_2 \left[E_2 + \left(\frac{\ell_s - s}{\ell_{d2}} \right) \right] \tag{5.2}$$

where $[E_2 + (\ell_s - s)/\ell_{d2}] \leq 1.00$,

and R_t = Ratio of tensile strength of all bars in connection with splices that are less than 100% effective to the strength of all bars where none are spliced

R_1 = Ratio of the area of group 1 bars to the area of all bars

R_2 = Ratio of the area of group 2 bars to the area of all bars

E_1 = Effectiveness in tension of splices in group 1 bars

E_2 = Effectiveness in tension of splices in group 2 bars

s = Spacing between individual group 1 and group 2 bars, if in excess of 6 in.

ℓ_s = Stagger distance between splices in group 1 and group 2 bars (not less than 24 in.)

ℓ_{d2} = Development length in tension of group 2 bars.

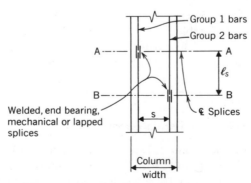

FIGURE 5.5 Elevation of one face of column.

If splices in group 1 and group 2 bars are different, the effectiveness of the connection should also be checked at section B-B.

The Code is silent on requirements for noncontact lapped splices in columns; however, it is reasonable to allow a 6-in. spacing as for flexural members (Code 12.14.2.3) and add to the required lap length when the spacing is greater than 6 in. In most four-bar columns, the spacing between corner bars will exceed 6 in. The term $(\ell_s - s)$ in Eq. 5.2 makes allowance for bar spacing greater than 6 in.

Equation 5.2 was used to develop Table A5.3.

Example 5.1 What is the effectiveness in tension of column bars in one face of a 12- × 12-in. column with 4 #10 bars and $f_c' = 5$ ksi when all bars are spliced with end-bearing splices and alternate bars are staggered one tension development length?

Solution The effectiveness in tension of end-bearing splices is zero. From Table A5.1, $\ell_d = 43''$ for #10 bars and group 1 and 2 bars are each 50% of the total. Bars are spaced about $7''$ c/c; therefore, the development length $= 0.80\ell_{db} = 0.8 \times 43 = 35''$. Substituting in Eq. 5.2 yields

$$R_t = 0.5 \times 0 + 0.5 \left[\frac{0 + (35 - 7)}{35} \right] = 0.40$$

Example 5.2 Using the same column as in Ex. 5.1, what is the minimum stagger required to provide tensile strength in each face of the column equal to 25% the area of vertical reinforcement multiplied by f_y?

Solution Inserting the data from Ex. 5.1 into Eq. 5.2 and dropping unnecessary factors,

$$R_t = 0.25 = 0.5 \left[\frac{\ell_s - 7}{35} \right]$$

$$\ell_s = \frac{35 \times 0.25}{0.5} + 7 = 25''$$

Because $25''$ is greater than the minimum lap length of $24''$, the computed lap length is acceptable.

Although end-bearing splices in all four vertical bars meet Code requirements, the arrangement would not be satisfactory if a diagonal moment could cause tension in one corner of the column.

Splices in Flexural Reinforcement

Splices in beams and slabs are much less common than in columns; however, the following situations normally require splices. To facilitate placing, lap splices are usually used rather than welded, end-bearing, or mechanical splices.

1. Bottom bars at a support required for compression reinforcement.
2. Bottom bars at a support when the flexural member is part of a primary lateral load-resisting system.
3. Top bars made continuous because negative moment exists over the entire span.
4. Bottom and truss bars in long spans where the required length of bar exceeds the available stock length or the maximum length for shipping or placing.

Situation 1 Bottom bar compression reinforcement spliced at a support. The length of lap required can be computed the same as for columns (Table A5.4), except that lap length may be reduced 17% if the area of stirrups in the lap area equals $0.0015b_w s$, where b_w is the width of the beam and s is the spacing of stirrups. (An area of $0.0015b_w s$ is almost twice the minimum stirrup area required.) In addition, lap length may be reduced proportionally if more steel is provided than is required.

Example 5.3 What lap is required for 2 #9 bars on each side of a support in which only 60% of the steel is needed for compression reinforcement? Assume $f_y = 60$ ksi, $f'_c = 3$ ksi, normal weight concrete and stirrups near the support exceed $0.0015b_w s$.

Solution From Table A5.4, compression lap length for #9 bars is $2' - 10'' = 34''$. Lap length can be reduced to 83% because of stirrups and reduced to 60% for excess reinforcement. The required lap length is

$$34'' \times 0.83 \times 0.60 = 17'' = 1' - 5''$$

Situation 2 Bottom bar tensile reinforcement spliced at a support. Wherever lateral loads (or other loading conditions) cause a reversal of normal moments and a positive moment exists at the support, tensile stress in the bottom bars must be developed. In addition, when the flexural member is part of a primary lateral load-resisting system, the Code requires that at least one-third the positive moment reinforcement in simple members and one-fourth the positive moment reinforcement in continuous members extend into the support and be fully anchored in tension at the face of the support. Full anchorage is required even though no computed tensile stress exists (Code 12.11.2). This provision does not apply if the flexural members contribute to the lateral load resistance but are not the primary system. Development of the tensile stress and anchorage of the bottom bars may be accomplished by lapping with bottom bars from the adjacent span or by extending the bars into the support a sufficient distance.

Example 5.4 What lap length or anchorage is required for bottom bars in beams of a primary lateral load-carrying system in which no computed tensile stress exists? At the given support, 2 #8 bars from one span and 2 #9 bars from the other span are required, and beam width is such that the bar spacing is greater than 6 in. and there is more than 3 in. clear to the bars on each side. Assume $f_y = 60$ ksi, $f'_c = 3$ ksi, and normal weight concrete.

Solution $A_s$prov./$A_s$reqd. is greater than 2, but both bars (100%) are spliced at the same location; therefore, a Class B splice ($1.3\ell_d$) is required (see Table 5.1).

From Table A5.1, tension ℓ_{db} = 34.6″ for #8 bars and 43.8″ for #9 bars. These lengths can be reduced to 80% for the wide bar spacing.

$$\text{Required lap length} = 34.6 \times 1.3 \times 0.8 = 36″ \ (3′-0″) \text{ for #8 bars}$$
$$= 43.8 \times 1.3 \times 0.8 = 46″ \ (3′-10″) \text{ for #9 bars}$$

The larger bar (#9) must be anchored behind the face of the support (43.8 × 0.8 = 35″), as shown in Fig. 5.6, or the longer lap length for the #9 bar must be used.

At the exterior support, bottom bars would probably need to end in a standard hook (see Section 5.5) because supoprts are rarely as wide as the required anchorage length.

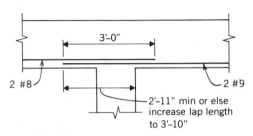

FIGURE 5.6

Situation 3   Top bar tension reinforcement spliced at mid-span. When continuous top bars are required, they are usually spliced at mid-span where the moment and stress in the bars are lowest. The required lap length is the basic development length multiplied by applicable factors to obtain the development length and by another factor for the class of splice required.

Example 5.5   What lap length is required for 2 #6 continuous top bars in a 12-in.-wide by 16-in.-deep T-beam shown in Fig. 5.7? Both bars are lapped at mid-span where capacity of the two bars is 1.8 times that required by analysis. Assume f_y = 60 ksi, f'_c = 4 ksi, and normal weight concrete.

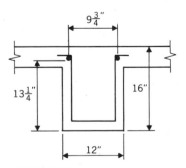

FIGURE 5.7

Solution   From Table A5.1, tension $\ell_{db} = 18''$ for #6 bars. From Table 5.1, a Class C splice is required.

Necessary modifying coefficients are as follows:

Top reinforcement	1.4
Spacing > 6″ c/c	0.8
$A_s$reqd./$A_s$prov.	1/1.8
Class C splice	1.7

Combining all coefficients,

$$\text{Lap length} = 18 \times 1.4 \times 0.8 \times \frac{1.7}{1.8} = 19'' = 1'-7''$$

Situation 4   Bottom and truss bar spliced at random locations. Occasionally, long spans or other special conditions require flexural bars that are longer than stock lengths or longer than can be shipped or placed conveniently. In such cases, the engineer should select splice locations where the moment and steel stress are lowest. These locations are likely to have the fewest number of bars, thus minimizing the number that need be spliced and affording more space for the laps of those that do need splicing. Required lap length can be computed from the basic development length multiplied by appropriate factors.

Example 5.6   What lap length is required for #7 slab bottom bars spaced 10 in. c/c where only 50% are spliced within the lap length and $A_s$prov./$A_s$reqd. = 1.4? Assume $f_y = 60$ ksi, $f_c' = 3.5$ ksi, and sand-lightweight concrete is used.

Solution   Interpolating from Table A5.1, tension $\ell_{db} = 26.3''$ for #7 bars and $f_c' = 3.5$. From Table 5.1, a class B splice is required. Modifying coefficients are as follows:

$$f_c' = 3.5 \qquad\qquad \sqrt{\frac{3}{3.5}} = 0.93$$

$$\text{Sand-lightweight concrete,} \quad 1.18$$

$$\text{Spacing} > 6'' \text{ c/c,} \qquad\quad 0.8$$

Combining all coefficients,

$$\text{Lap length} = 26.3 \times 0.93 \times 1.18 \times 0.8 \times 1.3 = 30'' = 2'-6''$$

Splices in Temperature Reinforcement

Assuming that no calculated stress exists on temperature reinforcement, that all bars are spaced 6″ c/c or more, and that all bars are spliced at one section, a Class B splice is required. (Splice lengths are tabulated in Table A5.5.)

If all splices are staggered at least one lap length, a class A splice may be used, and lap lengths in Table A5.5 divided by a factor of 1.3. A minimum length of 1 ft is still required. If concrete strength higher than $f'_c = 3$ ksi is used, no reduction in lap length for #3, #4 or #5 bars is permitted, and lap length for larger bars is reduced only slightly. For example, for $f'_c = 4$ ksi, lap length of #6 bars is reduced 1 in. and lap length of #7 bars is reduced 3 in.

Splices in Welded Wire Fabric

In welded wire fabric (WWF) there is a large variety of wire sizes and spacings possible for both directions. Wires can be smooth or deformed. Thus, tabulation of development lengths and lap splice lengths of WWF is voluminous. It is more efficient for the engineer to compute development and splice lengths for each case encountered, especially because cost and manufacturing considerations will limit the use of WWF to projects with large quantities of one style. CRSI has tabulated some of the more common splice requirements for WWF [3].

Requirements for development length are discussed in Section 5.2. Lap splice length of deformed WWF must equal $1.7\ell_d$, at least 8 in., and the outermost cross wires of each fabric sheet must overlap at least 2 in. The Code assumes implicitly that a cross wire on deformed fabric will serve as an anchor to develop a stress of 20 ksi in the wire. For smooth wire, at least two cross wires are needed to fully develop the wire under consideration because ASTM specifications require that each intersection of wires has an average shear strength of 35 ksi times the area of the larger wire. Because no bond strength on smooth wires is assumed, two cross wire intersections are needed to develop full strength of a smooth wire. The outermost cross wires of each fabric sheet in smooth WWF must overlap at least one spacing of cross wires plus 2 in.

5.4 ANCHORAGE AND BAR CUTOFF IN FLEXURAL MEMBERS

The computed stress (either tension or compression) in reinforcement at each section must be developed on each side of that section. Development can be by embedment length, end anchorage, or both. The bend in truss bars effectively anchors the bar in both negative and positive moment regions. In addition, reinforcement must extend beyond the point at which it is no longer needed. This extension must be equal to the effective depth d or 12 bar diameters d_c, whichever is greater (Code 12.1 and 12.10).

Tests and experience show that a bar ending in a tension zone tends to cause cracks near the end of the bar. Because these cracks reduce the shear capacity of the member, the ACI Code requires additional precautions for shear near the end of the terminated bar (Code 12.10.5). Unless there is some very good reason for doing so, it is best not to terminate bars in tension zones as the construction cost savings will be small or even nonexistent and design costs will be substantial.

Further requirements for bar anchorage and cutoff are discussed separately for negative moment and positive moment reinforcement.

Top Bars

In addition to general requirements for anchorage of reinforcement, the cutoff point of top bars (for negative moment) is determined as follows:

1. Each bar must extend a distance equal to its development length ℓ_d beyond the point of its maximum stress.

2. Each bar must extend beyond the point where it is no longer needed to resist flexure, a distance at least equal to the effective depth d of the number, or 12 bar diameters d_b, whichever is greater, except at ends of free cantilevers.

3. At least one third of the total tension reinforcement* must extend beyond the point of inflection, a distance equal to the effective depth of the member d, $12d_b$, or $\ell_n/16$, whichever is greater.

Top bar cutoff requirements are summarized in Fig. 5.8, which shows three different lengths of top bars for purposes of illustration. Note that the moment capacity curve is sloped to the left of the bar cutoff because a bar cannot be fully stressed at its end. The horizontal distance of the slope is equal to the development length of the bar.

Although computation of locations of the point of inflection and points at which bars are no longer needed is straightforward, it is tedious and time consuming. Fortunately, cutoff points for the large majority of top bars can be determined by simple rules of thumb. The bar cutoff diagram shown in Fig. A5.1 will suffice for most typical conditions. Bars cut off according to the diagram will meet requirements of the Code provided the following conditions are met:

1. The span-to-depth ratio equals $\ell_n/d \geqslant 16$. If a beam or slab does not meet this limit all top bars should be extended an additional distance equal to $d - \ell_n/16$ or the cutoff point calculated from basic principles.

2. Top bars are not larger than those given in Table A5.6 for the span of the beam or slab. In some cases, larger bars might still meet Code requirements because conservative assumptions were made in producing Table A5.6 and Fig. A5.1; however, additional computations would be required. If larger bars are used, it is simpler to extend the bars to $\ell/3$ and $\ell/4$, where ℓ is the span length given in Table A5.6 for the bar size used.

3. Distance from face of support to point of inflection $\leqslant \ell_n/4$. In most cases, this limit will be met if loads are uniformly distributed, the larger of two adjacent spans is not greater than the shorter by more than 20%, and negative moments do not exceed the approximate moments given by coefficients in Table A13.2.†

---

*The bar area, not the number of bars, is the significant parameter. Thus, one third of the top bars is the number of bars necessary to equal or exceed one third the required steel area.

†It is not necessary to compute moments by coefficients given in Table A13.2, but computed moments must not exceed moments computed by coefficients. This will be reasonably assured if the structure falls within the limitations for computing moments using Table A13.2.

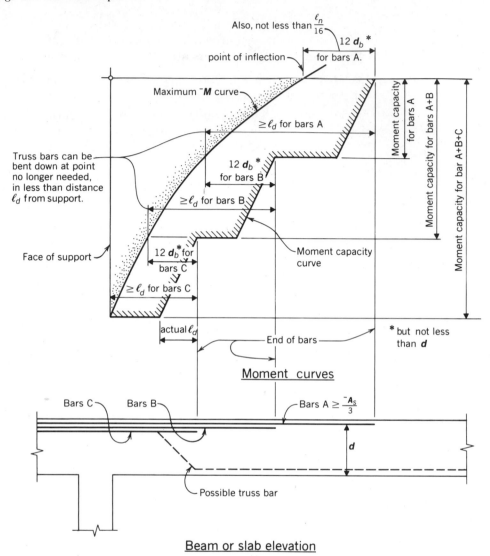

FIGURE 5.8 Code requirements for length of negative moment reinforcement in flexural members.

Points of inflection may be farther from the support than assumed for Fig. A5.1 or negative moment may exist across the entire span if these conditions are not met as, for example, if one span is loaded more heavily or is much longer than the adjacent span. In such cases, it is best to compute the location of the point of inflection and extend at least one half of the bars beyond it, as shown in Fig. 5.8. The point at which remaining steel is no longer needed must still be computed or estimated, or all steel must be extended beyond the point of inflection. However, unless the spans are long, the bars are heavy, and many are involved, the construction cost saving in using shorter bars will be small and not worth the design effort.

As suggested by Fig. A5.1, it is best to stagger the cutoff of top bars to avoid an abrupt change in negative moment capacity. Bars cut off in tension zones may require additional stirrups and added length of the continuing bars (Code 12.10).

Short or lightly loaded spans compared to adjacent spans may have the points of inflection farther from the support than under normal conditions, or even have moment reversal at mid-span. Engineers must be alert for such conditions, as shown in Fig. 5.9, and extend top bars as necessary.

At exterior supports, top bars required to resist negative moment must be anchored into the support. Bars used solely for other purposes, such as stirrup support, need not be anchored. Anchorage will require a hook if the support is not wide enough to accept the straight development length of the top bar (see Chapter 10, Joints).

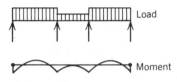

FIGURE 5.9 Load conditions creating negative moment at mid-span.

Example 5.7 What extension beyond the face of support should #8 top bars have in a 12-in.-deep continuous beam with a 13-ft clear span? Assume $f'_c = 3$ ksi and $f_y = 60$ ksi.

Solution From Table A5.6, a #8 bar is acceptable for a 13-ft span. Therefore, required extensions of #8 bars are as follows:

Longest bar: $13/3 = 4' - 4''$
Shortest bar: $13/4 = 3' - 3''$

As a cross-check, from Table A5.1, $\ell_{db} = \ell_d = 34.6''$ for a #8 bottom bar ($< 12''$ of concrete cast below the bar). Because $34.6'' < 3' - 3''$, the typical cutoff point is acceptable.

Bottom Bars

Based on a long tradition of successful usage, the ACI Code requires that at least one-third the positive moment reinforcement in simple span members and one-fourth the positive moment reinforcement in continuous members be extended into the support. In beams, such bars must extend at least 6 in. into the support. In slabs, the ACI Code has no minimum extension of bottom bars into supports. However, conservative engineers will be generous with embedment of all bottom bars in the support.

Bottom bars not extended into the support nor bent up as truss bars can be terminated beyond the point where they are no longer needed. However, it is advisable to extend all slab and beam bars into the support or just short of the support when the support is a column. Bars should be terminated in a region of compressive stress because bars terminating in a zone of flexural tension reduce shear strength of the member and additional stirrups may be needed to compensate (Code 12.10.5). Procedures are available for computing the length of bottom bars cut off within the span. The procedures result in laborious calculations, save little steel, and risk misplacement of bars in the field. For these reasons, cutoff within the span is not recommended and calculation procedures are not discussed further in this text.

Beam bars extended into a column may result in a heavy congestion of steel that prevents proper placement of the steel or the concrete, or both. In such cases, only the minimum number of bars (as discussed above) need be extended into the column.

The ACI Code limits the size of bottom bars so as to ensure that the stress at each section can be developed. Because the positive moment curve is rounded, the critical section is usually not at mid-span where moment and stress are largest. At some distance from mid-span, the stress is reduce slightly, but the available distance to develop the bar is reduced substantially. The Code accomplishes the development objective indirectly by requiring that at simple supports and at points of inflection, positive moment tension bars must be limited to a diameter such that

$$\ell_d \leq \frac{M_n}{V_u} + \ell_a \qquad (5.3)$$

where
ℓ_d = development length (see Section 5.2)
M_n = nominal moment strength, assuming all reinforcement at the section to be stressed to the specified yield strength f_y,
V_u = factored shear force at the section.

At a support,

ℓ_a = embedment length beyond the center of support;

At a point of inflection,

ℓ_a = effective depth of member d, or $12d_b$, whichever is greater.

The distance M_n/V_u can be increased 30% if the ends of the bottom bars are confined by a compressive reaction, as shown in Fig. 5.10.

Use of Eq. 5.3 can be visualized by reference to Fig. 5.11.

Use of Eq. 5.3 can be avoided in most cases if bars are made no larger than those given in Table A5.6. Table A5.6 is conservative for assumed conditions.* It is also conservative for most conditions not meeting the assumptions. Additional computa-

---

*Although Table A5.6 is frequently conservative by one or two bar sizes, there is usually little practical effect because bar size is usually limited by other factors, such as spacing to meet limits of crack control.

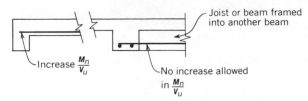

FIGURE 5.10 Bottom bar anchorage confinement.

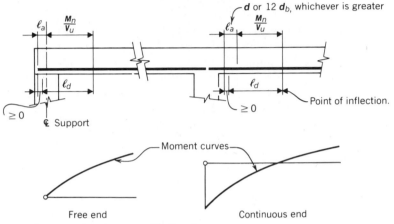

FIGURE 5.11 Development of bottom bars.

tions would be needed to determine whether larger bottom bars or shorter spans than those given by Table A5.6 would still meet Code requirements. For conditions in which Table A5.6 might not be conservative (see assumptions listed in the table), the maximum span can easily be computed by a coefficient given in Table A5.6 footnotes. Table A5.6 does not apply to the case of high negative moment with small positive moment and consequent small area of positive reinforcement required. The bar sizes for such cases can be conservatively estimated by limiting the bottom bar size to the size required for the span between inflection points.

Spans for bottom bars in Table A5.6 were computed for the case of the free end (no negative moment) of an end span of a continuous flexural member in which:

$$\text{positive moment, } \boldsymbol{M}_u = \boldsymbol{w}_u \ell_n^2 / 11 = 0.9 \boldsymbol{M}_n$$

$$\text{and shear at free end, } \boldsymbol{V}_u = 0.426 \boldsymbol{w}_u \ell_n$$

Substituting these values for $\boldsymbol{M}_n$ and $\boldsymbol{V}_u$ into Eq. 5.3,

$$\ell_d \leq \frac{\boldsymbol{M}_n}{\boldsymbol{V}_u} + \ell_a = \frac{\boldsymbol{w}_u \ell_n^2}{0.9 \times 11 \times 0.426 \boldsymbol{w}_u \ell_n} + \ell_a$$

or

$$\ell_n \geq 4.22(\ell_d - \ell_a) \tag{5.4}$$

Using the same procedure, for simple spans,

$$\ell_n \geqslant 3.6(\ell_d - \ell_a) \tag{5.5}$$

Using the same procedure, for continuous spans with equal end moments,

$$\ell_n \geqslant 5.1(\ell_d - \ell_a) \tag{5.6}$$

Because Eq. 5.5 yields lower values than Eq. 5.4, it is evident that the simple span case does not control. Conservatively assuming $\ell_a = 0$ for simple spans and for the free end of end spans, and assuming $\ell_a = 12d_b$ for continuous ends, it is evident that the case of continuous spans does not control. Therefore, minimum span lengths for bottom bars in Table A5.6 were computed using Eq. 5.4.

Table A5.6 might not be conservative for members with one end continuous, the other end free, and the maximum positive moment less than $w_u \ell_n^2/11$.

In simple spans the maximum bar size for a given span can be exceeded if the bar is anchored beyond the support centerline with a standard hook or equivalent mechanical anchorage. However, this option may present other problems, such as lack of space to embed the hook or anchorage.

Example 5.8   What is the maximum size bottom bar that can be used in a 15-ft simple span slab without end anchorage or confinement? Assume $f'_c = 4$ ksi, all lightweight concrete, and other conditions as listed in the assumptions for Table A5.6.

Solution   From Table A5.6, use #8 bar, or, from Eq. 5.5,

$$\ell_s \leqslant \ell_n/3.6 = 15 \times 12/3.6 = 50''$$

From Table 5.1, use #10 bar, $\ell_d = 48.2.''$

A larger bar size was justified because $f'_c = 4$ ksi in this example, and Table A5.6 is based on $f'_c = 3$ ksi and because Eq. 5.5 for a simple span was used in this example; whereas Eq. 5.4 for an end span was used for Table A5.6. In most practical conditions, even a #8 bar will be larger than needed in a 15-ft span.

Example 5.9   What is the maximum size bottom bar permissible in a 22-ft clear span with $^+M_u = w_u \ell_n^2/24$? Assume bottom bars are extended into, and therefore confined by, the support and that other conditions are those assumed by Table A5.6.

Solution   Distance between points of inflection can be computed as the square root of the ratio of moment coefficients

$$\sqrt{8/24} = 0.577$$

Minimum equivalent simple span = $22' \times 0.577 = 12.7'$
Divide the minimum span from Table A5.6 by 1.3 to account for confinement at the support.

From Table A5.6, select a #9 bar or smaller.

$$\text{Minimum clear span} = 15.5/1.3 = 11.9 \qquad \underline{\text{OK}}$$

Alternatively, using Eq. 5.5 for a simple span and, assuming $\ell_a = 0$,

$$\ell_d \leq \frac{0.577 \times (22 \times 12) \times 1.3}{3.6} = 55''$$

A #9 bar with $\ell_d = 43.8''$ is satisfactory.

A bottom bar bent up and made continuous with negative moment steel (i.e., a truss bar) need not be anchored in the area of the bend but only at each free end of the bar. Of course, even at the bend, the bar must extend beyond the point at which it is no longer needed to resist moment in both the negative and positive moment regions.

Two-Way Slab Bars

For two-way slabs without beams, the ACI Code specifies minimum bend point locations and bar extensions (Code 13.4.8). Code requirements are shown in Fig. A5.2, omitting reference to truss bars. As noted in Chapter 4, truss bars should not be used because an intricate placement order is required. The following modifications to these requirements are recommended. The Code permits straight bottom bars of three different lengths in any one panel; however, just one length (the longest length) should be used in many cases. Some additional material will be needed, but bar fabrication and placement will be greatly simplified. The possibility of misplacing short bars where longer bars are required will be decreased. Splicing bottom bars with a tension lap at the column centerline should be considered as an added precaution against progressive collapse.

Top bars should also be the same length for a given panel to simplify fabrication and placement. The engineer might compromise and specify one length for column strip top bars and one length for middle strip top bars.

Regardless of the above recommendations, the engineer might consider straight top and bottom bars of two or three lengths as permitted by the Code when the area of slab exceeds about 30,000 ft^2 and bay sizes and reinforcing are repeated many times. In this case, workers will have time to thoroughly understand the engineer's placement instructions with less chance for error. Also, quantities of each bar length should be sufficient to gain the economy of repetition.

Stirrup Bars

Rather than compute the development of stirrup bars, it is easier to compute the depth of a beam necessary to develop the stress in stirrups for typical situations in the engineer's practice. This has been done for common conditions in Table A5.7. For other conditions, the engineer can prepare similar tables (if the conditions are repeated frequently) or compute the required stirrup size and type (if the conditions are met only occasionally). Fig 5.12 shows the Code requirements for developing stirrups (Code 12.13) and the dimensional assumptions used in preparing Table A5.7.

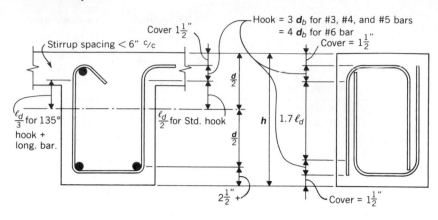

(a) Open stirrups

(b) Lapped U-shaped stirrups and ties.

FIGURE 5.12 Development of stirrup bars.

Other Anchorages

Other commonly encountered situations requiring anchorage of reinforcing bars are as follows:

1. Column dowels. Required minimum embedment of column dowels into the foundation is the basic development length in compression (see Table A5.1). Dowels must also transmit computed tension forces to the foundation. If tension forces are large, the development length in tension may exceed the development length in compression.

2. Dowels in construction joints to anchor brackets, walls, and other construction to be built later.* If the dowels have no computed stress, they may be considered to be acting in shear and need only extend into the concrete on each side of the joint a nominal distance, say 8 in. or 1 ft. Dowels carrying a computed stress (tension, compression, or shear) should be designed to develop the stress on each side of the construction joint. The most common problem with dowels is that there is not enough space to extend the bar straight into the member on one or both sides of the construction joint. In such cases, hooks or other anchorages must be used. Mechanical connectors are sometimes used to avoid cutting formwork to pass the dowel (also see Chapter 10, Joints).

3. Exterior supports for beams and slabs. If a support provides moment resistance to the end of a flexural member, flexural reinforcing in the member (top bars in particular) must be anchored into the support (see Chapter 10, Joints).

---

*"Monolithic" is a highly inaccurate term to describe cast-in-place concrete as construction almost always proceeds in stages for the convenience of the builder. Dowels are frequently used to tie the first piece of concrete to the next piece to help assure behavior resembling monolithic action.

5.5 HOOKS

When bars are bent in a 90° to 180° hook, the tension force in a bar can be developed in a shorter distance than that required for straight bar embedment. ACI Code requirements for such bends, called "standard hooks," are summarized in Fig. 5.13. Standard hooks are usually necessary for negative moment bars at the discontinuous ends of beams and are required at the ends of stirrup and tie bars. They are also useful in other situations. In detailing reinforcement, the total width of the hook as given in Table A5.8 must fit within concrete outlines after allowing for concrete cover on sides and ends.

Hooks are not effective for developing compression force in bars.

Minimum bend diameters are established for two reasons. The first is to prevent fracture of the bar due to bending. Bars manufactured to ASTM Standards are required to meet bend tests that are only slightly more restrictive than the minimum bends permitted by the ACI Code.

The second reason is to limit high local bearing pressures at the inside of the bend. Referring to Fig. 5.14, the force T in a bar is $\pi d_b^2 f_s / 4$ and the unit-bearing pressure p on concrete under the inside radius of the bend is

$$p = \frac{2T}{K d_b^2} = \frac{\pi f_s}{2K}$$

where K is the ratio of bend diameter to bar size. For example, the pressure on the inside of the hook of a bar with a $6d_b$ diameter bend is $\pi f_s / (2 \times 6) \simeq f_s / 4$. Not only

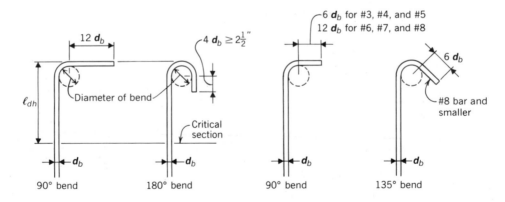

(a) For bars other than stirrups and ties;   (b) for stirrups and ties.

Bar size	Min. inside diam. of bend
For stirrups and ties	
#3, #4, & #5,	$4\,d_b$
#6, #7, & #8	$6\,d_b$
For other bars:	
#3 through #8	$6\,d_b$
#9, #10, & #11	$8\,d_b$
#14, & #18	$10\,d_b$

FIGURE 5.13 Standard hooks.

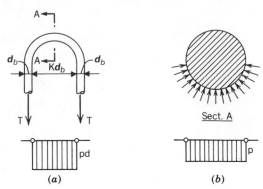

FIGURE 5.14   Bearing stresses at bend in bars.

do pressures approach or exceed cylinder strength of concrete but radial pressures on the bar tend to split the concrete in the plane of the hook [see Fig. 5.14(b)].

The ACI Code gives the basic development length ℓ_{hb} of a standard hooked bar in tension with $f_y = 60$ ksi, measured from the critical section to the outside end of the hook, as

$$\ell_{hb} = 1200 d_b / \sqrt{f'_c}$$

where f'_c is in psi.

The required development length ℓ_{dh} is the basic development length multiplied by factors for yield strength of bar, excess reinforcement, concrete cover, lateral ties, and lightweight concrete [4, 5]. Development lengths of standard hooks and modifying factors are summarized in Table A5.9.

PROBLEMS

1. Using ACI Code requirements, calculate the tension development length required for #9 top bars with $f_y = 60$ ksi in all-lightweight concrete with $f'_c = 4.5$ ksi when the spacing is 8 in. c/c and 94% of the area of bars is required by analysis.
2. For a 20-in. square column with 8 #11 vertical bars placed symmetrically, what length of lap splice is required if $f_y = 60$ ksi, $f'_c = 5$ ksi, and there is no tension on any face of the column? For the same column, select an acceptable arrangement for end-bearing splices.
3. For a 20-in. square column with 8 #6 vertical bars placed symmetrically, what length of lap splice is required if $f_y = 60$ ksi, $f'_c = 3$ ksi, and tension on one face of the column stresses the bars on that face to 85% of their capacity in tension?
4. Two beams that are members of the primary lateral load-carrying system frame into a column from opposite sides. One beam is reinforced with 2 #11 plus 3 #10 bottom bars and the other beam is reinforced with 3 #9 plus 2 #8 bottom bars. There is no calculated flexural tension in the bottom of the beam under any loading condition. Which bars should extend past the face of the column and how far should they extend? Assume $f_y = 50$ ksi and $f'_c = 4$ ksi.
5. Continuous top bars, #5 at 10 in. c/c, in an 8-in. slab are spliced at mid-span where $A_s$furn. is 25% more than $A_s$reqd. What length of lap splice is required if $f_y = 60$ ksi and $f'_c = 4$ ksi?

6. In a 36-in.-wide by 20½-in.-deep beam with interior exposure, uniformly loaded, spanning 27 ft-6 in. c/c between 18-in. square columns, negative moment at each end of the face of the column is 596 ft-kips and the corresponding positive moment at mid-span is 127 ft-kips. Assuming f_y = 60 ksi and f'_c = 3 ksi, select top bars to resist negative moment and determine their minimum length.

7. A 14-in.-wide by 24-in.-deep simple span T-beam supported by 18-in. square columns spaced 17 ft.-4 in. c/c, carries a factored load of 17k/ft. Assume end negative moments are negligible so that the point of inflection is at the face of the support. Can #11 bottom bars be anchored in the span? Assume f_y = 60 ksi and f'_c = 3 ksi.

8. Can #4 stirrups be anchored in a 16-in.-deep beam if f_y = 40 ksi and f'_c = 4 ksi? If not, what is the maximum usable yield strength for the stirrup bar?

Selected References

1. C.O. Orangun, J.O. Jirsa, and J.E. Breen. "A Reevaluation of Test Data on Development Length and Splices," *ACI Journal, Proc.*, V.74, pp 114–122, March 1977.
2. Phil M. Ferguson. "Small Bar Spacing or Cover—A Bond Problem for the Designer," *ACI Journal, Proc.*, V.74: pp 435–439, September 1977.
3. *Reinforcement Anchorages and Splices.* 2nd ed. Concrete Reinforcing Steel Institute, Chicago, 1984.
4. ACI Committee 408. "Suggested Development, Splice, and Standard Hook Provisions for Deformed Bars in Tension," *Concrete International Design and Construction*, V.1, No.7, pp 44–46, July 1979. See also *ACI Manual of Concrete Practice, Part 3, 1985.*
5. James O. Jirsa, LeRoy A. Lutz, and Peter Gergely. "Rationale for Suggested Development, Splice, and Standard Hook Provisions for Deformed Bars in Tension," *Concrete International Design and Construction*, V.1, No.7, pp 47–61, July 1979.

CHAPTER 6

SHEAR AND TORSION IN BEAMS

6.1 OBJECTIVES

In most beams, shear strength is as important a consideration as flexural strength. Safety must be assured in either mode, but if failure occurs in shear, it is more likely to be sudden and brittle than when it occurs in flexure. Shear strength of a beam or slab depends in part on tensile strength of concrete, which is more erratic than compressive strength (see Section 1.6). Therefore, a larger factor of safety and more care should be taken in design to resist shear than flexure. Considering the relatively small weight of steel used in shear reinforcement, it is the most effective reinforcement used in a structure to ensure structural integrity.

Torsion or twisting of concrete members produces stresses similar to transverse shear. Torsional failure can be as sudden and brittle as a shear failure. If torsion is statically determinate, as in an isolated beam with eccentric loads on one side only, consequences of a torsional failure could be sudden and severe and must be avoided. If torsion is statically indeterminate, as in a spandrel beam, the torsional moment will be redistributed into flexural members after the first cracking of the concrete, and consequences of a torsional failure will be less severe.

Following are objectives of design for shear and torsion in linear flexural members, that is, beams and one-way slabs:

1. Determine whether stirrups, larger concrete outlines, or other precautions are needed to resist shear.
2. If stirrups are needed, compute the length of the member over which stirrups must be spaced.
3. Select stirrup type, size, and spacing.
4. Design other details required for brackets, deep beams, walls, and for shear friction.

Definitions of statically determinate torsion, statically indeterminate torsion, stirrups, and shear friction are given later in this chapter.

Anchorage of shear reinforcement is covered in Chapter 5. Discussion of design procedures in this chapter assumes that concrete outlines have been determined previously by methods given in Chapter 14, Preliminary Design.

6.2 SHEAR BEHAVIOR

For simplicity, shear design is normally based on simple span shear diagrams and full loading. Engineers must be alert to conditions resulting in higher shear over a portion of the span, such as the two conditions illustrated in Fig. 6.1. Full live load on only part of the beam causes higher shear near mid-span than does full live load over the entire span. Because full live load over part of the span is unlikely to occur, this situation is usually disregarded except in extreme cases with known conditions of partial live load such as might be found in a library, warehouse, or heavy industrial plant. Increase in shear takes place in the area of low shear. Minimum requirements for stirrup size and spacing frequently provide considerable shear capacity where increase in shear might otherwise be a problem.

If end moments are nearly equal, engineers usually disregard the effect of unbalanced moments. But end spans with a small negative moment restraint at the exterior support will have a significant increase in shear at the first interior support, as shown in Fig. 6.1(*b*). End spans are normally designed for this increase in shear. The ACI Code indirectly allows a simplification for normal conditions by specifying a 15% increase in shear at the face of the first interior support in end members and no increase at the face of all other supports (Code 8.3.3). For conditions not meeting

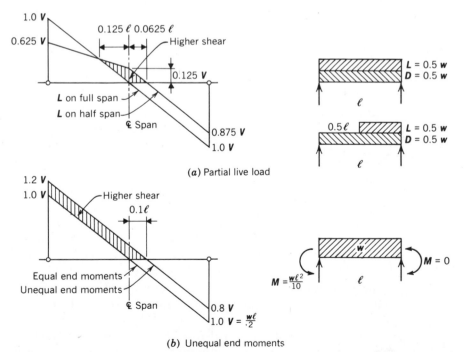

FIGURE 6.1   Conditions causing higher shear than caused by uniform load on simple span.

the limits of Section 8.3.3 of the ACI Code (see discussion of moment coefficients in Section 13.3), the increase in shear should be computed whenever there is a large difference in end moments, even in interior spans.

Procedures for design of reinforced concrete to resist shear are not entirely rational because an adequate theory does not yet exist to explain the observed shear behavior. Instead, shear design is based on successful longtime experience and fitting design equations and procedures to the best test evidence, as well as being based on rational analysis [1].

Consider a beam of homogeneous, isotropic material shown in Fig. 6.2(a). A small square element below the neutral axis is subjected to forces shown in Fig. 6.2(b), which can be resolved into diagonal tensile forces, as shown in Fig. 6.2(c). Shear forces **v** on each side of the element are equal, but magnitude of tensile force **t** relative to the shear force will vary according to the location of the element in the beam. Thus, magnitude of the diagonal tensile force t_1 as well as its angle of slope, will vary with location of the element in the beam. Force t_1 is called "diagonal tension." Strength of concrete in pure shear is almost as high as its compressive strength, but practical application of shear always causes diagonal tension. Despite misleading terminology, shear design is actually design to resist diagonal tension forces caused by shear.

Tensile stress trajectories are shown in Fig. 6.2(d). Note that the trajectory is horizontal at mid-span at the bottom of the beam where flexural reinforcement resists tensile stress. The trajectory crosses the neutral axis at a 45° angle and is the principal concern in design of shear reinforcement. At the top of the beam, the trajectory is vertical, but tensile stresses are usually small and therefore of minor concern. In long, thin beams, a diagonal shear crack will propagate into the compression zone and become nearly horizontal, thus giving graphic evidence of vertical tensile stress.

Typical cracks resulting from tensile stresses are shown in Fig. 6.2(e). In shallow beams, vertical or sloping shear reinforcement must cross the diagonal cracks. In deep beams and walls, horizontal, as well as vertical, shear reinforcement is required to ensure that all potential cracks are crossed by reinforcement.

Before a concrete member is cracked, it is stressed approximately as shown in Fig. 6.2(a) to 6.2(d). Shear reinforcement is stressed only in proportion to the stress in the surrounding concrete, that is, with very low stress. As load on the member is increased, probably the first cracks to form will be those at mid-span and at the support where flexural stresses are highest. When the load is increased still further, diagonal cracks in areas of high shear will form. Cracks start at the tension face of the member and propagate diagonally toward the compression face. If no shear reinforcement is present, cracks may continue through the compression zone, cutting the member in two, causing immediate failure. Sometimes diagonal cracks stop at the compression zone and the member will carry additional load before failure, but such behavior is unpredictable. For this reason, the ACI Code assumes a low value for shear strength provided by concrete and requires reinforcement for shear in excess of this assumed strength.

After the occurrence of the first diagonal tension crack caused by shear, it is possible for concrete to continue to carry as much shear as caused the first crack

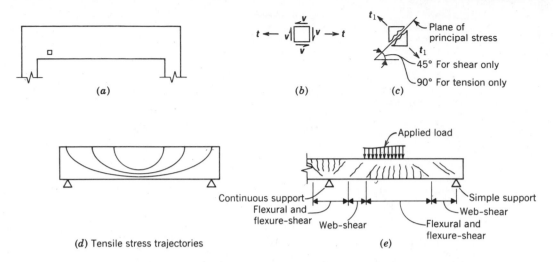

FIGURE 6.2 Diagonal tension and cracks in concrete beams. (Courtesy of the American Concrete Institute.)

if the crack itself does not become too wide. Concrete continues to resist shear in four ways:

1. Aggregate interlock across the crack.
2. Shear in the compression zone.
3. Dowel action of flexural reinforcement.
4. Arch action, especially in deep beams. The vertical component of the compressive force in an arch reduces the shear force in concrete.

Figure 6.3 illustrates shear mechanisms after cracking. The ACI Code assumes that concrete can carry some shear regardless of the magnitude of external shearing force and that shear reinforcement must carry the remainder. Thus,

$$V_u \leq \phi V_n = \phi(V_c + V_s) \tag{6.1}$$

where V_u = factored shear force
V_n = nominal shear strength of member
V_c = nominal shear strength provided by concrete
V_s = nominal shear strength provided by shear reinforcement
and ϕ = strength reduction factor = 0.85 for shear and torsion.

After the first diagonal tension crack, if shear reinforcement is present, it increases shear capacity primarily in four ways:

1. Most importantly, shear reinforcement provides tensile strength across the crack.

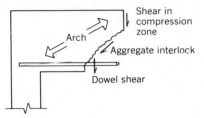

FIGURE 6.3   Postcracking shear mechanisms.

2. By keeping cracks small, it preserves the shear strength due to aggregate interlock.

3. By keeping cracks small, it also prevents propagation of cracks into the compression zone, thus reducing or destroying its shear capacity.

4. It prevents the splitting of concrete by flexural reinforcement and maintains shear strength provided by dowel action.

Computations can be made in terms of shear force V or in terms of unit shear stress v. Stress is easier to compare with allowable or limiting values, giving the engineer a better frame of reference, which helps to reduce chances of error. Shear stress is the shear force divided by the area of the beam web $b_w d$.

$$v_u = \frac{V_u}{b_w d} \qquad (6.2)$$

Usable shear stress capacities of a member are the nominal shear strengths multiplied by the strength reduction factor and divided by the area of the beam web.

$$v_n = \frac{\phi V_n}{b_w d}$$

$$v_c = \frac{\phi V_c}{b_w d}$$

$$v_s = \frac{\phi V_s}{b_w d}$$

Dividing Eq. 6.1 through by $b_w d$ gives a comparable equation in terms of shear stress.

$$v_u \leqslant v_n = v_c + v_s \qquad (6.3)$$

The nominal shear strength of concrete can be stated conservatively as

$$V_c = 2\sqrt{f'_c}\,b_w d$$

(6.4)

Just as with flexural reinforcement (see Section 2.2), it is necessary to have a minmum amount of shear reinforcement to prevent sudden failure on first cracking. The ACI Code requires a minimum area (Code 11.5.5.3)

$$\min A_v = \frac{50 b_w s}{f_y}$$

(6.5)

where: $\min A_v$ = area of shear reinforcement within a distance s along the axis of the member
and b_w = web width.

When shear reinforcement perpendicular to the axis of the member is used, nominal shear strength provided by shear reinforcement V_s is

$$V_s = \frac{A_v f_y d}{s}$$

(6.6)

Combining Eqs. 6.5 and 6.6,

$$\min V_s = 50 b_w d$$

(6.7)

Or usable ultimate unit shear strength stress provided by minimum shear reinforcement, regardless of steel strength,

$$\min v_s = \frac{\phi \min V_s}{b_w d} = \frac{\phi 50 b_w d}{b_w d} = 42 \text{ psi}$$

When factored shear stress at the critical section exceeds ultimate unit shear strength provided by concrete v_c by less than 42 psi, the only additional calculation needed is to determine the length of the member to receive shear reinforcement.

Discussion in the following sections is based on normal weight concrete. If lightweight concrete is used, its average splitting tensile strength f_{ct} (see Section 1.6) is a better measure of its shear strength than is its compressive strength f'_c. Therefore, provisions for strength of concrete in shear and torsion should be modified by inserting the value of $f_{ct}/6.7$ for $\sqrt{f'_c}$ (but not to exceed $\sqrt{f'_c}$) or by multiplying $\sqrt{f'_c}$ by 0.75 for all-lightweight concrete or 0.85 for sand-lightweight concrete. Linear interpolation may be used when partial sand replacement is used (Code 11.2).

6.3 SHEAR AND AXIAL LOADS

Magnitude and direction of diagonal tension is affected by the presence and magnitude of axial stress (See Fig. 6.2(b) and 6.2(c) and discussion in Section 6.2). For axial tension, shear capacity of concrete drops rapidly as tensile stress increases. Under axial compression, shear capacity of concrete increases more slowly and is affected by several variables. Shear stress resisted by concrete under axial load can be visualized by reference to Fig. A6.1.

To make design easier, shear strength provided by concrete under significant axial tension can safely be assumed to be equal to zero, and shear strength provided by concrete under axial compression need not be increased over that permitted for members without compression. For greater precision, shear strength of members subject to axial compression can be increased by a factor $(1 + N_u/2A_g)$, where N_u is the factored axial load in kips and A_g is the gross area of section, so that

$$V_c = 2 \left(1 + \frac{N_u}{2A_g}\right) \sqrt{f'_c} b_w d \tag{6.8}$$

Shear capacity of concrete under axial compression can be further increased when moment is low and large quantities of flexural tension reinforcement are present (Code 11.3.2.2). These provisions are not discussed in this text because of their limited applicability. When applicable, the cost of design effort needed to take advantage of higher shear stress is frequently more than the savings in construction cost.

When the average axial tensile stress (N_u/A_g) is less than 0.5 ksi, shear strength of concrete in members subject to axial tension can be computed as

$$V_c = 2 \left(1 + \frac{N_u}{0.5A_g}\right) \sqrt{f'_c} b_w d \tag{6.9}$$

where N_u is negative for tension.

When average axial tensile stress is greater than 0.5 ksi, no shear strength is assigned to concrete and shear reinforcement must be designed to carry all shear.

The ACI Code refers to "members subject to significant axial tension" without defining "significant." As a guide, engineers should consider axial tension in tension members and in other members where tension is part of the load-carrying system or where construction causes a tensile force that can be calculated. For example, prestressed members will shorten and may cause tension in adjacent nonprestressed members. In structures of ordinary proportions, tension need not be computed when it is due to creep and shrinkage or thermal shortening unless the member is restrained. For borderline cases, an engineer can safely ignore tension less than about 5% of 500 psi, or about 25 psi.

6.4 SLABS AND SHALLOW BEAMS

Shallow beams are those with a span-to-depth ratio ℓ_n/d equal to 5 or more, where ℓ_n is the clear span measured face to face of supports and d is the effective depth. In design to resist shear, slabs are distinguished from beams only in that slabs rarely have shear reinforcement.

To make the best fit with test data, the critical section for shear is at a distance **d** from the face of support. Sections less than distance **d** from the face of support may be designed for the same shear V_u as that computed at distance **d**, provided that

1. Support reaction in the direction of applied shear introduces compression into the end regions of the member.
2. No concentrated load occurs between the face of support and the critical section.

If the above conditions are not met, the critical section is at the face of support (see Fig. 6.4 for examples of location of the critical section). Other critical sections may occur where member cross-sectional area is reduced or load combinations cause higher shear forces. When large concentrated loads occur near the support, the portion of the beam between support and load should be reinforced as a deep beam by using horizontal as well as vertical shear reinforcement (see Section 6.5).

Shear strength provided by concrete of members subject to shear and flexure only is given by Eq. 6.4. Multiplying the strength by $\phi/b_w d$ gives the usable strength in terms of stress:

$$v_c = \phi 2\sqrt{f_c'} = 1.7\sqrt{f_c'}$$

Equation 6.4 gives shear strength provided by concrete in areas of high moment. When moment is low in relation to shear, the ACI Code permits a somewhat higher shear stress on the concrete (Code 11.3.2.1). The maximum increase is 75% and, for $f_y = 60$ ksi, this occurs when the flexural reinforcement ratio is about twice the value permitted for a rectangular beam; that is, when $\rho_w/0.75\rho_b$ is about two.* A T-beam is required for such heavy reinforcement. Figure 6.5 illustrates application of these Code provisions in a typical set of conditions. Concrete shear capacity varies along the length of the member and varies with changing ratios of moment to shear, span to depth, and ρ_w. Most engineers find it expedient to disregard these provisions and use the simpler Eq. 6.4, except for large concentrated loads near the support or for a member accidentally built with a shear stress on concrete higher than $1.7\sqrt{f_c'}$.† Provisions of ACI Code Section 11.3.2 are not discussed in this text because of their limited applicability.

Shear reinforcement is required whenever shear stress exceeds $\phi 2\sqrt{f_c'}$ (or that given by more detailed methods mentioned above and described in ACI Code Section 11.3.2.1). As protection against sudden shear failure, the ACI Code also

---

*For those who are curious, the minimum ratio $\rho_w/0.75\rho_b$ at which maximum shear stress occurs varies from 2.2 for $f_c' = 3$ ksi to 1.8 for $f_c' = 5$ ksi when $f_y = 60$ ksi and from 1.26 for $f_c' = 3$ ksi to 1.03 for $f_c' = 5$ ksi when $f_y = 40$ ksi. Note that these high flexural reinforcement ratios occur where moment is low and much of the reinforcement is not required for flexure.

†Also excepting those few cases where the engineer has an academic interest in leisurely examining all stress conditions in the member, probably without being paid for the effort.

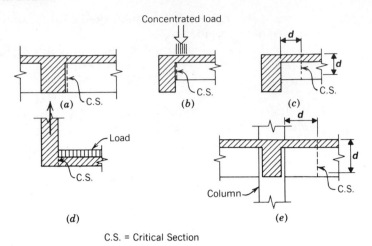

C.S. = Critical Section

FIGURE 6.4   Location of critical section for shear for typical conditions.

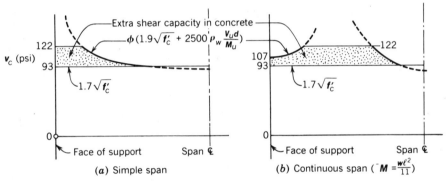

FIGURE 6.5   Extra shear capacity using special ACI Code provisions for $f'_c = 3$ ksi, $\rho_w = 1.6\%$, $\ell_n/d = 10$.

requires minimum shear reinforcement whenever factored shear stress $\boldsymbol{v}_u$ exceeds one-half the usable shear stress capacity of concrete $\boldsymbol{v}_c$, except for the following:

1. Slabs and footings.
2. Joists meeting ACI Code limits (see Section 3.3).
3. Shallow beams 10 in. or less in total depth.
4. Beams twice as wide as they are deep, or more ($\boldsymbol{b/h} \geqslant 2$).
5. Beams with flange thickness equal to 40% of the beam depth or more (Code 11.5.5.1) (see Fig. 6.6).

In slabs, footings, and wide beams, potential diagonal tension cracks must follow a longer, more difficult transverse path through the member than in narrow beams, thus reducing chances of a sudden failure. Beams with thick flanges contain extra

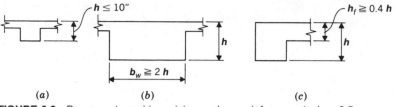

FIGURE 6.6 Beams not requiring minimum shear reinforcement when $0.5v_u < v_u \leq v_c$.

concrete to carry shear not accounted for by normal design procedures. Slabs and joists can transfer load laterally, thus reducing the risk of a sudden failure. In addition, the ACI Code allows a 10% higher shear stress on concrete in joists at least 4 in. wide, with a depth not more than 3½ times the minimum web width and spacing not more than 30 in. clear between webs (Code 8.11). Skip-joists (discussed in Chapter 3) are not permitted a 10% higher shear stress.

Example 6.1   Given a 7-in. slab spanning 16 ft clear between beams and supporting a 100-psf live load, determine whether the slab is satisfactory in shear, without shear reinforcement. Assume $f'_c = 3.5$ ksi and $d = 6''$.

Solution

Dead load: $87.5 \times 1.4 = 122$ psf

Live load: $100 \times 1.7 = \underline{170}$

$\qquad w_u \qquad\qquad = 292$ psf

The critical section is at $6''$ ($0.5'$) from face of support.

$$V_u = 0.292 \left(\frac{16}{2} - 0.5 \right) = 2.19 \text{k/}'$$

$$v_u = \frac{2.19 \times 1000}{6 \times 12} = 30 \text{ psi} < v_c$$

where $v_c = 0.85 \times 2\sqrt{3500} = 100$ psi   OK

Example 6.2   Consider a $12 + 4$ pan slab with 5-in. joists at 35 in. c/c, with tapered end pans as in Ex. 3.7, with $w_u = 0.95$ k/f and a 28-ft clear span. Are joists satisfactory in shear? If not, what precautions are necessary? Supporting beams are the same depth as the joists. Assume $f'_c = 3$ ksi.

Solution   For joists,

$$v_c = 0.85 \times 2 \times 1.1\sqrt{3000} = 102 \text{ psi}$$

Critical sections for shear are at the face of support (because the support does not introduce compression into the end region of the joists) and at the end of the tapered pan (because the width is substantially narrower than at the face of support). From Ex. 3.7, $d = 14.9''$. The average width of the joists is $1''$ wider than the minimum width because both sides of the joists taper at a slope of 1 in 12. The average width is computed as in Fig. 6.7.

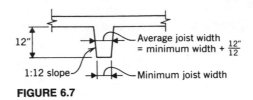

FIGURE 6.7

At the face of support,

$$V_u = 0.95 \times \frac{28}{2} = 13.3 \text{ k}$$

Average $b_w = 10 + 1 = 11''$

$$v_u = \frac{13.3 \times 1000}{11 \times 14.9} = 81 \text{ psi} < 102 \text{ psi}$$ OK

At the end of the tapered pan,

$$V_u = 0.95 \left(\frac{28}{2} - 3 \right) = 10.4 \text{ k}$$

Average $b_w = 5 + 1 = 6''$

$$v_u = \frac{10.4 \times 1000}{6 \times 14.9} = 117 \text{ psi} > 102 \text{ psi}$$ NA

Try $6''$-wide joists:

Average $b_w = 6 + 1 = 7''$

Extra load — Joists, 12.5 psf × 1.4   = 17#/'
              Live load, 0.08 × 100 psf × 1.7 = 14
              Total   = 31#/'

New $w_u = 0.95 + 0.03 = 0.98 \text{ k/}'$

$$V_u = 0.98(28/2 - 3) = 10.8 \text{k}$$

$$v_u = \frac{10.8 \times 1000}{7 \times 14.9} = 103 \text{ psi} = 102 \times 1.01$$ OK

The engineer should return to Ex. 3.7 and verify that the flexural reinforcement selected will still be adequate for increased load of $6''$ joists instead of $5''$ joists.

Shear Reinforcement

When factored shear stress exceeds shear strength provided by concrete (i.e., when $v_u > v_c$), shear reinforcement must be used. The upper limit permitted by the ACI Code on shear stress resisted by reinforcement is

$$\boxed{\max v_s = \phi 8\sqrt{f_c'}} \tag{6.10}$$

Table A6.1 lists the maximum shear stress provided by concrete v_c and reinforcement v_s.

Three types of shear reinforcement are used with cast-in-place concrete (Code 11.5.1).

1. *Stirrups perpendicular to the axis of the member, including spirals and rectangular helixes.* Spirals and helixes are rarely selected for shear reinforcement when that is the only reason for their use, but they can be used to resist shear when present for other reasons (e.g., in a spiral column). Welded wire fabric (WWF) is frequently used in precast joists and beams for shear reinforcement but is rarely used in cast-in-place construction. The latter usually lacks the repetition necessary to justify a sufficient volume of fabric for production.

2. *Longitudinal reinforcement with bent portion making an angle of 30° or more with longitudinal tension reinforcement.* These are commonly called truss bars (see Section 2.6 for a discussion on use of truss bars). They are usually bent at a 45° angle but sometimes at a 30° angle. The upper limit on shear stress resisted by a single bar or a group of parallel bars all bent up at the same distance from the support is $\phi 3\sqrt{f_c'}$.

3. *Inclined stirrups.* These are rarely used because placement may cost more than the savings in steel and because of the high risk of misplacement (e.g., stirrups placed at the wrong angle or at a reverse angle, i.e., sloping toward mid-span instead of toward the support).

The spacing of vertical shear reinforcement must not exceed $d/2$ nor 24 in. c/c, and truss bars must be so spaced that every 45° line, representing a potential crack, extending toward the reaction from mid-depth of member $d/2$ to longitudinal tension reinforcement, is crossed by at least one line of shear reinforcement (Code 11.5.4). Only the center three fourths of the inclined portion of a truss bar is effective in resisting shear. Maximum spacing for truss bars varies from $d/2$ to d depending on the angle of slope of the bent portion of the bar (see Fig. 6.8). When v_s exceeds $\phi 4\sqrt{f_c'}$, maximum spacings must be reduced by one-half.

The most common type of shear reinforcement is stirrups placed vertically (i.e.,

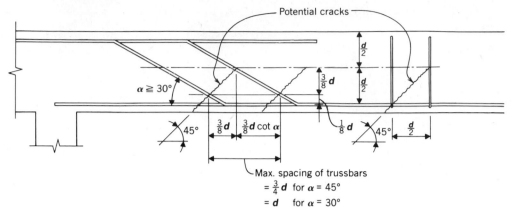

FIGURE 6.8   Maximum spacing of shear reinforcement.

perpendicular to the axis of the member). Criteria for selecting size of stirrups are the following:

1. Stress in bars must be developed within the depth of the beam. If stirrups are fully stressed, their size is limited by beam depth (see Table A5.7). Larger stirrups can be used, but they will not be fully stressed; hence, they are generally uneconomical.

2. Bars should be small enough so that steel is not wasted meeting maximum spacing limits of $d/2$ and $d/4$.

3. Bars should be large enough so that minimum spacing is 3 in. or more.

4. Bars should be large enough so that excessive numbers of stirrups are not required to meet minimum steel requirements (see Eq. 6.5). Table A6.2 lists the maximum spacing for various beam widths that will meet minimum requirements for shear reinforcement. For example, in a 24-in.-wide by 36-in.-deep beam ($d \approx 33$ in.), the maximum spacing is $d/2 = 16\frac{1}{2}$ in., but #3 stirrups cannot be spaced more than 11 in. c/c for minimum reinforcement. A better selection would be #4 stirrups, which can be spaced up to 20 in. c/c for minimum reinforcement but are limited to the $16\frac{1}{2}$-in. spacing.

Simple U-shaped stirrups with hooks at the ends are usually used. In rectangular beams, hooks must be turned in, but they are usually turned out if the beam has a flange of sufficient size to permit it. This allows longitudinal bars to be placed in the beam more easily. Closed ties are used only when necessary (e.g., for torsion) because they make placement of longitudinal bars more difficult. Hooks bent at 135° or 180°, with a longitudinal bar in the hook, may be desirable in shallow beams (see Table A5.7). For very wide beams, multiple stirrups or galloping stirrups are used to provide sufficient area of shear reinforcement and to place shear reinforcement near the center of the beam as well as the sides. Multiple stirrups are preferred over galloping stirrups. Figure 6.9 illustrates various stirrup types.

Force in one stirrup can be visualized as the volume of the portion of shear stress

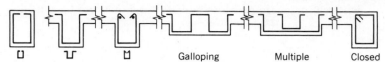

FIGURE 6.9 Stirrup types. Multiple stirrups are preferred over galloping stirrups.

wedge that is concentric with the stirrup. If vertical stirrups perpendicular to the axis of the member are used, the nominal shear strength of stirrups is

$$V_s = \frac{A_v f_y d}{s}$$

and inserting the capacity reduction factor ϕ,

$$A_v = \frac{V_s s}{\phi f_y d} \tag{6.11}$$

Because the maximum spacing of stirrups is $d/2$, their shear capacity at that spacing can be computed by inserting $d/2$ for s into Eq. 6.11, giving

$$V_s = 2\phi A_v f_y$$

Thus, the shear capacity of stirrups using steel $f_y = 60$ ksi, spaced at $d/2$, is about

22 kips for #3 U-shaped stirrups ($A_v = 0.22$ in.)^2
40 kips for #4 U-shaped stirrups ($A_v = 0.40$ in.)^2
62 kips for #5 U-shaped stirrups ($A_v = 0.62$ in.)^2

If steel $f_y = 40$ ksi, the shear capacity of stirrups spaced at $d/2$ is 15, 27, and 42 kips for #3, #4, and #5 bars, respectively.

These values are easily remembered. Stirrup shear capacity at other spacings can be computed by ratio of the spacing to $d/2$.

If shear reinforcement consists of a single truss bar or a group of parallel truss bars all bent up at the same distance from the support, the nominal shear strength of truss bars is $V_s = A_v f_y \sin \alpha$ and inserting the capacity reduction factor ϕ,,

$$A_v = \frac{V_s}{\phi f_y \sin\alpha} \tag{6.12}$$

where α = the angle between truss bar and longitudinal bars.

If inclined stirrups or parallel truss bars are located at different distances from the support, the nominal shear strength of a single bar is

$V_s = A_v f_y(\sin \alpha + \cos \alpha)d/s$ and inserting the capacity reduction factor ϕ,

$$A_v = \frac{V_s s}{\phi f_y d(\sin\alpha + \cos\alpha)} \tag{6.13}$$

where $\mathbf{s}$ = spacing of inclined stirrups or truss bars in direction parallel to longitudinal reinforcement.

The minimum total number of stirrups is the total volume of the unit shear stress wedge divided by the capacity of one stirrup. Uniformly loaded beams always require more stirrups than the minimum due to the necessity of meeting maximum spacing requirements.

Referring to Fig. 6.10, the volume of the unit shear stress wedge is the shaded area of the total shear diagram divided by the effective depth of the beam $\boldsymbol{d}$.

$$\frac{\boldsymbol{\phi V_s}}{\boldsymbol{b_w d}} \times \frac{\boldsymbol{\ell_s b_w}}{2} = \frac{\boldsymbol{\phi V_s \ell_s}}{2\boldsymbol{d}}$$

Volume of the shear stress wedge can easily be computed by geometry after ultimate shear $\boldsymbol{V_u}$ and shear strength provided by concrete $\boldsymbol{\phi V_c}$ have been computed.

For a prismatic beam (one with a constant cross section throughout its length) the size, type, number, and spacing of stirrups should be the same at each end of the beam to minimize stirrup placement errors in the field. Of course, stirrup selection is based on the end of the beam with the largest shear force or most critical conditions. For a nonprismatic beam where placement errors cannot occur, stirrup selection at one end can be different from that at the other end. Referring to Fig. 6.10, the procedure for determining the number and spacing of stirrups in a uniformly loaded beam can be summarized as follows:

1. Compute usable concrete shear capacity $\boldsymbol{\phi V_c}$ at the critical section and compare it to the factored shear $\boldsymbol{V_u}$. If $\boldsymbol{\phi V_c}/2 > \boldsymbol{V_u}$ (or $\boldsymbol{\phi V_c} > \boldsymbol{V_u}$ for certain

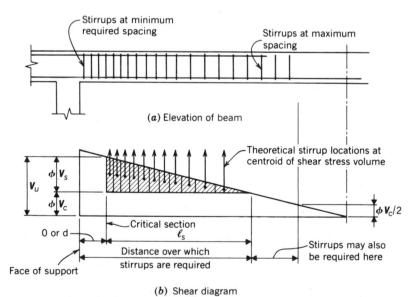

FIGURE 6.10   Spacing of stirrups.

members), no stirrups are required. If $5\phi V_c < V_u$, the cross section must be enlarged. For intermediate values, shear reinforcement is required.

2. By trial and error, select a stirrup size and compute the maximum stirrup spacing (see criteria listed above for selecting size of stirrups).

3. Compute stirrup shear capacity at maximum spacing and at a few appropriate smaller spacings, including a spacing that will result in adequate capacity at the critical section.*

4. On the shear diagram for the member, compute the length of the beam for which each stirrup spacing is adequate. Alternatively, sketch the shear diagram to scale and measure the length on the sketch.

5. Summarize the number of stirrups required at each spacing. The first stirrup is usually spaced one-half the computed minimum spacing but not less than 3 in. from the face of support under conditions illustrated in Fig. 6.4(b), 6.4(c) and 6.4(e) and at the face of support under conditions illustrated in Fig. 6.4(a) and 6.4(d).

6. Check the accuracy of calculations by comparing the total length of the beam covered by stirrups to the distance that stirrups are required.

Example 6.3   Consider a 24- × 17-in. beam (shown in Figs. 2.9 and 2.11) with service dead load = 2.7 k/′ and service live load = 2.2 k/′. What stirrups are required? Assume $f'_c = 4$ and $f_y = 60$.

Solution $d = 17 - 2.7 = 14.3'' = 1.2'$

Critical section for shear is 14.3″ from the face of support.

Dead load: $2.7 \times 1.4 = 3.8 \text{k}/′$
Live load: $2.2 \times 1.7 = \underline{3.7}$
$$w_u = 7.5 \text{k}/′$$

At face of support, $V_u = \dfrac{7.5 \times 25}{2} = 94.0 \text{ k}$

At critical section, $V_u = 7.5 \left(\dfrac{25}{2} - 1.2 \right) = 85.0 \text{ k}$

$v_c = 0.85 \times 2\sqrt{4000} = 107 \text{ psi}$,

$\phi V_c = 0.107 \times 24 \times 14.3 = 36.7 \text{ k}$

Maximum stirrup spacing = $d/2 = 7.15''$; say 7″ c/c

Try #3 U-shaped stirrups, $A_v = 0.22 \text{ in.}^2$

---

*Stirrups can be spaced precisely by a simple slide rule procedure: Place the cross hair on distance ℓ_s on the D scale, and move the slide to place the number of stirrups required N_s on the B scale under the cross hair. Then, moving the cross hair successively to 0.5, 1.5, 2.5, and so forth, less than N_s, read the location of each stirrup on the D scale. By successive subtractions, read the required stirrup spacings.

#3 stirrup shear capacity:

$$\phi V_s = \frac{0.85 \times 0.22 \times 60 \times 14.3}{7''} = 22.9 \text{ k at } 7'' \text{ c/c}$$

$$= 32.1 \text{ k at } 5'' \text{ c/c}$$
$$= 53.5 \text{ k at } 3'' \text{ c/c}$$

Total shear capacity with stirrups spaced at 3″ c/c at the critical section is

$$\phi V_c + \phi V_s = 36.7 + 53.5 = 90.2 \text{ k} > V_u = 85.0 \text{ k} \qquad \underline{\text{OK}}$$

The shear capacity of concrete and of stirrups at 3″, 5″, and 7″ spacings is plotted on the shear diagram in Fig. 6.11. The length of the beam to be covered by each stirrup spacing is computed by ratio of congruent triangles. Stirrups must extend to the point where $V_u \leq \phi V_c/2$. Stirrup spacing for each end of the beam can then be selected as

26 #3 U,* 13 at 3″/3 at 5″/10 at 7″

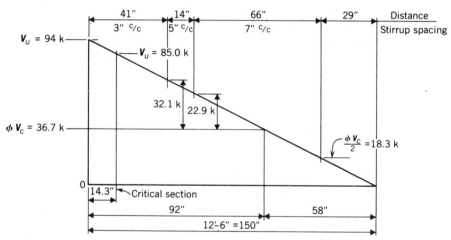

FIGURE 6.11   Stirrup spacing for Ex. 6.3.

As a check, stirrups cover a length of beam = 124″ (13 × 3 + 3 × 5 + 10 × 7), which is slightly more than the required length of 121″.

Skip-Joists   A skip-joist slab system described in Chapter 3 will probably require shear reinforcement because the wide joist spacing puts more load on individual joists and because a 10% higher shear stress is not permitted as it is for standard pan slab joists.

---

*Engineers should always sketch the type of stirrup contemplated when listing number and spacing of stirrups in calculations—a habit that may save costly mistakes and a red face (the engineer's own) during construction.

One method of providing shear reinforcement is to use a galloping stirrup (see Fig. 6.12) placed in the center of each joist, extending along the length of the joist as far as shear reinforcement is required. Adequate reinforcement will be provided in most cases by a #3 or #4 bar with multiple vertical legs at the maximum spacing of **d**/2. Minimum cover should be provided on the stirrups at both top and bottom faces of the skip-joists. If legs are spaced at least 6 in. c/c, the minimum depth of a joist in which a stirrup bar can be anchored is 15 in. for #3 bars, 18 in. for #4 bars and 21 in. for #5 bars. Strictly speaking, a galloping stirrup does not meet the ACI Code because the bends do not enclose a longitudinal bar. WWF can also be used for shear reinforcement, but much repetition is required to obtain a minimum production run of fabric.

6.5 DEEP BEAMS

Deep beams are beams with a clear span to effective depth ratio ℓ_n/d of 5 or less. Deep beams are usually much narrower than they are deep. When such beams are loaded on the top edge, as in Fig. 6.13(a), they exhibit significantly different structural behavior from shallow beams and must be designed using special rules [2]. The ACI Code is silent on rules for design of members with deep beam proportions but loaded on the bottom edge, as in Fig. 6.13(b), or loaded throughout their depth by intersecting beams, as in Fig. 6.13(c). In such cases, it would be satisfactory to design the member using shallow beam procedures and place additional shear reinforcement as required for deep beams.

Transfer girders to carry column loads over large spans, basement walls distributing foundation loads, floor diaphragms, some shear walls, short and deep beams carrying heavy loads, and storage bins are designed as deep beams. Although not required by the Code, a portion of a shallow beam between a support and a nearby heavy concentrated load should also be reinforced as a deep beam when the size is large enough to permit it.

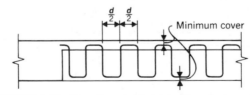

FIGURE 6.12   Elevation of skip-joist with galloping stirrup shear reinforcement.

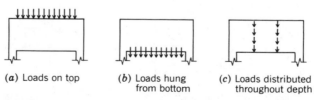

(**a**) Loads on top   (**b**) Loads hung   (**c**) Loads distributed
                              from bottom   throughout depth

FIGURE 6.13   Loading conditions for deep beams.

Deep beams tend to carry some load directly to the support by a diagonal strut, as shown in Fig. 6.14(a). This strut, combined with compression force near the top and tensile reinforcement near the bottom, acts as a tied arch. One implication of the tied arch is that shear strength provided by concrete is larger and more dependable than that in a shallow beam. Another implication is that flexural tension reinforcement is more nearly fully stressed for the full length of the span as it would be in a tied arch, rather than stressed in proportion to the moment diagram. Thus, flexural reinforcement should be anchored at the end of the span or beyond the span supports to assure full development of design stresses. The portion of flexural reinforcement that need be so anchored is proportional to the portion of load carried by a diagonal strut. It is convenient to think of this anchorage as the connection at the end of a tied arch between the compression member and the tension member. To avoid secondary stresses due to joint eccentricity, the centerline of concentrated loads, compressive force, and diagonal struts at the top of the beam, and the centerline of reactions, tension tie and diagonal strut at the bottom of the beam should intersect at common points. This can be checked most easily by drawing a sketch of intersections to scale.

As beams become deeper, flexural tension tends to become distributed over a larger height of the beam near the bottom. When beams exceed overall depth-to-clear-span ratio of 2/5 for continuous spans or 4/5 for simple spans ($\ell_n/h < 2.5$ or 1.25), the ACI Code requires that nonlinear distribution of strain and buckling be taken into account (Code 10.7.1). As an approximation, the effective depth **d** can be taken as the lesser of $0.8\ell_n$ or $0.8h$. Tensile reinforcement should then be spread over a portion of the beam symmetrical with the assumed centroid of tensile reinforcement.

Because of stout proportions of deep beams and difficulty in defining the exact location of maximum diagonal tensile stresses, the ACI Code requires that shear be computed at one critical section, and shear reinforcement required at that critical section (if any) be used throughout the span. Shear reinforcement should be anchored at edges (ends, top, and bottom) of the beam. Due to tied arch behavior, shear reinforcement tends to be ineffective at low load levels, even up to service loads. However, shear reinforcement is essential to assure satisfactory behavior at service load and ductility at ultimate load.

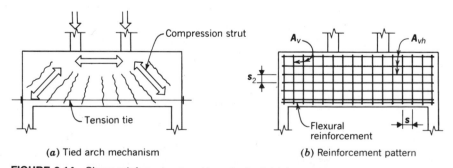

(a) Tied arch mechanism   (b) Reinforcement pattern

FIGURE 6.14 Shear reinforcement and transfer in deep beams.

Because shear need be computed at only one section in deep beams, it is much more convenient to take advantage of the long procedure, giving credit to increased shear capacity of concrete than is the case for shallow beams. The distance from the face of support to the critical section for shear in deep beams is equal to $0.15\ell_n$ for uniformly loaded beams or half the shear span $a/2$ for beams with concentrated loads, but not greater than d. The shear span a is the distance between concentrated load and face of support.

A beam with a small portion of the total load hung from the bottom edge, as in Fig. 6.13(b), can still be designed using deep beam procedures if the beam meets all other limitations for deep beam design (span-to-depth ratio and load location). However, steel to support the hung load should be added to the required vertical shear reinforcement.

Design Procedure

Design of deep beams can be completed efficiently by following this procedure.

Step 1 Is the beam span-to-depth ratio ℓ_n/d less than five and is it loaded on the top? If either one of these conditions is not true, the beam should use the lower shear strength assigned to concrete by shallow beam design procedures. If $\ell_n/d < 5$ but most of the load is not on the top, the beam should be provided with minimum horizontal shear reinforcement as required for deep beams.

Step 2 Compute maximum allowable shear stress max v_n by using Eq. 6.14 or by referring to Fig. A6.2. If $v_u > $ max v_n, enlarge the concrete outlines as necessary.

$$\text{max } v_n = \frac{\phi 2}{3}\left(10 + \frac{\ell_n}{d}\right)\sqrt{f_c'} \qquad (6.14)$$

when ℓ_n/d is less than 2, max $v_n = 8\sqrt{f_c'}$.

Step 3 The use of Step 3 is optional to save computational labor. Using Fig. A6.3, estimate the usable shear stress capacity of concrete, v_c. If ultimate shear stress $v_u < v_c/2$, then no shear reinforcement is needed and design for shear is complete.

If $v_u > v_c/2$, select the usable shear capacity of minimum shear reinforcement v_s from Table A6.3. If $v_c/2 < v_u < (v_c + \text{min } v_s)$, then minimum shear reinforcement is required. If v_u is close to or exceeds $v_c + \text{min } v_s$, computations should be made as described in Step 4.

If no shear reinforcement is required, the beam web should be reinforced as a wall if the distance between flexural reinforcement and the bottom of the compression flange (or the top of the beam if no flange exists) is more than 18 in. Also, the ACI Code requires that additional longitudinal flexural reinforcement be distributed in the zone of flexural tension near the side faces of the web if depth of the web exceeds 3 ft (Code 10.6.7). The area of additional reinforcement must be at least equal to 10% of the flexural tension reinforcement A_s but may not be included in strength computations unless a strain compatibility analysis is made to determine stress in each bar. Spacing must not exceed web width nor 12 in.

Minimum vertical shear reinforcement (perpendicular to flexural tension reinforcement) $A_v = 0.0015b_w s$, and spacing s must not exceed $d/5$ or 18 in. Minimum horizontal shear reinforcement (parallel to flexural tension reinforcement) $A_{vh} = 0.0025b_w s_2$ and spacing s_2 must not exceed $d/3$ or 18 in.

Step 4   Compute v_c using Eq. 6.15.

$$v_c = \frac{\phi V_c}{b_w d} = \phi \left(3.5 - 2.5 \frac{M_u}{V_u d} \right) \left(1.9 \sqrt{f_c'} + 2500 \rho_w \frac{V_u d}{M_u} \right) \qquad (6.15)$$

where M_u = factored moment occurring simultaneously with V_u at the critical section.

$$\text{The term } \left(3.5 - 2.5 \frac{M_u}{V_u d} \right) \text{ must not exceed 2.5}$$

$$\text{and } v_c \text{ must not exceed } \phi 6\sqrt{f_c'}$$

Using Eq. 6.17, compute the usable shear stress capacity of minimum shear reinforcement min v_s. For any quantity more than the minimum, the usable shear stress capacity provided by shear reinforcement v_s is

$$v_s = \phi \left[\frac{A_v}{s} \left(\frac{1 + \ell_n/d}{12} \right) + \frac{A_{vh}}{s_2} \left(\frac{11 - \ell_n/d}{12} \right) \right] \frac{f_y}{b_w} \qquad (6.16)$$

where A_v = area of shear reinforcement perpendicular to flexural tension reinforcement within a distance s.

A_{vh} = area of shear reinforcement parallel to flexural tension reinforcement within a distance s_2.

Substituting the values given above for minimum shear reinforcement into Eq. 6.16 gives

$$\text{min } v_s = \phi \left[0.0015 \left(\frac{1 + \ell_n/d}{12} \right) + 0.0025 \left(\frac{11 - \ell_n/d}{12} \right) \right] f_y \qquad (6.17)$$

Table A6.3 lists min v_s for f_y = 40 ksi and 60 ksi for permissible ranges of ℓ_n/d.

If $v_u < v_c/2$, no shear reinforcement is required and design for shear is complete. As noted above, the web may require reinforcement as a wall and may require additional flexural reinforcement. If $v_c/2 < v_u < (v_c + \text{min } v_s)$, then minimum shear reinforcement is required.

Step 5   If $v_u > (v_c + \text{min } v_s)$, compute the required area of shear reinforcement and select bars to meet the requirements. After minimum reinforcement is provided both horizontally and vertically, additional shear capacity required by the ACI Code could be satisfied with either additional horizontal or additional vertical reinforcement. It is apparent from Eq. 6.16 that horizontal reinforcement is more efficient than vertical reinforcement in meeting the Code. Horizontal bars are more nearly perpendicular to potential diagonal tension cracks and they provide some support by doweling. Nevertheless, a well-proportioned beam will include a reasonable amount of vertical reinforcement as well as horizontal reinforcement. The ratio of horizontal to vertical reinforcement areas should be in the range of 1 to 1.67, with the higher ratios reserved for small ratios of ℓ_n/d.

Solving Eq. 6.16 for A_v and A_{vh} and disregarding, first, the contribution of horizontal reinforcement and, then, the contribution of the vertical reinforcement, the amount of shear reinforcement can be computed directly.

$$A_v = \frac{12v_s s b_w}{\phi(1 + \ell_n/d)f_y} \tag{6.18}$$

$$A_{vh} = \frac{12v_s s_2 b_w}{\phi(11 - \ell_n/d)f_y} \tag{6.19}$$

where $v_s = v_u - v_c$, or that portion of shear stress not resisted by concrete.

Shear stress carried by shear reinforcement v_s must be resisted by either horizontal or vertical reinforcement but need not be resisted by both. Equations 6.18 and 6.19 can be used to compute required shear reinforcement in addition to the minimum by considering $v_s = v_u - v_c - \min v_s$ or that portion of the shear stress not resisted by concrete plus minimum shear reinforcement.

The above procedure is summarized in flowchart form in Fig. 6.15.

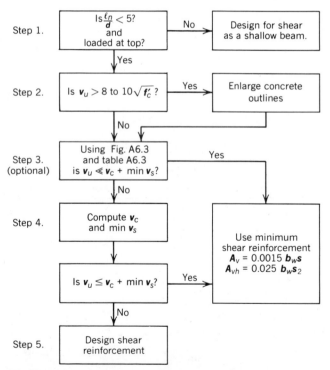

FIGURE 6.15 Flowchart for design procedure for shear in deep beams.

Example 6.4   Select reinforcement for a transfer girder supporting two columns and floor loads at top and bottom edges, as shown in Fig. 6.16. Factored column loads are P_u = 800 kips, and factored floor loads are 4 kips/ft each. Assume f'_c = 4 ksi and f_y = 60 ksi. (This, and subsequent examples relating to the girder, illustrate a major structural member that is not common in actual practice. However, design of all deep beams involves most of the same considerations, even though some will not require explicit calculations.)

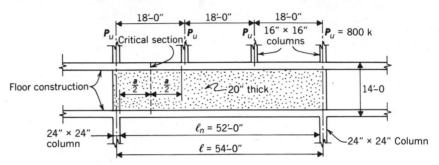

FIGURE 6.16   Transfer girder for Ex. 6.4.

Solution

Design of Flexural Reinforcement

Moment from concentrated loads:

$$800 \times 18 = 14{,}400'k$$

Moment from uniform loads:

$$w_u = 2 \times 4 + 14 \times 0.25 \times 1.4 = 12.9 \text{ k}/'$$

$$\ell = 54', \frac{w_u \ell^2}{8} = \frac{12.9 \times 54^2}{8} = \underline{4{,}700'k}$$

$$M_u = 19{,}100'k$$

Assume d = 14 − 1 = 13′, and $(\phi f_y j)$ = 4.0

$$A_s = \frac{19{,}100}{13 \times 12 \times 4.0} = 30.6^{\square''}$$

Use 20 #11 bars, $A_s$prov. = 31.2$^\square$

$$\rho_w = \frac{31.2}{13 \times 12 \times 20} = 1.00\% < 2.14\% = 0.75\rho_b \qquad\qquad \text{OK}$$

[From Fig. A2.2, the assumption for $(\phi f_y j)$ is safe.]

Design of Shear Reinforcement

$$\ell_n/d = 52.0/13 = 4.00 < 5$$

Use deep beam shear design procedures.

Distance to critical section from face of supporting column,

$$= a/2 = (18 - 1.67)/2 = 8.17' \text{ for concentrated load}$$
$$= 0.15\ell_n = 0.15 \times 52.00 = 7.80' \text{ for uniform load}$$

Use $a = 8.17'$ because the concentrated load predominates, but conservatively include the uniform load to a distance of 7.8' from the face of support.

Shear at critical section:

Column load = 800 k

Uniform load = $12.9 (27 - 7.8 - 1) =$ $\underline{235}$

$$V_u = 1035 \text{ k}$$

Using Eq. 6.14,

max $v_n = 0.85 \times 2/3(10 + 4.00) \sqrt{4000} = 502$ psi

$$v_u = \frac{1035 \times 1000}{156 \times 20} = 332 \text{ psi} < \text{max } v_n \qquad\qquad \underline{\text{OK}}$$

Therefore, girder web thickness is adequate.

If web thickness were reduced to 14", $v_u = 474$ psi, and would still be satisfactory. More shear reinforcement would be required and there would be more congestion of reinforcing bars in the bottom of the girder with more difficulty in placing concrete. Little, if any, construction cost would be saved. This example will continue with a 20" web.

From Fig. A6.3, for $f'_c = 4$ ksi, $\rho_w = 1.0\%$, and $\ell_n/d = 4.0$ for uniform loads,

$$v_c \simeq 230 \text{ psi}$$

From Table A6.3, min $v_s = \underline{106}$

$$v_c + v_s = 336 \text{ psi} > v_u = 332 \text{ psi}$$

Minimum shear reinforcement may be adequate, but detailed computations are needed to confirm this tentative conclusion. Note that even if the above estimate of shear conditions is confirmed, minimum web reinforcement is required and some horizontal reinforcement is required near the bottom of the girder because it is a beam exceeding 3 ft in depth.

At the critical section,

$$M_u = 800(1.00 + 8.17) \qquad\qquad\qquad = 7336'\text{k}$$
$$+ 12.9 \times 27(9.17) - \frac{12.9(9.17)^2}{2} = \underline{2652}$$
$$M_u = 9990'\text{k}$$

$$V_u = 800 + 12.9(27 - 9.17) = 1030 \text{ k}$$

$$\frac{M_u}{V_u d} = \frac{9990}{1030 \times 13} = 0.746$$

From Eq. 6.15,

$$v_c = 0.85(3.5 - 2.5 \times 0.746)\left(1.9\sqrt{4000} + \frac{2500 \times 0.01}{0.746}\right) = 214 \text{ psi}$$

Using Eq. 6.17,

$$\min v_s = 0.85\left[0.0015\left(\frac{1+4}{12}\right) + 0.0025\left(\frac{11-4}{12}\right)\right]60{,}000 = 106 \text{ psi}$$

$$v_c + \min v_s = 214 + 106 = 320 \text{ psi} < v_u = 332 \text{ psi}$$

Therefore, slightly more than minimum shear reinforcement is required. Results of these calculations are inconsistent with the values from Table A6.3b because locations of the critical section for concentrated loads and for uniform loads are not identical.

Vertical shear reinforcement:

$$A_v = 0.0015 \, b_w s = 0.0015 \times 12 \times 20 \qquad = 0.36 \text{ in.}^2$$

Tension steel to hang lower floor:

$$A_v = \frac{4}{0.9 \times 60} \qquad\qquad = \underline{0.07^{\square''}/'}$$

Total required A_v $\qquad\qquad = 0.43^{\square''}/'$

Use #5 at 16" c/c vertical each face, A_s prov. $= 0.46^{\square''}/'$

Horizontal shear reinforcement:

$$A_{vh} = 0.0025 b_w s_2 = 0.0025 \times 12 \times 20 = 0.60^{\square''}/'$$

Required extra A_{vh} using Eq. 6.19

$$A_{vh} = \frac{12(332 - 320)20}{0.85(11 - 4)60{,}000} \qquad = \underline{0.01}$$

$$\text{Total } A_{vh} \qquad\qquad = 0.61^{\square''}/'$$

In beams with a depth exceeding 3 ft, reinforcement is required in the zone of flexural tension equal to 10% of longitudinal reinforcement. This zone may be taken as the bottom half of the beam because $a/d = 0.38$ to 0.44 from Table A2.1 and the neutral axis is located a little distance below the compressive stress block. For this example, the minimum area of horizontal reinforcement required in the web is

$$\frac{0.10 A_s}{0.5 d} = \frac{0.1 \times 31.2}{0.5 \times 13} = 0.48^{\square''}/' < 0.61^{\square''}/'$$

Therefore, requirements for shear reinforcement control.

Use #5 at 12" c/c horizontal each face, $A_s$prov. = 0.62$^{□"}$/'

After making bar selections for flexural and shear reinforcement, the engineer's design is no more than 50% complete. Success of the structure will depend on proper detailing of these bars, as well as selection and placement of secondary reinforcement. Figures 6.17, 6.18, and 6.19 illustrate how bars might be placed in a structure and some of the considerations involved. The following comments are coded to numbers shown in Fig. 6.17.

1. Referring to Figs. 6.17 and 6.18, only four bars per layer of tension reinforcement can pass the column cage even though minimum spacing would allow more than four bars per layer. Furthermore, extra space between bars will help ensure that concrete can be placed and consolidated around bars. For this latter reason, 4" vertical spacing (2½" clear) is also specified. In addition, heavier than normal bar supports must be used to carry the weight of steel in the bottom of the girder. By inspection, or by computation, the assumption of 1' from bottom of the beam to centroid of longitudinal reinforcement is conservative.

2. Check the connection of diagonal strut to tension tie by assuming that the increase in shear stress in concrete from 108 to 214 psi is due to the diagonal strut. Referring to Fig. 6.19, the diagonal strut compression force is

$$800 \times \frac{21.63}{12} \times \frac{(214 - 108)}{214} = 714 \text{ k}$$

and the tension tie force required to resist the diagonal strut compression is

$$714 \times \frac{18}{21.63} = 594 \text{ k}$$

Using $\phi = 0.9$ for flexure, the required number of #11 bars in a tension tie is

$$\frac{594}{0.9 \times 60 \times 1.56} = 8$$

Anchor all 8 bars below the floor line. Stress in the diagonal strut is

$$\frac{714}{20 \times 28} = 1.28 \text{ ksi}$$

This is less than the permitted ultimate stress on concrete, as will be seen in Chapter 9, Columns. To assure anchorage of tension tie bars below the floor line, extend eight bars beyond the inside face of support a distance equal to the basic development length. Anchorage of the bars presents an interesting problem of a compression member (see Ex.9.1).

3. Depending on confinement provided by floor construction, ties may be required to assure development of #11 bars (see Ex.9.1).

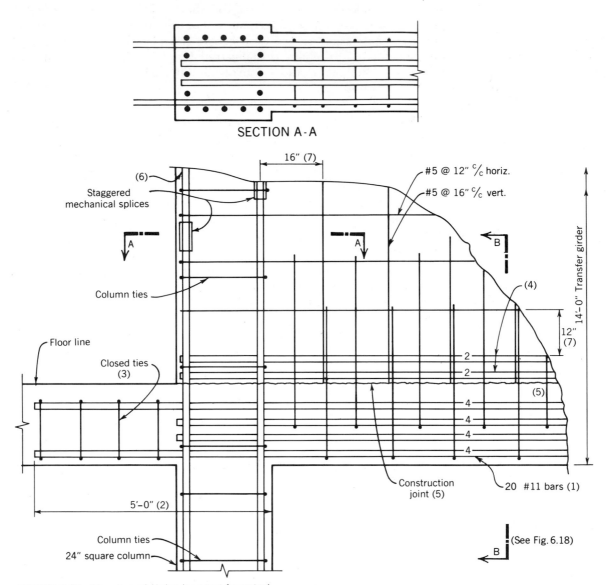

SECTION A-A

FIGURE 6.17 Elevation of girder (see text for notes).

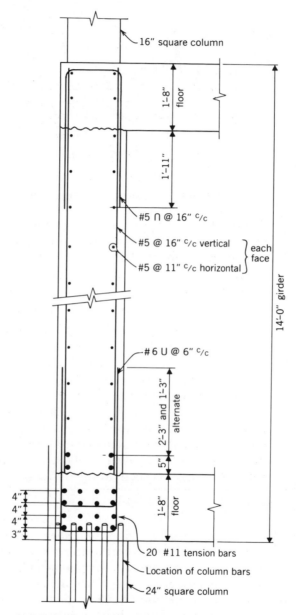

16" square column

1'-8" floor

1'-11"

#5 ∩ @ 16" c/c

#5 @ 16" c/c vertical } each face
#5 @ 11" c/c horizontal }

14'-0" girder

#6 ∪ @ 6" c/c

2'-3" and 1'-3" alternate

5"

1'-8" floor

4"
4"
4"
3"

20 #11 tension bars

Location of column bars

24" square column

FIGURE 6.18   Section B (see text for notes).

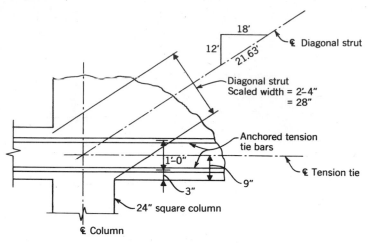

FIGURE 6.19   Tied arch joint at support column for Ex. 6.4.

4. To provide more space for placing concrete and cleaning the construction joint just before placing concrete, the top four bars are placed only on the outside face of the girder.

5. The builder may want to use a construction joint at the top of the lower floor construction and at the bottom of the upper floor construction. Extra dowels are needed to assure composite action (see Ex. 6.5).

6. Loading the column at each end of the transfer girder through shear requires special considerations (see Ex. 9.1).

7. Spacing of shear reinforcement can start from the nearest column bar or flexural bar. Horizontal shear reinforcement can be anchored by extending bars through the columns at each end because the development length of a #5 bar (15″) is less than the width of the column. Vertical shear reinforcement is anchored at the top by extending to the top of the beam because the development length is less than the depth of the compressive stress block. At the bottom, bars should be anchored by lapping with U-shaped stirrups (without hooks) wrapped around tension reinforcement.

6.6 SHEAR FRICTION

Up to this point, nearly all provisions for shear have been concerned with diagonal tension and reinforcement to resist it. In contrast, shear friction is concerned with slippage of one member across another along a plane. Planes of slippage can occur, for example, between a precast floor member and its topping, between two pieces of concrete cast at separate times but intended to act together structurally, between concrete and a steel plate, and at an imaginary plane through monolithic concrete. In the latter case, an imaginary plane can become real if unexpected forces crack the concrete and then allow shearing forces to cause a slippage. Shear-friction concepts are widely used in precast concrete construction but also find application in cast-in-place concrete in brackets and construction joints [3].

Conceptually, shear force in a case of shear-friction is resisted by friction between the two members being sheared, or sliding against each other. The shear-friction strength equals the force perpendicular to the plane multiplied by the coefficient of friction. When shear-friction reinforcement is placed perpendicular to the shear plane, nominal shear strength V_n is

$$V_n = A_{vf}f_y\mu \qquad (6.20)$$

where A_{vf} = area of shear-friction reinforcement

and μ = coefficient of friction, which depends on roughness of contact surfaces, and is given in Table A6.4 (Code 11.7.4).

The ACI Code limits the specified yield strength of shear-friction reinforcement f_y to 60 ksi or less.

Equation 6.20 has been compared to test results and found conservative up to a limit of $V_n = 0.2f'_cA_c \leq 0.8 A_c$, where A_c is the contact area of the slippage plane resisting shear transfer. The 0.8-ksi limit corresponds to $f'_c = 4$ ksi.

For design purposes, inserting $V_u/\phi = V_n$ into Eq. 6.20 and solving for A_{vf},

$$A_{vf} \geq \frac{V_u}{\phi f_y \mu} \qquad (6.21)$$

Additional reinforcement should be added to resist net tension across the plane. If the engineer is absolutely certain there will be a net compression across the shear plane under all circumstances, the compression force may be used to reduce the force provided by shear-friction reinforcement.

Angled reinforcement is somewhat more efficient than perpendicular reinforcement if it is slanted in the direction of the shear force, as in Fig. 6.20. Individual pieces of slanted shear-friction reinforcement are not recommended because of the risk of misplacement in the field. Slanted reinforcement so fabricated as to prevent misplacement is acceptable.

When shear-friction reinforcement is inclined to shear plane, as in Fig. 6.20, the area of shear-friction reinforcement required is

$$A_{vf} = \frac{V_u}{\phi f_y(\mu \sin \alpha_f + \cos \alpha_f)} \qquad (6.22)$$

where α_f = angle between shear-friction reinforcement and the shear plane.

FIGURE 6.20 Angled shear-friction reinforcement. (Courtesy of the American Concrete Institute.)

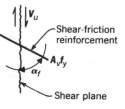

If no moment acts across the shear plane, shear-friction reinforcement should be uniformly distributed along the shear plane to minimize crack widths. If a moment acts across the shear plane, it is desirable to distribute the shear-friction reinforcement so that most of it lies in the flexural tension zone.

Example 6.5   Assume that the builder of the transfer girder in Ex. 6.4 must cast the girder in three sections divided at the top of the lower floor and the bottom of the upper floor. What reinforcement is needed across construction joints?

Solution

Lower Construction Joint

Shear on the construction joint for shear friction can be conservatively taken as the unit shear stress in the girder (332 psi from Ex. 6.4) multiplied by the ratio of the number of flexural reinforcing bars below the construction joint to the total number of flexural reinforcing bars. Conservatively disregard the compression force across the construction joint imposed by end columns.

$$v_u = 332 \times \frac{16}{20} = 266 \text{ psi} < 800 \text{ psi} \qquad \underline{\text{OK}}$$

Using Eq. 6.21, and a coefficient of friction of 1.0,

$$A_{vf} = \frac{0.266 \times 20 \times 12}{0.85 \times 60 \times 1.0} = 1.25^{\square''}/'$$

Add steel to support the hanging floor,

from Ex. 6.4,	$\underline{0.07}$
Total A_{vf}reqd.	$= 1.32^{\square''}/'$

Use #6 U bars at 8″ c/c, $A_s$prov. $= 1.32^{\square''}/'$

Bars can be anchored in the lower section by bending around flexural reinforcement. Anchorage is obtained in the upper section by embedment equal to development length. Because bars are spaced more than 6″ c/c, development length is $0.8 \times 18'' = 15''$ (see Table A5.1). It is prudent to avoid an abrupt change in reinforcement and to reinforce potential shear planes above the construction joint, especially near the ends of the girder. A conservative engineer would alternate U bars extending 15″ above the flexural reinforcement and U bars extending an additional 12″ into the girder web.

Due to anchorage stresses, the main flexural reinforcement might split the beam. To help prevent splitting, one-half the #6 U bars could be placed to enclose all #11 tension bars, and the other half only the top two layers of #11 tension bars, as shown in Figs. 6.17 and 6.18.

Note in Fig. 6.18 that lower corner bars of flexural reinforcement must be moved upward slightly due to the radius of bend in #6 U bars.

To justify the coefficient of friction used, the construction joint must be intentionally roughened to a full amplitude of approximately ¼". Some engineers require a layer of grout placed on the construction joint just prior to placing concrete to help assure a honeycomb-free joint.

If no construction joint were used, the coefficient of friction μ would be 1.4, $A_{vf} = 1.25/1.4 = 0.89^{\square"}/'$ and #6 U bars at 12" c/c could be used.

Upper Construction Joint

Instead of a more rigorous analysis, it is sufficiently conservative to assume that the shear friction force is the compressive force in concrete above the shear plane at ultimate load and that it must be developed between the end of the girder and the point of maximum moment, or at the interior column.

The height of the compressive stress block is

$$a = \frac{A_s f_y}{0.85 f'_c b} = \frac{31.2 \times 60}{0.85 \times 4 \times 20} = 27.5"$$

The floor thickness is 20", thus,

$$\text{Shear force } V_u = 31.2 \times 60 \times \frac{20}{27.5} = 1360 \text{ k}$$

The required shear-friction force using a coefficient of friction $\mu = 1.0$,

$$V_n = \frac{V_u}{\phi \mu} = \frac{1360}{0.85 \times 1.0} = 1600 \text{ k}$$

The column load can be deducted from this force because the principal load on the girder is from the columns. The design force then becomes $1600 - 800 = 800$ k and $A_{vf} = 800/60 = 13.3$ in.^{2}. Dividing by 18 ft between reaction and column,

$$A_{vf} \text{reqd.} = \frac{13.3}{18} = 0.74^{\square"}/'$$

Vertical web reinforcement is #5 at 16" c/c
Add #5 ∩ at 16" c/c
Total A_{vf}prov. $= 0.31 \times 2 \times 12/8 = 0.93^{\square"}/'$

The upper construction joint must also be intentionally roughened to a full amplitude of approximately ¼ in. (see Fig. 6.18 for illustration of placement of shear-friction bars).

6.7 TORSION

Sometimes a concrete member must carry torsion as well as shear and bending moment. Torsion is the twisting of a member around its axis, as in the crankshaft of an automobile. If the magnitude of torsion can be computed from statics, it is statically determinate and called "equilibrium torsion." A reinforced concrete member must resist such torsion or the structure will fail. Examples of equilibrium torsion are shown in Fig. 6.21.

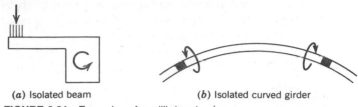

(a) Isolated beam   (b) Isolated curved girder

FIGURE 6.21 Examples of equilibrium torsion.

If the torsion cannot be computed from statics, and a rotation or twist of a member is required for compatibility with the deflection or rotation of other elements of construction such as slabs or other beams, the torsion is statically indeterminate and called "compatibility torsion." Examples of compatibility torsion are shown in Fig. 6.22. If a member fails to resist compatibility torsion, the structure will not necessarily fail because loads can be carried in some other manner, that is, compatibility torsion is redundant. Torsional failure may adversely affect serviceability by opening large cracks or allowing excessive deflections; therefore, members should be designed to resist at least a portion of compatability torsion.

Torsional stiffness is an important factor in determining the magnitude of compatibility torsion. Before cracking, torsion is resisted by plain concrete, and torsional stiffness is relatively high. After cracking, torsion is resisted by both concrete and reinforcement, but torsional stiffness is much lower. Thus, a large torsional moment may crack a member after which both stiffness and moment are greatly reduced. For this reason, there is an upper limit on compatibility torsion for which a member must be designed. This limit is intended to provide sufficient reinforcement to ensure ductility and to keep torsion cracks small and well spaced.

An isolated, rectangular beam subject to torsion but not to bending moment, transverse shear, or axial load will rotate about its axis, as in Fig. 6.23(a). At its centroid, the beam will not deflect. As the beam rotates, each side, top, and bottom must change shape from a rectangle to a shape approximating a parallelogram, as shown in Fig. 6.23(b), 6.23(c), and 6.23(d). One diagonal on each side will lengthen, causing tension in the concrete. The beam will crack when torsional shear stress exceeds concrete strength. The other diagonal on each side will tend to shorten, causing compression in the concrete. Because concrete is strong in compression, little shortening in the diagonal can occur. Instead, the beam will tend to lengthen so that the compression diagonal of the parallelogram is nearly the same length as the compression diagonal of the original rectangular side of the beam. Thus, torsion causes axial tension in a member that must be resisted by reinforcement.

The behavior of beams with flanges is similar to that described above. Each rectangular component of the beam can be considered to respond independently to applied torsion. For simplicity, the torsional resistance of a T-beam is considered equal to the sum of the torsional resistances of the rectangular components of the T-beam.

Bending moment and transverse shear are usually present at the same time as torsion. Sometimes axial load is also present. The effects of these forces and moments may increase or decrease the effects of torsion. For example, torsional and

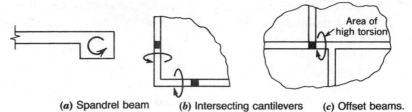

(a) Spandrel beam   (b) Intersecting cantilevers   (c) Offset beams.

FIGURE 6.22 Examples of compatibility of torsion.

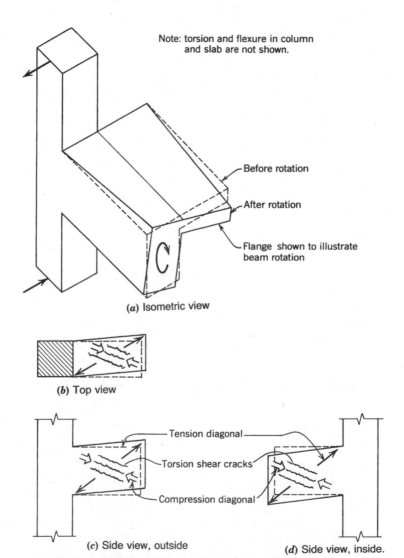

Note: torsion and flexure in column
and slab are not shown.

Before rotation

After rotation

Flange shown to illustrate
beam rotation

(a) Isometric view

(b) Top view

Tension diagonal

Torsion shear cracks

Compression diagonal

(c) Side view, outside   (d) Side view, inside.

FIGURE 6.23 Spandrel beam subject to torsion.

transverse shear are additive on one side of a beam but subtractive on the other side.

Torsional shear stresses are largest in the middle of the longest side and reduce to zero at corners and at the center of the section, as shown in Fig. 6.24(a). Torsion behaves somewhat like shear and causes diagonal tension, which cracks concrete when stress reaches the strength of concrete. Torsion cracks spiral around a member, as shown in Fig. 6.24(b). As torsion increases, failure finally occurs by crushing concrete along a narrow line connecting diagonal cracks. Because failure can occur suddenly at first cracking or at slightly higher loads, torsion strength assigned to concrete is a conservative fraction of its average ultimate strength. Torsion in excess of assumed concrete torsional strength must be carried by reinforcement.

As with shear, design for torsion is not yet completely rational but depends upon conforming to many test results. For summaries of test results and their application to design, see Refs. 4, 5, and 6.

To be consistent with shear design, the critical section for torsion is taken at a distance from the face of support equal to the effective depth d. Factored torsional moment T_u at a section is resisted by the sum of nominal torsional moment strengths provided by concrete T_c and by torsional reinforcement T_s, reduced by capacity reduction factor ϕ.

$$T_u = \leq \phi T_n = \phi(T_c + T_s) \tag{6.23}$$

where T_n = total nominal torsional moment strength.

and $\phi = 0.85$ for torsion.

Research has determined that the relationship between torsional moment strength of concrete and shear strength of concrete is best defined by a circular interaction diagram.

$$\left(\frac{V_c}{V_o}\right)^2 + \left(\frac{T_c}{T_o}\right)^2 = 1 \tag{6.24}$$

where V_c and T_c are the shear capacity and torsion capacity, respectively, under combined loading

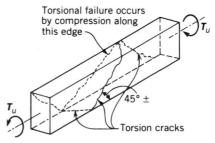

(a) Torsion stresses   (b) Pattern of torsion cracks

FIGURE 6.24 Torsion in a rectangular beam.

and V_o and T_o are the cracking shear capacity and torsional moment capacity of the member when subject, respectively, to transverse shear and torsional moment alone.

Equation 6.24 is shown graphically in Fig. 6.25.

For design purposes it is necessary to have separate expressions for shear and torsional moment strength of concrete. Solving Eq. 6.24 first for V_c and then for T_c gives

$$V_c = \frac{V_o}{\sqrt{1 + \left(\dfrac{V_o}{T_o}\right)^2 \left(\dfrac{T_c}{V_c}\right)^2}} \tag{6.25}$$

$$T_c = \frac{T_o}{\sqrt{1 + \left(\dfrac{T_o}{V_o}\right)^2 \left(\dfrac{V_c}{T_c}\right)^2}} \tag{6.26}$$

From laboratory tests, useful ultimate strengths of concrete can be conservatively taken as $T_o = 0.8\sqrt{f'_c}\,\Sigma x^2 y$ for torsion and $V_o = 2\sqrt{f'_c}\,b_w d$ for shear (Eq. 6.4). Using these values for T_o and V_o,

$$\frac{T_o}{V_o} = \frac{0.4}{C_t} \tag{6.27}$$

$$\text{Where } C_t = \frac{b_w d}{\Sigma x^2 y} \tag{6.28}$$

and where x = shorter overall dimension of the rectangular part of the cross section and y = longer overall dimension of the rectangular part of the cross section.

Component rectangles of flanged sections should be selected to yield the largest value of $\Sigma x^2 y$ but should not overlap. In general, this will be achieved when component rectangles are most nearly square. To agree with the data, the overhanging flange width used in design must not exceed 3 times the flange thickness (see Fig. 6.26 for illustration of typical conditions in the calculation of $\Sigma x^2 y$).

Values of C_t can be easily visualized by reference to Table A6.5, which has been prepared for rectangular beams. For beams with flanges of normal proportions, C_t is lower than for rectangular beams. The reduction in C_t can be quickly estimated by dividing C_t for rectangular beams given in Table A6.5 by the ratio $\Sigma x^2 y / x^2 y$ from Fig. A6.4.

FIGURE 6.25 Interaction of concrete shear and torsion strengths.

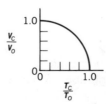

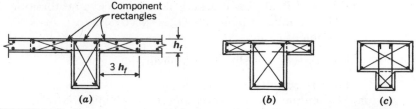

FIGURE 6.26   Component rectangles for the calculation of $\Sigma x^c y$. (Courtesy of the American Concrete Institute.)

It is safe to assume that

$$\frac{V_c}{T_c} = \frac{V_u}{T_u}$$

Substituting this identity, Eq. 6.27, and values given above for V_o and T_o into Eqs. 6.25 and 6.26 yields the following design equations:

$$V_c = \frac{2\sqrt{f_c'}b_w d}{\sqrt{1 + \left(2.5 C_t \dfrac{T_u}{V_u}\right)^2}} \tag{6.29}$$

$$T_c = \frac{0.8\sqrt{f_c'}\,\Sigma x^2 y}{\sqrt{1 + \left(\dfrac{0.4 V_u}{C_t T_u}\right)^2}} \tag{6.30}$$

Because of the circular interaction curve (see Fig. 6.25), low levels of torsion can be tolerated without serious reduction in shear and moment capacity. The ACI Code requires that the effects of torsion be included with shear and flexure only where factored torsional moment T_u exceeds $\min T_u$ where

$$\boxed{\min T_u = \phi 0.5\sqrt{f_c'}\,\Sigma x^2 y} \tag{6.31}$$

Torsional moment below which torsion may be disregarded, $\min T_u$, is given in Table A6.6 for rectangular beams of normal size and proportions. Torsonal moment in this table may be increased for flanges by coefficients taken from Fig. A6.4.

The maximum factored torsional moment T_u for which a member need be designed for compatability torsion in a statically indeterminant structure, where reduction in torsional moment can occur due to redistribution of internal forces, is

$$T_u = \frac{\phi 4\sqrt{f_c'}\,\Sigma x^2 y}{3} \tag{6.32}$$

By dividing Eq. 6.32 by Eq. 6.31, it can easily be seen that the maximum T_u for which a member subjected to compatability torsion need be designed is $\min T_u$ multiplied by 2.67, where $\min T_u$ is given by Eq. 6.31.

Torsion Reinforcement

Tests have shown that both longitudinal and transverse reinforcement are necessary to properly resist torsion. Transverse reinforcement or stirrups must be closed (see Fig. 6.9) because torsional shear stress occurs on all sides of the member. Because all four sides of a stirrup are stressed, ends of stirrup bars must be anchored to resist full stress, preferably by hooking into the concrete core enclosed by the reinforcement cage or by extending into flanges. Potential spiral torsion cracks on two adjacent sides must be crossed by at least four stirrups or ties. Maximum stirrup spacing is $(x_1 + y_1)/4$ but not more than 12 in where x_1 is the shorter center-to-center dimension of a closed rectangular stirrup, and y_1 is the longer center-to-center dimension of a closed rectangular stirrup. This is equivalent to the requirement for maximum spacing of stirrups resisting transverse shear (see Fig. 6.8).

To ensure that all potential spiral cracks are crossed by reinforcement, stirrups for torsion must be provided at least a distance $(b_t + d)$ beyond the point theoretically required, where b_t is the width of that part of the cross section containing stirrups. Stirrups need not extend beyond the point where torsional moment is less than half the torsional moment strength provided by concrete. Figure 6.27 explains stirrup spacing limits for torsion reinforcement.

Required torsion reinforcement is in addition to reinforcement required to resist shear, flexure, and axial forces. Individually required areas can be combined for bar selection, provided the most restrictive requirements for spacing and placement are met.

Two legs of a stirrup resist shear, but torsion design procedures assume that only one leg of a stirrup resists torsion at failure. When combining shear and torsion reinforcement requirements, the total area of stirrups for two legs of each stirrup is $A_v + 2A_t$. Stirrups provided for torsion can be combined with stirrups provided for

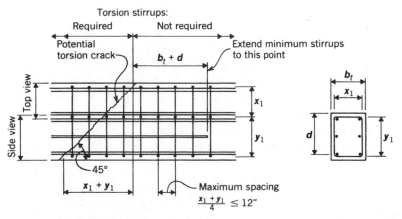

(a) Expanded view of beam reinforcement   (b) Beam cross section

FIGURE 6.27   Torsion stirrup spacing limits.

shear to meet requirements for minimum area of stirrups. Thus, minimum area of stirrup reinforcement is

$$\min (A_v + 2A_t) = \frac{50b_w s}{f_y} \qquad (6.33)$$

where A_t = area of one leg of a closed sitrrup resisting torsion and f_y is in psi.

To determine the area of torsion stirrups required, solve Eq. 6.23 for T_s.

$$T_s = \frac{T_u - \phi T_c}{\phi} \qquad (6.34)$$

By referring to Fig. 6.24(b), it is apparent that one leg of a torsion stirrup will cross a diagonal tension crack and the other leg will be near the diagonal line of compression at failure. By considering the effectiveness of stirrups in resisting torsional moment, the area of one closed stirrup leg resisting torsion A_t can be computed as

$$A_t = \frac{T_s s}{\alpha_t x_1 y_1 f_y} \qquad (6.35)$$

where $\alpha_t = 0.66 + 0.33(y_1/x_1)$ but not more than 1.50.

Nominal torsional moment strength provided by torsion reinforcement T_s shall not exceed $4T_c$.

Equation 6.35 can be considered as the torsional moment per stirrup $T_s s$ divided by the combined effective depths $x_1 y_1$, the yield strength of steel f_y, and a factor to account for efficiency of the section α_t.

As noted above, longitudinal reinforcement is required to resist torsional moment. One bar should be placed in each corner of closed stirrups and other bars at a spacing not to exceed 12 in. between corner bars. Ends of longitudinal bars used as torsion reinforcement must be anchored inside the face of supports and beyond the point in the span where torsion reinforcement is no longer required. Extension beyond this point is the same as for stirrups. Extension must be at least equal to $(b_t + d)$ but not less than the bar tension development length.

Tests indicate that the volume of longitudinal reinforcement must at least equal the volume of transverse reinforcement (stirrups) not including hooks, laps, or other forms of anchorage. Thus, the area of longitudinal reinforcement A_ℓ can be computed as

$$A_\ell = 2A_t \left(\frac{x_1 + y_1}{s} \right) \qquad (6.36)$$

As a further precaution, the ACI Code requires that A_ℓ not be less than

$$A_\ell = \left[\frac{0.4\, xs}{f_y} \left(\frac{T_u}{T_u + \dfrac{V_u}{3C_{t/}}} \right) - 2A_t \right] \left(\frac{x_1 + y_1}{s} \right) \tag{6.37}$$

where x = shorter overall dimension of rectangular part of cross section.

If Eqs. 6.36 and 6.37 are set equal to each other and solved for A_t, it can be seen that Eq. 6.36 governs for determining A_ℓ when

$$A_t > \frac{0.1\, xs}{f_y} \left(\frac{T_u}{T_u + \dfrac{V_u}{3C_{t/}}} \right) \tag{6.38}$$

When A_t is less than the value given in Eq. 6.38, then Eq. 6.37 governs for determining A_ℓ. For further simplification, the engineer can make a preliminary computation disregarding the term in parenthesis in Eq. 6.38 $[T_u/(T_u + V_u/3C_t)]$ as it will always be equal to or less than one.

Design Procedures

Design for torsion can be completed efficiently by following this procedure.

1. Compute the factored torsional moment T_u at the critical section and compare it to the torsional moment for the section below which torsion need not be considered, min T_u (given by Eq. 6.31). If T_u is less than minT_u, design for torsion is not required.

2. If torsion is statically indeterminant, that is, if torsion is compatibility torsion, limit T_u for design to 2.67 times minT_u.

3. Design the member for transverse shear by computing v_u, v_c and, if required, A_v.

4. Design the member for torsion by computing ϕT_c, ϕT_s and, if required, A_t.

5. Combine A_v and A_t. Select stirrup number, size, type, and spacing.

6. Compute the required area of longitudinal torsion bars A_ℓ. Select the number and size of longitudinal bars where required at locations not coincident with flexural reinforcement.

7. Where longitudinal torsion bars are in the same face of the beam as flexural reinforcement, add the reinforcement required for torsion to that required for flexure and select longitudinal reinforcing bars.

Example 6.6 Given an isolated beam with a 24-ft clear span, shown in Fig. 6.28, with a superimposed service dead load of 10 psf and service live load of 30 psf, select necessary torsion reinforcement. Assume $f'_c = 3$ ksi and $f_y = 60$ ksi.

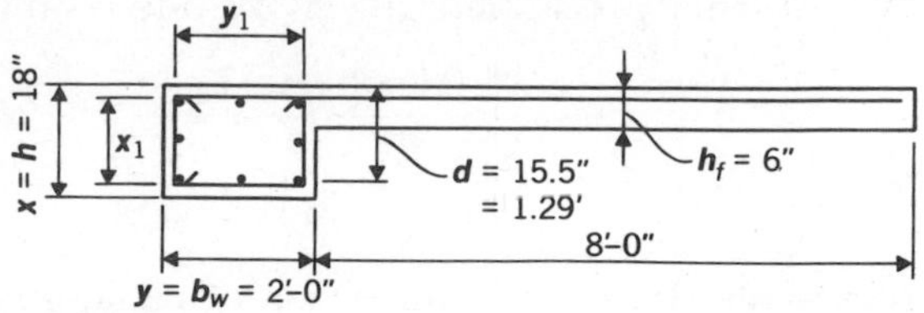

FIGURE 6.28

Solution

Slab loading: $(75 + 10)1.4 \times 8 = 0.95$ k/'

$30 \times 1.7 \times 8 = \underline{0.41}$

$w_u = 1.36$ k/'

Beam loading: $(225 + 10)1.4 \times 2 = 0.66$ k/'

$30 \times 1.7 \times 2 = \underline{0.10}$

$w_u = 2.12$ k/'

Torsional moment at critical section, distance d from the support is as follows:

$$T_u = 1.36(4 + 1)\left(\frac{24}{2} - 1.29\right) = 72.8'\text{k}$$

$$V_u = 2.12\left(\frac{24}{2} - 1.29\right) = 22.7 \text{ k}$$

$$v_u = \frac{22.7}{24 \times 15.5} = 61 \text{ psi}$$

Compute the minimum factored torsional moment below which torsion can be safely disregarded. From Fig. A6.4, using $y/x = 24/18 = 1.33$ and $h_f/y = 6/24 = 0.25$, $\Sigma x^2 y/x^2 y \simeq 1.08$. Referring to Table A6.6 and interpolating,

$\min T_u \simeq 15.25 \times 1.08 = 16.5'\text{k}$

Or by computation using Eq. 6.31,

$\Sigma x^2 y = 18^2 \times 24 + 6^2 \times 18 = 8420$

$$\min T_u = \frac{0.85 \times 0.5\sqrt{3000} \times 8420}{12,000} = 16.3'\text{k} < T_u = 72.8'\text{k}$$

Therefore, the beam must be designed for torsion. An experienced engineer would make this decision by inspection without computing $\min T_u$.

Compute the capacity of concrete in shear:
Using Eq. 6.28,

$$C_t = \frac{24 \times 18}{8420} = 0.0513$$

$$\text{and } 2.5 C_t \frac{T_u}{V_u} = 2.5 \times 0.0513 \times \left(\frac{72.8 \times 12}{22.7} \right) = 4.94$$

Using Eq. 6.29,

$$v_c = \frac{0.85 \times 2\sqrt{3000}}{\sqrt{1 + (4.94)^2}} = 19 \text{ psi} < v_u = 61 \text{ psi}$$

Therefore, shear stirrups are required in the length of the beam shown in Fig. 6.29(a).

At the critical section, for a unit length of 1' (12"),

$$A_v = \frac{(61 - 19)12 \times 24}{60,000} = 0.20^{\square''}/'$$

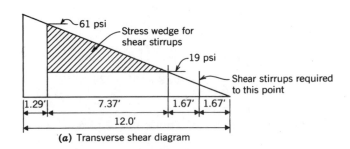

(a) Transverse shear diagram

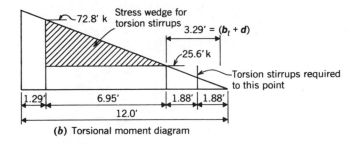

(b) Torsional moment diagram

FIGURE 6.29

Compute the capacity of concrete in torsion using Eq. 6.30 and values previously computed for $\Sigma x^2 y$ and $2.5 C_t T_u/V_u$.

$$\phi T_c = \frac{0.85 \times 0.8\sqrt{3000} \times 8420}{12,000\sqrt{1 + (1/4.94)^2}} = 25.6'\text{k} < T_u = 72.8'\text{k}$$

Therefore, torsion stirrups are required in the length of the beam shown in Fig. 6.29(b).

Design of Torsion stirrups:
At the critical section,

$$4T_c = \frac{4 \times 25.6}{0.85} = 120.5'\text{k}$$

Using Eq. 6.34,

$$T_s = \frac{72.8 - 25.6}{0.85} = 55.5'\text{k} < 4T_c \qquad\qquad \underline{\text{OK}}$$

At the critical section, assuming a unit length of 1' (12") and using Eq. 6.35,

$$x_1 = 18 - 1.75 \times 2 = 14.5", \qquad y_1 = 24 - 1.75 \times 2 = 20.5"$$

$$y_1/x_1 = 1.41, \qquad\qquad \alpha_t = 0.66 + 0.33 \times 1.41 = 1.13$$

$$A_t = \frac{55.5 \times 12 \times 12}{1.13 \times 14.5 \times 20.5 \times 60} = 0.39^{\square}\text{"}/'$$

Minimum stirrups for shear and torsion are as follows:

$$\min(A_v + 2A_t) = \frac{50 b_w s}{f_y} = \frac{50 \times 24 \times 12}{60,000} = 0.24^{\square}\text{"}/'$$

Minimum stirrup area does not control because the required area of shear and torsion stirrups

$$A_v + 2A_t = 0.20 + 2 \times 0.39 = 0.98"$$

Maximum torsion stirrup spacing $= \dfrac{x_1 + y_1}{4}$

$$= \frac{14.5 + 20.5}{4} = 8.75" \text{ c/c}$$

Maximum shear stirrup spacing $= \dfrac{d}{2} = \dfrac{15.5}{2} = 7.75" \text{ c/c}$

Use maximum stirrup spacing $= 7\frac{1}{2}"$ c/c (8" c/c might be used, but as will be seen below, it is convenient to use $7\frac{1}{2}"$.)

Use #4 stirrups at $4\frac{1}{2}"$ c/c, $A_t\text{prov.} = 0.40 \times 12/4.5 = 1.07^{\square}\text{"}/'$

At maximum stirrup spacing,

#4 stirrups at 7½" c/c, $A_t$prov. $= 0.40 \times 12/7.5 = 0.64^{□''}/'$

The selected stirrup spacing in the diagram in Fig. 6.30 shows shear and torsion stirrup requirements.

Some engineers would choose to add six stirrups in the beam and use 4" and 7" stirrup spacing to avoid spacings measured in ½" increments.

Design the cantilever slab.

$M_u = 1.36 \times 4 = 5.4'k/'$

Assume $(\phi f_y j) = 4.2$ and $d = 6 - 1.75 = 4.25"$

$$A_s = \frac{5.4}{4.2 \times 4.25} = 0.31^{□''}/'$$

Use #4 7½" c/c (same as maximum stirrup spacing).

$A_s$prov. $= 0.30^{□''}/'$

The stirrup top bar and the slab top bar can be combined as the same bar. Their required areas need not be added before bar selection.

Compute the required longitudinal reinforcement using Eq. 6.36,

$A_t = 0.39^{□''}/'$ and $s = 12"$

$$A_\ell = 2 \times 0.39 \frac{(14.5 + 20.5)}{12} = 2.28^{□''}$$

To check, use Eq. 6.38.

$$\frac{0.1 \, xs}{f_y} = \frac{0.1 \times 18 \times 12}{60} = 0.36^{□''} < 0.39^{□''}$$

Therefore, the above computation using Eq. 6.36 controls area of longitudinal torsion reinforcement required.

Eight longitudinal bars are required to maintain a maximum spacing of 12", each with an area of $2.28/8 = 0.29^{□''}$ or more.

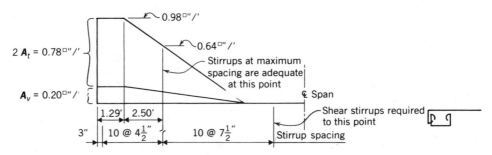

FIGURE 6.30

Use 1 #5 at mid-height on each side, and add
$3 \times 0.29 = 0.87^{□''}$ to the top flexural reinforcement required and $0.87^{□''}$ to the bottom flexural reinforcement required.

This example shows a structure of extreme proportions. Deflection at the tip of the slab would be substantial from bending of the slab, torsional rotation of the beam, and flexure of the beam. It should be used only for a roof canopy or other purpose not supporting or attached to walls or construction likely to be damaged by large deflections. Deflection is likely to be annoying to pedestrians on the slab. The engineer should consider cambering the cantilever slab upward more than the anticipated deflection to avoid an aesthetically objectionable sag (see Problem 8.10).

6.8 SHEAR WALLS

Design for shear forces perpendicular to the face of the wall is performed in the same manner as for one- or two-way slabs, depending on the manner of load application and support of the wall. Shear walls resist horizontal forces, such as wind and earthquake forces, in the plane of the wall and must meet provisions given in ACI Code Section 11.10 [7].

Design of shear walls in structures of normal proportions is usually controlled by flexure and overturning rather than shear, especially for those walls with a large height-to-length ratio. Unless a strain compatability analysis is made, effective depth d shall be taken equal to $0.8\ell_w$, where ℓ_w is the horizontal length of the wall.

Maximum factored shear force V_u permitted on a wall with minimum reinforcement is the sum of the nominal shear strength provided by concrete and the nominal shear strength provided by shear reinforcement, both reduced by the capacity reduction factor ϕ.

Assuming $d = 0.8\ell_w$ and $V_c = 2\sqrt{f_c'}\,hd$, the nominal shear strength provided by concrete V_c,

$$V_c = 1.6\sqrt{f_c'}\,h\ell_w \tag{6.39}$$

where h = thickness of the wall.

Substituting s_2 for s in Eq. 6.11, omitting the capacity reduction factor ϕ, and solving for V_s,

$$V_s = \frac{A_v f_y d}{s_2} \tag{6.40}$$

where A_v = area of horizontal shear reinforcement.
and s_2 = spacing of horizontal reinforcement in the wall.

Spacing s_2 must not exceed $\ell_w/5$, $3h$, nor 18 in., where h is the wall thickness.

The minimum area of horizontal shear reinforcement permitted by the ACI Code minA_v is

$$\min A_v = 0.0025 h s_2$$

Substituting this value and $0.8\ell_w$ for d into Eq. 6.40 gives the minimum shear strength provided by shear reinforcement.

$$\min V_s = 0.002 f_y h \ell_w \tag{6.41}$$

Combining Eqs. 6.39 and 6.41, the minimum shear strength ϕV_n of a shear wall is

$$\min \phi V_n = \phi h \ell_w (1.6\sqrt{f_c'} + 0.002 f_y) \tag{6.42}$$

Values for Eq. 6.42 are tabulated for typical conditions in Table A6.7. The maximum factored shear permitted by the ACI Code on a shear wall is $\phi 10\sqrt{f_c'}hd$ and is also tabulated in Table A6.7, assuming $d = 0.8\ell_w$.

When more horizontal shear reinforcement A_v is required than the minimum, the amount can be computed by using Eq. 6.11.

$$A_v = \frac{V_s s_2}{\phi f_y d}$$

where $V_s = \dfrac{V_u - \phi V_c}{\phi}$

and A_v = the area of horizontal shear reinforcement within a distance s_2.

The required vertical shear reinforcement depends on the amount of horizontal shear reinforcement required. Ratio ρ_n of vertical shear reinforcement area to gross concrete area of the horizontal concrete section shall not be less than

$$\rho_n = 0.0025 + 0.5 \left(2.5 - \frac{h_w}{\ell_w}\right)(\rho_h - 0.0025) \tag{6.43}$$

where $\rho_h = A_v/h s_2$ or the ratio of the horizontal shear reinforcement area to the gross concrete area of the vertical section

and h_w = total height of the wall from base to top
when $h_w/\ell_w > 2.5$, $\rho_n = 0.0025$
when $h_w/\ell_w < 0.5$, $\rho_n = \rho_h$.

Influence of ρ_h on ρ_n, as given by Eq. 6.43, can be easily visualized by referring to Fig. A6.5.

As with beams, the increase in shear strength provided by concrete can be conservatively disregarded when the wall is in compression and can be taken as zero when the wall is in tension.

Ordinary walls that do not resist significant shear can be reinforced with slightly less vertical steel than required for shear walls (see Chapter 12). If shear force is less than about one fourth of the shear strength of walls with minimum reinforcement given in Table A6.7 for steel with $f_y = 60$, the lower vertical reinforcement requirements for ordinary walls would be sufficiently conservative. The factored shear force would be about two thirds of nominal shear strength provided by concrete alone.

Example 6.7   What length is required for two symmetrically placed shear walls, as shown in Fig. 6.31, designed to resist earthquake loads, in a 200-ft square warehouse with a roof weighing 100 psf and seismic coefficient of 0.2? Assume $f'_c = 3$ and $f_y = 60$.

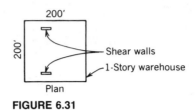

200'

200'

Shear walls

1-Story warehouse

Plan

FIGURE 6.31

Solution   Total factored horizontal shear is

$$V_u = 200 \times 200 \times 0.100 \times 0.2 \times 1.7 = 1360 \text{ k}$$

Using the shear capacities from Table A6.71,

$$\text{length of 8'' wall} \ = \frac{1360}{2 \times 16.9} = 40.2' \text{ maximum}$$

$$= \frac{1360}{2 \times 35.8} = 19.0' \text{ minimum}$$

$$\text{length of 10'' wall} = \frac{1360}{2 \times 21.1} = 32.2' \text{ maximum}$$

$$= \frac{1360}{2 \times 44.7} = 15.2' \text{ minimum}$$

$$\text{length of 12'' wall} = \frac{1360}{2 \times 25.4} = 26.8' \text{ maximum}$$

$$= \frac{1360}{2 \times 53.6} = 12.7' \text{ minimum}$$

Problems of delivering shear force to the walls and designing the walls to resist overturning are more critical than designing these walls for shear.

Example 6.8   If shear walls beside two symmetrically located stairwells, 15½ ft long, are placed in a four-story building with a total height and length of 55 ft and 150 ft respectively, as shown in Fig. 6.32, and the walls must resist all horizontal wind forces of 30 psf, how thick should the walls be to resist wind shear with minimum reinforcement? Assume $f'_c = 3$ ksi and $f_y = 60$ ksi.

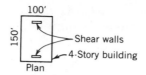

FIGURE 6.32

Solution   Total factored horizontal shear is

$$V_u = 150 \times 55 \times 0.030 \times 1.7 = 421 \text{ k}$$

$$\text{Shear per ft} = \frac{421}{2 \times 15.5} = 13.6 \text{ k/}'$$

Referring to Table A6.7, an 8″ wall with minimum reinforcement is more than adequate. A 6½″ wall is adequate for stress because $(6.5/8)16.9 = 13.7\text{k/}' > 13.6\text{k/}'$, but is not recommended due to construction difficulties in placing concrete in such a thin wall.

Reinforcement required in an 8″ wall is

$$0.0025 \times 8 \times 12 = 0.24^{\square''}/'$$

Use one layer of #4 bars at 10″ c/c or #5 bars at 15 c/c in the center of the wall, each way.

$$A_s\text{prov.} = 0.24^{\square''}/'$$

The preferable selection is #5 bars because there will be fewer pieces of steel to handle in the shop and in the field.

Design for flexure and overturning is more critical than design for shear resistance.

6.9 BRACKETS AND CORBELS

Very short cantilevers are called "brackets." As used in the ACI Code, "corbel" is a synonymous term. When shear span-to-depth ratio **a/d** is one or less, diagonal tension cracks are nearly vertical. Brackets frequently support loads that exert on the bracket horizontal tensile forces which also cause nearly vertical cracks. Thus, bracket design emphasis is on horizontal reinforcement. Unless engineers are certain that there will be no tensile forces under any circumstances, they should design brackets for some tension.

Physical requirements for brackets are shown in Fig. 6.33. A bracket is similar to a deep beam with a compression strut meeting tension bars near the outside end of the bars. Because tension bars may be fully stressed to the outside end, they must be anchored at the front face of the bracket by one of three methods.

1. Weld tension bars to a cross bar of a size at least equal to the tension bars. The weld must develop full specified yield strength f_y of the tension bars.

2. Bend tension bars in a horizontal loop back into the support.

3. Anchor tension bars mechanically, such as by welding to a stiff steel angle exposed on the edge of the bracket.

To prevent splitting of the face shell covering reinforcement, the bearing area of the load on the bracket must not project beyond the interior face of the cross bar anchor, the straight portion of looped tension bars, or the interior face of the mechanical anchorage.

The ends of tension bars must also be anchored in the support by embedment or hooks or other means (see Chapter 5).

Because bracket behavior is controlled by shear, a single value of $\phi = 0.85$ must be used for all design conditions including flexure and tension.

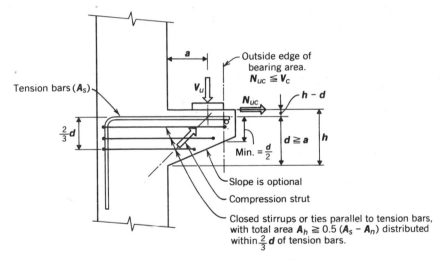

(a) Bracket with tension bar anchor, Method 1.

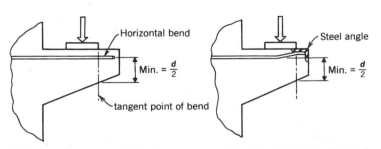

(b) Tension bar anchor, Method 2.   (c) Tension bar anchor, Method 3.

FIGURE 6.33   Bracket details.

Moment M_u on a bracket is the sum of moments caused by vertical load and by horizontal force,

$$M_u = V_u a + N_{uc}(h - d)$$

where V_u = load or shear force on the bracket [see Fig. 6.33(a)]

N_{uc} = factored tensile force applied at the top of the bracket acting simultaneously with V_u, to be taken as positive for tension.

The area of reinforcement A_f to resist moment can be computed as for beams but using $\phi = 0.85$ instead of $\phi = 0.9$. The value of $(\phi f_y j)$ from Table A2.3 or Fig. A2.2 must be decreased 5%. Flexural compression does not control the design of brackets.

The area of reinforcement A_n to resist tension is $N_{uc}/(\phi f_y)$, where N_{uc} is at least equal to $0.2V_u$ unless special provisions are made to avoid tensile forces. In determining N_{uc}, a load factor of 1.7 must be used regardless of the source of the horizontal force.

The shear V_u must be resisted by shear friction using procedures described in Section 6.6. For normal weight concrete, the maximum shear strength in kips of a bracket is the lesser of

$$0.2f'_c b_w d \quad \text{or} \quad 0.8b_w d$$

where b_w = width of the bracket.

For lightweight concrete, the maximum shear strength in kips of a bracket is the lesser of

$$\left(0.2 - 0.07\,\frac{a}{d}\right) f'_c b_w d \quad \text{or} \quad \left(0.8 - 0.28\,\frac{a}{d}\right) f'_c b_w d$$

where a and d are defined in Fig. 6.33(a).

The total area of tension bars can then be computed as the greater of

$$A_s = A_f + A_n$$

$$\text{or } A_s = \frac{2A_{vf}}{3} + A_n$$

$$\text{or } A_s = \rho_{min} b_w d$$

where A_{vf} = the area of shear-friction reinforcement
and $\rho_{min} = 0.04f'_c/f_y$.

In addition to tension bars, closed stirrups or ties parallel to the tension bars must be uniformly distributed within two thirds of the effective depth adjacent to the tension bars. Area A_n of the additional ties shall be not less than $0.5(A_s - A_n)$.

PROBLEMS

1. What is the shear stress in concrete in a 6-in. slab resisting a shear force of 1.2 klf? Assume ¾ in. cover on tension reinforcement.

2. How much factored shear force is resisted by concrete in a 36-in.-wide by 20½-in.-deep beam? Assume $f'_c = 4$ ksi and 2 in. cover is used on reinforcement.
3. Select the type, size, and spacing of stirrups for the beam in Problem 2 if the uniformly distributed factored load is 9.6 klf and the clear span is 29 ft between columns. Use $f_y = 40$ ksi for stirrups.
4. A simple span, 20 × 36-in. beam (shown in Fig. 6.34) supports a heavy piece of equipment weighing 100 tons near one end. Columns are 16 in. square. In addition, the beam supports a uniformly distributed dead load of 2 klf and a live load of 3 klf. Design shear reinforcement for the beam between the load and the nearest column. Use $f'_c = 4$ ksi and $f_y = 40$ ksi for shear reinforcement.

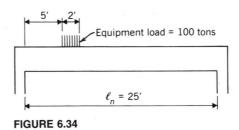

FIGURE 6.34

5. How much transverse shear can be resisted by shear friction across a construction joint at a point where the moment is 50% of the maximum moment in a 7-in. slab with flexural tension reinforcement of #5 bars at 10 in. c/c? Assume that the joint is smooth and perpendicular to the axis of the member and $f_y = 50$ ksi.
6. A connection angle is welded to a steel plate exposed on the face of a concrete wall, as in Fig. 6.35. A 24-in.-deep steel beam with a 75-kip service load end reaction will be bolted to the angle. What size studs are required and how many studs are needed? Show the location of studs on the plate. Assume $f'_c = 4$ ksi and $f_y = 36$ ksi for steel studs.

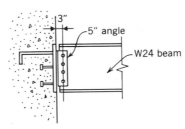

FIGURE 6.35

7. If the wall service dead load is 0.96 klf, what is the maximum torsional moment for which the isolated beam in Fig. 6.36 need be designed? Clear span between columns is 24 ft. If $f'_c = 3$ ksi and $f_y = 60$ ksi, are stirrups required? If so, select the number, type, size, and spacing of the stirrups.
8. Consider the same beam as in Problem 7, but with a wall dead load of 1.3 klf and a live load on the wall of 0.8 klf. What is the maximum design torsional moment? Are stirrups

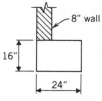

FIGURE 6.36

required? If so, select the number, type, size, and spacing of stirrups and longitudinal torsion bars.

9. Design a concrete bracket to carry a steel beam with an unfactored 120-kip end reaction bearing on a 12 × 12-in. steel plate. The bracket is on the face of a 16-in.-wide column. Assume $f'_c = 4.5$ ksi, $f_y = 60$ ksi and no welding may be used in the bracket. The depth of the bracket must not be more than 22 in. for architectural reasons.

Selected References

1. ACI-ASCE Committee 426. "The Shear Strength of Reinforced Concrete Members," (ACI 426-74)(Reaffirmed 1980), Chapters 1 to 4, *Proceedings, ASCE*. V.99, No.ST6, pp 1148–1157, June 1973. See also, *ACI Manual of Concrete Practice, Part 4*, American Concrete Institute, Detroit, Michigan, 1985.
2. H.A.R. dePaiva and C.P. Siess. "Strength and Behavior of Deep Beams in Shear," *Proceedings, ASCE*, V.91, No.ST5, Part 1, pp 19–41, October 1965.
3. R.F. Mast. "Auxiliary Reinforcement in Precast Concrete Connections," *Proceedings, ASCE*, V.94, No.ST6, pp 1485–1504, June 1968.
4. A.H. Mattock. "How to Design for Torsion," *Torsion of Structural Concrete*, SP-18, American Concrete Institute, pp 469–495, Detroit, Michigan, 1968.
5. ACI Committee 438. "Tentative Recommendations for the Design of Reinforced Concrete Members to Resist Torsion," *ACI Journal, Proceedings*, V.66, No. 1, pp 1–8, January 1969. See also Discussion of "Tentative Recommendations for the Design of Reinforced Concrete Members to Resist Torsion," by ACI Committee 438, *ACI Journal, Proceedings*, V.66, No. 7, pp 576–588, July 1969.
6. T.T.C. Hsu and C.S. Hwang. "Torsional Limit Design of Spandrel Beams," *ACI Journal, Proceedings*, V.74, No.2, pp 71–79, February 1977.
7. Alex E. Cardenas, John M. Hanson, W. Gene Corley, and Eivind Hognestad. "Design Provisions for Shear Walls," *ACI Journal, Proceedings*, V.70, No.3, pp 221–230, March 1973. See also, *Research and Development Bulletin* RD028D, Portland Cement Association, Skokie, Illinois.

CHAPTER 7

SHEAR AND TORSION IN TWO-WAY SLABS

7.1 OBJECTIVES

This chapter deals with shear in two-way slab systems. Except for a few special situations covered in Chapter 11, shear in two-way footings is the same as shear in two-way slabs. In both types of members, load funnels in toward a narrow perimeter around the concentrated load or support (usually a column), a perimeter that is small and highly stressed. Because shear failure can be sudden and catastrophic, it must be avoided.* ACI Code requirements for shear are more conservative than for flexure.

Assuming that slab thickness and column size are given, the following are objectives of shear design in two-way slabs:

1. Check shear strength provided by concrete and select a method for providing additional shear capacity, if required.
   Depending on the method selected for providing additional shear capacity,
2. Determine size and shape of the column capital.
3. For shear reinforcement consisting of bars or wires, select the bar size, number, bending, anchorage, and location.
4. For shear reinforcement consisting of steel I− or channel-shaped sections, select the steel size, length, number of arms, fabrication details, and location.

7.2 SHEAR STRENGTH OF PLAIN CONCRETE

Shear strength of slabs and footings in the vicinity of concentrated loads and reactions is limited by the more severe of two conditions. In the first condition, the

---

*The sight of slightly cracked floor slabs stacked on top of each other (maybe with a few paper-thin construction workers in between) and columns poking through several stories high, like naked tree trunks, is enough to cure most engineers of any latent tendency to cut corners on slab shear design.

critical section is located at a distance **d** from the face of the concentrated load or reaction and extends across the entire width of the member. This is called one-way action or beam action and is discussed in Chapter 6.

In the second condition, the critical section surrounds the concentrated load or reaction area, is perpendicular to the plane of the slab, and is located so that its perimeter b_o is a minimum but need not approach closer than $d/2$ to the perimeter of the concentrated load or reaction area (Code 11.11.1.2). This is called two-way action or punching shear and is discussed in this chapter (see Fig. 7.1 for typical examples of critical section for punching shear).

Punching shear stress is computed simply by dividing the shear force by the area of the critical section perimeter.

$$v_u = \frac{V_u}{b_o d} \tag{7.1}$$

Shear failure in two-way action occurs by punching out a cone of concrete through the slab, as shown in Fig. 7.2. The slope of the cone is assumed to be about 45°, but the critical section is located $d/2$ from the support because it gives the best correlation with hundreds of tests to failure [1]. When the column (loaded area) is round or nearly square, ultimate shear stress on plain concrete at the critical section may be conservatively taken as $\phi 4\sqrt{f'_c}$. As the column becomes more elon-

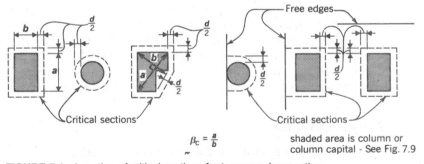

FIGURE 7.1   Location of critical sections for two-way shear action.

$$\beta_c = \frac{a}{b}$$

shaded area is column or column capital - See Fig. 7.9

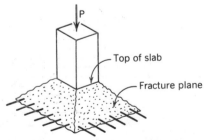

FIGURE 7.2   Punching shear failure for footing slab.

gated, more shear is resisted at corners and along the short side. On the long side, ultimate usable shear stress approaches that allowable in beams, or one-way action, that is, $\phi 2\sqrt{f'_c}$. To account for this degradation of shear capacity in rectangular columns, the ACI Code permits a usable stress based on the nominal strength provided by concrete,

$$v_c = \phi\left(2 + \frac{4}{\beta_c}\right)\sqrt{f'_c} \tag{7.2}$$

where β_c is the ratio of long side to short side of the concentrated load or reaction area and v_c is not greater than $\phi 4\sqrt{f'_c}$. Although Eq. 7.2 is applicable for all ratios of β_c, a shear strength v_c greater than $\phi 2\sqrt{f'_c}$ is appropriate only where negative moment over the column acts in two perpendicular directions. Figure 7.3 illustrates usable shear capacity of concrete for elongated columns based on Eq. 7.2.

In computing shear stress, average effective depth for steel in two directions may be used and V_u need not include shear force or loads inside the critical section perimeter. For most slabs, this reduction in shear force is small and is usually disregarded. In footings, the reduction in shear force for loads inside the shear perimeter is usually considered as it is significant. For example, a 9-ft-square by 2-ft-deep footing supporting a 18-in. square column will have a shear perimeter 3 ft-2 in. on a side. The area inside the shear perimeter is 10.0 ft.^{2} or 12% of the total footing area. For higher soil-bearing pressures requiring thicker footings and for larger columns, the area inside the shear perimeter will be a higher percentage of the total.

Equation 7.2 and punching shear stress of $\phi 4\sqrt{f'_c}$ apply only to critical shear perimeter immediately adjacent or near to the column (loaded area). Shear stress at the edge of drop panels should be limited to $\phi 2\sqrt{f'_c}$.

As with beam shear, ACI Code philosophy is to assume that concrete can carry some shear regardless of magnitude of the shear force and that shear reinforcement must carry the remainder. Thus, Eq. 6.1 also applies to two-way shear.

Openings in Slabs

Holes or openings in the slab near the column reduce the shear capacity of concrete. Horizontal holes or voids in the slab, such as those caused by electrical

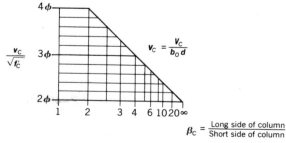

FIGURE 7.3 Punching shear strength.

conduit, can be disregarded if they are no closer to the column than two times the slab thickness (2**h**) and are not larger than a width equal to the slab thickness and a depth equal to one-third the slab thickness. In any case, horizontal voids should not be permitted within a shear perimeter where the shear stress is equal to or greater than that permitted on concrete $\phi 2\sqrt{f'_c}$. Conservative engineers will require a somewhat greater clear distance around columns.

Vertical holes passing through the slab reduce shear strength if located within intersecting column strips or if located closer than 10 times the slab thickness to the column. Referring to Fig. 7.4, reduction in shear strength is conservatively approximated by considering as ineffective that part of the perimeter of the critical section that is enclosed by straight lines projecting from the centroid of the column to edges of the openings. In addition, engineers should be alert to situations in which the location of holes prevents two-way shear action, as in Fig. 7.5. For such situations, the shear perimeter is reduced and shear stress permitted on concrete is $\phi 2\sqrt{f'_c}$. In general, holes near columns in flat slab or flat plate structures should be avoided.*

*Prudent engineers make allowance for the "plumber factor." During construction, workers will put pipes through slabs at the place most convenient for them, that is, near columns.

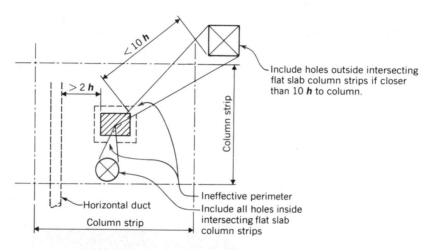

FIGURE 7.4 Effective perimeter b_o when holes are near the column.

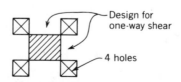

FIGURE 7.5 Holes that prevent two-way shear action.

Waffle Slabs

Shear in waffle slabs should be computed in at least two locations: one at critical shear perimeter b_o around columns, as for a flat slab or flat plate, and the other in joists around drop panels. Joist shear can be computed in the same manner as for one-way joists by using a shear force that excludes loads on the drop panel and by summing the widths of all joists framing into a drop panel (see Fig. 4.29). Due allowance should be made for the diagonal width of two joists framing into a corner, as shown in Fig. 7.6.

Joists near the centerline of the column may carry more moment than joists at the edge of the drop panel. [See Fig. 4.9(b) for lateral distribution of moment in a two-way slab system.] It seems reasonable that the same joists would also carry higher shear. Because the ACI Code allows a 10% higher shear capacity on joists than on beams, it is prudent to recognize the uneven distribution of shear on joists. This can easily be done by using only 80% to 90% of the web width of the joists framing directly, or nearly directly, into columns. Assuming such joists are about one-half the total, a conservative engineer would use only 90% to 95% of total joist web width around a drop panel in computing shear capacity of joists in a waffle slab.

7.3 TRANSFER OF MOMENT BETWEEN SLAB AND COLUMN

When loading conditions from any source cause transfer of moment M_t between slab and column, a portion γ_f is transferred by flexure $M_{tf} = \gamma_f M_t$, where γ_f is given by Eq. 4.1 and listed in Table A4.1. The remainder, $M_{tv} = \gamma_v M_t$, is transferred by eccentric shear around the critical section perimeter b_o where $\gamma_v = 1 - \gamma_f$ (Code 11.12.2).

Shear stresses due to moment transfer vary linearly about the centroid of the critical section in a manner similar to flexural stresses in a homogeneous beam.

Referring to Fig. 7.7, the shear stress on line AB is

$$v_u = \frac{V_u}{b_o d} + \frac{M_{tv} c_{AB}}{J_c} \qquad (7.3a)$$

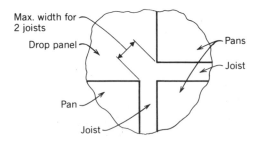

FIGURE 7.6   Plan at corner of drop panel in a waffle slab.

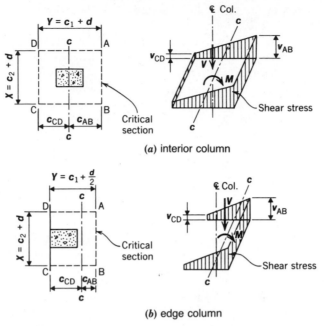

FIGURE 7.7 Stress distribution from moment transfer by shear. (Courtesy of the American Concrete Institute.)

and the shear stress on line CD is

$$v_u = \frac{V_u}{b_o d} - \frac{M_{tv} c_{CD}}{J_c} \qquad (7.3b)$$

where c_{AB} and c_{CD} are the distances from the centroidal axis c-c of the critical section to lines AB and CD, respectively, and J_c is a property of the critical section analogous to polar moment of inertia.

$$J_c = \frac{dY^3}{6} + \frac{Yd^3}{6} + \frac{dXY^2}{2} \qquad (7.4)$$

The first term in Eq. 7.4 is the moment of inertia of the two planes on lines DA and CB about the axis c-c.

The second term in Eq. 7.4 is the moment of inertia of the two planes on lines DA and CB about a horizontal axis at depth $d/2$ below the top of the slab.

The third term in Eq. 7.4 is the moment of inertia of slab areas on lines CD and AB transferred to axis c-c, and X and Y are the dimensions of the critical section.

The centroidal axis c-c is a horizontal line through the middle of the effective slab depth, located laterally to coincide with the center of gravity of the critical section perimeter b_o. The location of the centroidal axis may be computed by simple statics when b_o is asymmetrical.

Because many slab-column connections transfer at least small amounts of moment, it is convenient to have a simple, short method for computing eccentric shear stress.* Because only the maximum shear stress is normally needed, J_c can be divided by the distance from the centroidal axis to the extreme fiber of the critical section perimeter to obtain the property of the critical section analogous to the polar section modulus S_c. For a symmetrical interior rectangular column, dividing Eq. 7.4 by $Y/2$ gives

$$S_c = Y^2 d \left(\frac{1}{3} + \frac{X}{Y} \right) + \frac{d^3}{3} \tag{7.5}$$

The analogous section modulii for other common cases, illustrated in Fig. 7.8, are given below. Figure 7.8(d) explains computation of J_c for Case IV. Computations for other cases are similar.

*Unless, of course, the engineer likes to spend nights and weekends producing reams of well-reasoned computations.

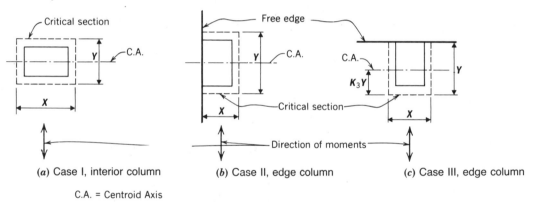

(a) Case I, interior column   (b) Case II, edge column   (c) Case III, edge column

C.A. = Centroid Axis

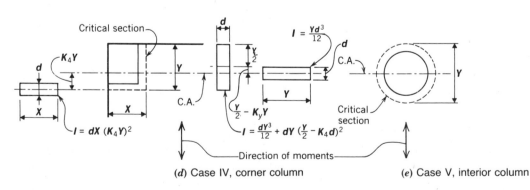

(d) Case IV, corner column   (e) Case V, interior column

FIGURE 7.8   Common cases of moment transfer at beam-column connections.

Case II. Rectangular edge column with moments parallel to the edge.

$$S_c = Y^2 d \left(\frac{1}{6} + \frac{X}{Y} \right) + \frac{d^3}{6} \tag{7.6}$$

Case III. Rectangular edge column with moments perpendicular to the edge.

$$S_c = Y^2 d \left[\frac{2}{3K_3} - 1 \right] + \frac{d^3}{6K_3} \tag{7.7}$$

where $K_3 Y$ is the distance from the centroidal axis to the inner perimeter of the critical section,

$$K_3 = \frac{Y}{X + 2Y}$$

Case IV. Rectangular corner column.

$$S_c = Y^2 d \left[\frac{1}{3K_4} - \frac{1}{2} \right] - \frac{d^3}{12K_4} \tag{7.8}$$

where $K_4 Y$ is the distance from the centroidal axis to the inner perimeter of the critical section,

$$K_4 = \frac{Y}{2(X + Y)}$$

Case V. Round interior column.

$$S_c = \frac{\pi Y^2 d}{4} + \frac{d^3}{3} \tag{7.9}$$

To visualize the variation in S_c as the shape of the critical section perimeter changes, refer to Fig. A7.1. In preparation of this graph, the last term in Eqs. 7.5 through 7.8 has been conservatively disregarded. It makes only a small contribution to the total in most cases.

The ACI Code provides no limits to the ratio β_c, but there is some question about the applicability of this procedure to critical sections of extreme proportions because tests were performed mostly on square columns and a few columns twice as long as they are wide ($\beta_c = 2$). An important assumption in developing the procedure is that plane sections remain plane after bending. In this case, the neutral axis is a horizontal plane at mid-depth of the slab. The plane of the neutral axis probably does not remain plane after bending when the column aspect ratio is very large or very small, thus invalidating the procedure for slabs where columns have extreme proportions.

When the aspect ratio is larger than 3 (i.e., $0.33 > Y/X > 3$), the analogous section modulus should be conservatively computed for a column in which the longest dimension of the critical section is limited to three times the shortest dimension. For example, if the critical section of a column is a rectangle 17×60 in., the analogous section modulus S_c should be computed in each direction for a critical section 17×51 in.

In Cases III and IV it has been assumed that shear stress caused by moment transfer increases shear stress caused by shear force. This occurs at the inside edge of the critical section perimeter. Where large moments caused by lateral loads are transferred and shear forces are small, it is possible that the critical shear stress will occur at the free edge of the slab rather than at the inside edge of the critical section perimeter. In such cases, it is best to work from fundamental principles using Eqs. 7.3 and 7.4.

Example 7.1   Check shear in the slab around columns in the flat plate structure considered in Ex. 4.1 (see Fig. 4.17).

Solution

By inspection, assume that shear in the slab around column C2 is most severe.

Shear Stress from Shear Forces

$V_u = 15 \times 15 \times 0.207 = 46.6$ k

$d = 4.25''$ from Ex. 4.1

$b_o = 2(12 + 4.25 + 16 + 4.25) = 73''$

$v_u = \dfrac{46.6}{73 \times 4.25} = 150$ psi

Shear from N-S Moments on Line 2

(See Fig. 4.18 from Ex. 4.1 calculations for North-South direction, lines 3 and 4.)
Unbalanced moment $M_t = 83.5 \times 0.65 - 63.9 \times 0.70 = 9.5'$k
Dimensions of the critical section are as follows:

$X = 12 + 4.25 = 16.25''$

$Y = 16 + 4.25 = 20.25''$

$\dfrac{Y}{X} = \dfrac{20.25}{16.25} = 1.25$

From Table A4.1, $\gamma_t = 0.58$

Moment transferred by eccentric shear

$$M_{tv} = 9.5(1 - 0.58) = 4.0'\text{k}$$

Using Eq. 7.5,

$$S_c = (20.25)^2 4.25 \left(0.333 + \frac{16.25}{20.25}\right) + (4.25)^3/3 = 2000 \text{ in.}^3$$

Shear caused by transfer of N-S moment

$$v_u = \frac{4000 \times 12}{2000} = 24 \text{ psi}$$

Shear from E-W Moments on Line C

(See Fig. 4.22 from Ex. 4.1 calculations for East-West direction, lines 3A and 4A, conservatively assuming moments on line C are the same as moments on line B.)
Unbalanced moment $M_t = 80.7(0.70 - 0.65) = 4.0'k$

Dimensions of the critical section are as follows:

$$X = 20.25'', \quad Y = 16.25''$$

$$\frac{Y}{X} = \frac{16.25}{20.25} = 0.80$$

From Table A4.1, $\gamma_t = 0.63$

Moment transferred by eccentric shear

$$M_{tv} = 4.0(1 - 0.63) = 1.5'k$$

Using Eq. 7.5, where $X = 20.25''$ and $Y = 16.25''$,

$$S_c = (16.25)^2 4.25 \left(0.333 + \frac{20.25}{16.25}\right) + (4.25)^3/3 = 1800$$

E-W moment $v_u = 1500 \times 12/1800 = 10 \text{ psi}$

Shear Summary

Vertical shear stress $= 150 \text{ psi}$

N-S moment transfer shear stress $= \quad 24$

E-W moment transfer shear stress$= \quad \underline{10}$

Maximum total shear stress at
N-W corner of critical section $= 184 \text{ psi} < V_c$ <u>OK</u>

where $v_c = 0.85 \times 4\sqrt{3000} = 186 \text{ psi}$

Confirm the selection of column C2 as having the highest shear stress by checking shear at column B2. Compute shear stress at column B2 by ratio with the shear stress at column C2. Shear stress from shear force

$$150 \text{ psi} \times \frac{16}{15} = 160 \text{ psi}$$

N-S moment transfer shear stress

$$24 \text{ psi} \times \frac{83.5(0.70 - 0.65)}{9.5} = \quad 11$$

E-W moment transfer shear stress = __10 psi__

Total shear stress = 181 psi $< v_c$ OK

Although the slab in this example does meet Code requirements, there is little margin for error in construction tolerances and there is almost no allowance for holes through the slab within intersecting column strips or within 4'-7" (10 × 5.5") of column faces. A better design would use a 6" or 6½"-thick slab or column capitals. The thicker slab would have less deflection but would require more concrete and reinforcing steel. Column capitals would probably cost less than a thicker slab but would not reduce deflection nearly as much. Another option is discussed in Ex. 7.3.

Slab punching shear strength may be increased by increasing the strength provided by concrete or by providing shear reinforcement. Shear reinforcement is discussed in the next section. In most cases, increasing the strength provided by concrete is preferred as it is less expensive. Shear strength provided by concrete may be increased by increasing the strength of concrete. This may be an expensive option as the shear strength increases slowly as concrete strength increases. For example, a one-third increase in f'_c from 3 to 4 ksi increases shear strength 15%. Shear strength provided by concrete may be increased by increasing slab thickness, but this is not an attractive solution if slab thickness has been established for optimum cost, deflection, and other considerations. Shear strength provided by concrete may also be increased by increasing the size of columns. This too is not an attractive solution as it is usually more economical to establish a column size based on axial load and moment in the column.

Shear strength in the slab provided by concrete is increased most economically by providing column capitals if they do not interfere with other elements of construction. These can take a variety of forms to fit architectural and aesthetic requirements. Capitals are usually made symmetrical about column centerlines, except for capitals at exterior columns. Asymmetrical capitals cause moment at the joint by eccentric loading of the column. These moments should be allocated to columns and slabs in proportion to their relative stiffnesses and taken into account when designing both slabs and columns. In any case, no portion of a column capital or bracket shall be considered for structural purposes that lies outside the largest right circular cone, right pyramid, or tapered wedge whose planes are oriented no greater than 45° to the column (Code 13.1.2) (see Fig. 7.9 for a few examples of acceptable column capitals and location of the critical section for computation of slab shear strength).

For preliminary design, it is convenient to have a simple, direct method of determining column or column capital size necessary to limit the punching shear stress due to both shear force and moment transfer when the unbalanced moment is due to unequal spans or loading. For interior rectangular or round columns without nearby holes through the slab, this can be computed approximately by increasing the shear force by ratio R, where R is the ratio of punching shear stress caused by moment transfer M_{tv} to stress caused by shear force.

$$R = \frac{M_{tv}/S_c}{V_u/b_o d} \tag{7.10}$$

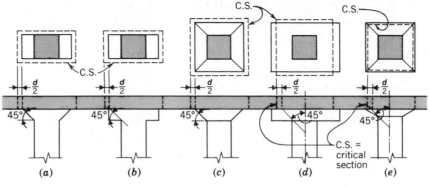

FIGURE 7.9 Acceptable column capitals.

where M_{tv} = the unbalanced moment transferred by shear due to unequal spans with uniform load,

$$M_{tv} = (1 - \gamma_f)0.65w_u\ell_2 \frac{[(r\ell_n)^2 - \ell_n^2]}{8}$$

r = ratio of the longest clear span to the shortest clear span, each adjacent to the column in the direction that moments are being determined.

S_c = analogous section modulus given by Eq. 7.5 for rectangular columns or Eq. 7.9 for round columns,

$$V_u = \frac{w_u\ell_1\ell_2(1 + r)}{2}$$

ℓ_n and ℓ_1 = shortest clear span and corresponding center-to-center span, respectively, adjacent to the column

$b_o = 2(X + Y)$ for the interior rectangular columns,

$b_o = \pi Y$ for the interior round columns,

[see Fig. 7.8(a) and 7.8(e)], and

$(1 - \gamma_f)$ = portion of the unbalanced moment transferred by shear, and γ_f is given by Eq. 4.1.

Inserting these parameters into Eq. 7.10 and dropping the last term in Eqs. 7.5 and 7.9 yields the following:

$$R = C_v \left(\frac{\ell_n}{Y}\right) (r - 1) \tag{7.11}$$

For rectangular columns,

$$C_v = \frac{0.975\sqrt{Y/X}}{1.5 + \sqrt{Y/X}} \times \frac{X + Y}{3X + Y} \times \frac{\ell_n}{\ell_1} \tag{7.12}$$

Values for C_v are shown graphically in Fig. A7.2.

For square columns,

$$C_v = 0.195 \frac{\ell_n}{\ell_1} \tag{7.13}$$

For round columns,

$$C_v = 0.26 \frac{\ell_n}{\ell_1} \tag{7.14}$$

For a further simplification, using a conservatively high value for ℓ_n/ℓ_1, for square columns, $C_v = 0.19$. For round columns, $C_v = 0.25$.

At the first interior column, with equal spans on each side, negative moment at the column from the end span is higher than negative moment from the interior span. Equation 7.11 can be conservatively adjusted for this imbalance by increasing the term $(r - 1)$ by 0.04 when the end span is the longest span and reducing the term $(r - 1)$ by 0.04 when the interior span is the longest span.*

Example 7.2   Using the direct method, how much does the condition of unequal spans in the North-South direction increase the shear stress at column C2 in Ex. 4.1?

Solution $\dfrac{Y}{X} = \dfrac{20.25}{16.25} = 1.25$ and $\dfrac{\ell_n}{\ell_1} = \dfrac{12.83}{14.0} = 0.92$

By computation, using Eq. 7.12, or from Fig. A7.2, $C = 0.20$.

Because the interior span is the longest, reduce $(r - 1)$ by 0.04,

$(r - 1) - 0.04 = 14.7/12.8 - 1.04 = 0.108$

Using Eq. 7.11,

$R = 0.19 \times \dfrac{154}{20.25} \times 0.108 = 0.16$

Checking the answer by reference to Ex. 7.1,

shear stress due to shear force = 150 psi

Shear stress due to moment transfer = 24 psi.

From this example, shear stress due to moment transfer
$$= 150 \times 0.16 = 24 \text{ psi} \qquad\qquad \text{OK}$$

7.4 SLAB SHEAR REINFORCEMENT

When shear reinforcement consisting of bars or wires is provided, shear stress permitted on concrete at all critical sections is $\phi 2\sqrt{f_c'}$ or as little as half that permitted

*In deriving Eq. 7.11, the factor $(r\ell_n)^2$ is either increased or decreased by 0.70/0.65 = 1.08, which is equivalent to increasing or decreasing r by $\sqrt{0.70/0.65} = 1.04$.

when no shear reinforcement is provided. Shear reinforcement consisting of steel sections is stiffer than bars or wires; therefore, no reduction in shear stress on concrete is needed. Maximum permitted nominal shear stress is $\phi 6\sqrt{f_c'}$ for bar or wire reinforcement and $\phi 7\sqrt{f_c'}$ for steel sections. If moment is being transferred between slab and column, maximum permitted shear stress on concrete at the smallest perimeter around the column for shear forces and moments combined is $\phi(2 + 4/\beta_c)\sqrt{f_c'} \leqslant \phi 4\sqrt{f_c'}$

Shear stresses permitted are summarized in Table 7.1.

Shear Reinforcement Consisting of Bars or Wires

When shear reinforcement consisting of bars or wires is used, critical sections are located (1) at distance $d/2$ from face of support as for a slab unreinforced for shear (the inner critical section or inner perimeter), (2) at distance $d/2$ beyond the last set of stirrups (the outer critical section or outer perimeter), and (3) at successive sections between the inner and outer perimeter.

Shear stress permitted on concrete at the inner perimeter is only $\phi 2\sqrt{f_c'}$ even though a stress as high as $\phi 4\sqrt{f_c'}$ is permitted when no shear reinforcement is required. The reduction in shear capacity of concrete is due to the flexibility of bar or wire shear reinforcement. Because shear reinforcement need not be considered unless shear stress on the inner perimeter exceeds $\phi 4\sqrt{f_c'}$ [or $\phi(2 + 4/\beta_c)\sqrt{f_c'}$ for rectangular columns], and because shear stress on concrete at the outer perimeter must be $\phi 2\sqrt{f_c'}$ or less, it is evident that the outer perimeter must be more than twice that of the inner perimter for round columns, square columns, or rectangular columns with an aspect ratio $\beta_c \leqslant 2.0$.

The practical effect of this ACI Code provision is to require many small, closely spaced stirrups for a significant area around the column (see Fig. 7.10). Engineers can exercise their ingenuity to reduce the number of stirrups required, such as in Fig. 7.10(*b*).

Truss bar shear reinforcement as shown in Fig. 7.11 has been proposed. It has the advantage of anchoring the shear bar easily, but many bars are still required to

TABLE 7.1   Allowable Punching Shear Stress with Use of Shear Reinforcement

	Shear Reinforcement Consisting of	
Critical Section	Bars or Wires	Steel Sections
At $d/2$ from face		
of support	$v_n = \phi 6\sqrt{f_c'}$	$v_n = \phi 7\sqrt{f_c'}$
With moment transfer[a]	$v_c = \phi(2 + 4/\beta_c)\sqrt{f_c'}$	$v_c = \phi(2 + 4/\beta_c)\sqrt{f_c'}$
	$\leqslant \phi 4\sqrt{f_c'}$	$\leqslant \phi 4\sqrt{f_c'}$
No moment transfer	$v_c = \phi 2\sqrt{f_c'}$	NA[b]
At perimeter of shear		
reinforcement[c]	$v_c = \phi 2\sqrt{f_c'}$	$v_c = \phi 4\sqrt{f_c'}$

[a] Moment transferred by eccentricity of shear (see Section 7.3).
[b] NA = not applicable.
[c] With or without moment transfer.

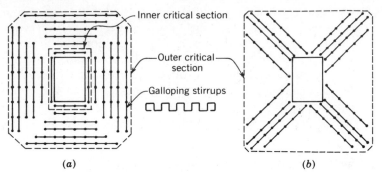

FIGURE 7.10   Bar or wire shear reinforcement.

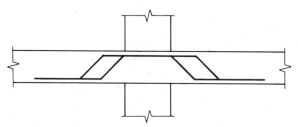

FIGURE 7.11   Truss bar slab shear reinforcement.

reinforce the slab to the outer shear perimeter. Congestion of truss bars over or near the column will discourage the use of this system of reinforcement.

Bars or wires for slab shear reinforcement are designed in the same manner as for beam shear reinforcement (see Chapter 6). Concrete cover on stirrups should be the same as the minimum required for flexural reinforcement. Due to shallow depth in most slabs, bars and wires must be small to enable them to be anchored properly. For this reason, stirrups should not be considered for slabs less than about 10 in. or 1 ft in thickness. Even then, fabrication and placement of stirrups will be measured in fractions of an inch, and tolerances must be tight and strictly observed.

Bar or wire shear reinforcement in slabs should be avoided wherever possible for the following reasons:

1. Proper anchorage of bars is difficult to obtain.

2. Careful layout and design is required to avoid interference with top and bottom slab flexural reinforcement. Otherwise, irreconcilable interference will occur and shear reinforcement will not be placed correctly.

3. Extreme accuracy is needed in placement.

4. Small size bars and wires are easily damaged and displaced by foot traffic during construction.*

*A normal, healthy young man can bend a #3 bar over his knee, and so can most young women. If you don't want shear reinforcement to look like a plate of spaghetti, provide construction workers with levitators so they can hover weightlessly over the slab while performing their chores. A supply of soup spoons for placing concrete around the bars might also be wise.

5. Reinforcement is inefficient as many small bars or wires are needed.

6. Small bars look inconsequential and, hence, are more likely to be left out of the slab.

7. Consequently, bar and wire shear reinforcement tends to be expensive.

Shear Reinforcement Consisting of Steel Sections

Slabs can be reinforced for shear using a "shearhead," consisting of structural steel I- or channel-shaped sections. (Rectangular bars or plates are not satisfactory because flanges are needed for bearing.) Two arms at right angles consisting of identical sections are required. More arms may be used. Where the steel sections intersect, full penetration welds are required to develop the bending and shear strength of each arm.

Figure 7.12(a) illustrates the physical limitations and terminology used in shearhead design. Although steel shearheads have been used successfully for over 50 years, research leading to design procedures was not conducted until the 1960s [2]. Design of shearheads can be completed efficiently by following this procedure.

1. Select a shearhead configuration, that is, the number of arms and their location. They must clear vertical reinforcing bars in the column. Arms should consist of standard structural steel sections. Because the smallest standard section is 3 in. deep and clearance is required above the bars to pass

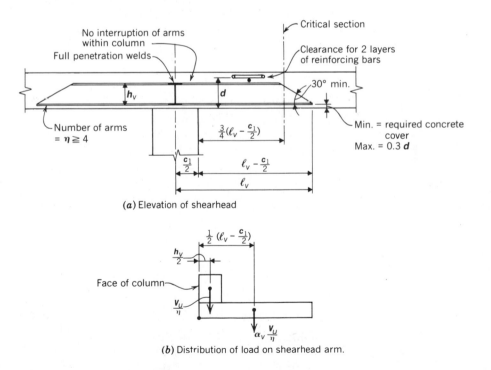

(a) Elevation of shearhead

(b) Distribution of load on shearhead arm.

FIGURE 7.12 Slab shearhead.

the slab top bars, and concrete cover is required below the shearhead, the thinnest slab in which a shearhead can be used is about 5½ in. If more than ¾ in. concrete cover is necessary, if larger than #4 top bars are used, if slab bars must pass under the shearhead as well as over it, or if a 3-in. section ultimately proves inadequate, then the minimum slab thickness will be more than 5½ in. In any case, the compression flange of the shearhead must be within $0.3d$ of the compression surface of the slab.

2. Assume that the deepest steel section that will fit within the slab will be selected. Compute the minimum web thickness of the section equal to the depth of the steel section divided by 70. The purpose of this step is to assure adequate shear strength in each structural steel arm of the shearhead.

3. Shear carried by the shearhead will be in proportion to α_v, its relative flexural stiffness compared to the surrounding concrete section,

$$\alpha_v = \frac{E_s I_s}{E_s I_s + E_c I_{cr}} \tag{7.15}$$

where $E_c I_{cr}$ is the flexural stiffness of the composite cracked slab section (disregarding the shearhead) surrounding the shearhead arm with a width of $(c_2 + d)$ and where c_2 is the width of the column perpendicular to the arm. (Refer to Chapter 8 for methods of computing $E_c I_{cr}$.)

Tests indicate that flexible shearheads do not contribute to shear strength of the slab when $\alpha_v < 0.15$. Therefore, solving Eq. 7.15 to obtain the minimum permissible I_s for each shearhead arm, gives

$$I_s \geqslant 0.176 \frac{(E_c I_{cr})}{E_s} \tag{7.16}$$

4. Given the shearhead configuration, compute the length of the arms such that the concrete shear stress at the critical section $v_c \leqslant \phi 4\sqrt{f_c'}$. The critical section must be located so that its perimeter b_o is a minimum but need not approach closer to the column than $d/2$ nor closer than the critical section on the shearhead arm shown in Fig. 7.12(a).

5. Using a trial-and-error procedure, assume a steel section for the shearhead arms, compute the relative stiffness α_v from Eq. 7.15, and then compute the required plastic moment strength M_p of the shearhead arm.

$$M_p = \frac{V_u}{\phi 2\eta} \left[h_v + \alpha_v \left(\ell_v - \frac{c_1}{2} \right) \right] \tag{7.17}$$

where h_v, ℓ_v, c_1, and η are defined in Fig. 7.12(a) and $\phi = 0.85$.

Select a structural steel section that will meet the required maximum depth limit, minimum web thickness limit, minimum I_s, and M_p.

If the moment of inertia I_s for the steel section selected is more than that assumed in computing α_v, it will be necessary to recompute α_v and M_p and either confirm the selection of steel section or select a new and stronger section. If I_s for the steel section selected is less than that assumed in computing α_v, recomputation of α_v and M_p are not necessary but might be desirable if a reduction in size of the steel section is possible.

The derivation of Eq. 7.17 is based on the assumed loading of a shearhead arm cantilevered from the face of the column, as shown in Fig. 7.12(b). Tests indicate that shear on the steel section is much higher near the column. As a simplification to a complex interaction of loads, shears, and moments, it can be conservatively assumed that an additional shear equal to the entire shear acts on the steel section within a distance h_v from the face of the column.

At the engineer's option, outer ends of arms may be cut at an angle not less than 30° with the horizontal, provided that the full depth of the steel section extends beyond the face of the column.

6. If unbalanced moments are considered, that is, if the shearhead must transmit some moment between the slab and the column, design the connection between the shearhead and the column to transmit the required moment. In most cases, shearheads embedded in columns extending above and below the slab will need no further anchorage to transmit the moment. When a column does not extend above the slab, some method may be needed to transmit the required moment. The most common method is to weld a short steel column section to the bottom of the shearhead and extend it into the concrete column below the slab. Extension of the steel column must be adequate to transmit moment to the concrete column. At the underside of the shearhead, the steel column and its connection to the shearhead arms must resist the entire transmitted moment.

7. At the engineer's option, reduce the moment resistance contributed by slab flexural reinforcement by an amount equal to the moment resistance provided by shearhead reinforcement M_v.

$$M_v = \frac{\phi\alpha_v V_u}{2\eta}\left(\ell_v - \frac{c_1}{2}\right) \qquad (7.18)$$

In many cases, such reduction will be small and not worth computing. In any case, it may be conservatively disregarded.

Example 7.3   Assume that the slab in Ex. 4.1 must have shearheads to carry at least one third of the shear force to allow passage of holes through the slab near columns. (also see Ex. 7.1). What size shearhead is required?

Solution

Step 1   Because #4 top bars are used, the maximum depth of the steel member is

$$5.5 - 0.75 \times 2 - 0.5 \times 2 = 3''$$

Use three 3″ I or channel sections crossed on the centerline of the columns, as shown in Fig. 7.13. (Assume for purposes of this example that column vertical bars will not interfere with the shearheads.)

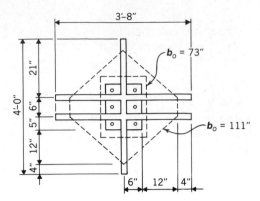

FIGURE 7.13

Step 2   Minimum web thickness = 3/70 = 0.04″

Step 3   If all arms on the shearhead project the same distance beyond the face of the column, the two E-W arms are critical because stiffness of the shearhead arms is larger compared to stiffness of the concrete slab than for the N-S arms.

$(c_2 + d) = 16 + 4.25 = 20.25″$

If 11 #4 bars are spaced 8″ c/c in the column strip (see Fig. 4.22 for Ex. 4.1),

$$\rho = \frac{0.20}{8 \times 4.25} = 0.59\%$$

From Fig. A8.2, $\beta_{cr} = 1.15$

$E_c I_{cr} = 1.15 \times 10^5 \times 20.25 \times (4.25)^2 = 4.2 \times 10^7$

Using Eq. 7.16,

$$\text{Minimum } I_s = \frac{0.176 \times 4.2 \times 10^7}{2.9 \times 10^7} = 0.26 \text{ in.}^4$$

Step 4 b_o at column = 73″ (see Ex. 7.1).

If the shearhead must carry 1/3 of the total shear, required b_o near the ends of the shearhead arms

$$b_o = \frac{73″}{0.67} = 109.5″$$

Referring to the sketch of the shearhead in Fig. 7.13,

$$b_o = 111″ \qquad\qquad\qquad \text{OK}$$

Step 5 $V_u = 46.6$ k   (from Ex. 7.1)

$h_v = 3″, \qquad \ell_v = 22″, \qquad c_1 = 12″$

Assume use of S 3×5.7, $I = 2.52$

$E_sI_s = 2.9 \times 10^7 \times 2.52 \times 2 = 14.6 \times 10^7$

$E_cI_{cr} = 4.2 \times 10^7$

$$\alpha_v = \frac{14.6}{14.6 + 4.2} = 0.78$$

Compute moment in in.-kips due to small member sizes. Using Eq. 7.17,

$$\text{Reqd.}M_p = \frac{46.6}{0.85 \times 2 \times 6}\left[3 + 0.78\left(22 - \frac{12}{2}\right)\right] = 70.5''k$$

Required plastic section modulus

$$Z_x = \frac{70.5}{36} = 1.96 \text{ in.}^3$$

Step 6   Use S 3×5.7, $Z_x = 1.95$ in.^{3}, $I_x = 2.52$ in.^{4},

and depth $= 3''$,   web thickness $= 0.170''$ OK

Similar computations for the N-S arm indicate that it is less critical than the two E-W arms. If arm(s) in one direction are much more critical than arm(s) in the other direction, the strength of the shearhead in each direction can be balanced by adjsuting the distance $(\ell_v - c_1/2)$.

PROBLEMS

1. In a two-way flat plate system, what is the maximum punching shear capacity of an 8-in. slab supported on an 18-in. square column, without column capital or drop panel if $f_c' = 4$ ksi and $f_y = 60$ ksi?
2. What is the maximum punching shear capacity of the slab in Problem 1 if the column is 10- × 32 in. in size?
3. What is the maximum punching shear capacity of the slab in Problem 2 if there are two 12-in. round holes along one long side of the column with 9 in. clear between the face of the column and the edge of the holes, and the holes are placed symmetrically with respect to the column centerline and with 8 in. clear between them?
4. Derive Eq. 7.7.
5. What size round column capital is needed for a 10-in. slab supporting a factored punching shear of 220 kips and an unbalanced factored moment transferred by shear of 30 ft-kips? Assume $f_c' = 3.5$ ksi.

Selected References

1. ACI-ASCE Committee 426, "The Shear Strength of Concrete Members," Chapter 5, *ACI Manual of Concrete Practice, Part 4*, American Concrete Institute, Detroit, Michigan, 1985.
2. W. Gene Corley and Neil M. Hawkins. "Shearhead Reinforcement for Slabs," *ACI Journal*, V. 65, pp 811–824, October 1968.

CHAPTER 8

DEFLECTION

8.1 OBJECTIVES

Laypersons intuitively believe that structures showing large deflections are unattractive and unsafe. There are good reasons for them to feel so because a well-designed, ductile structure will indeed deflect a large amount just before failure. Usually the deflection will be visible. Therefore, it is advisable for engineers to limit deflection of structures under working loads to tolerable amounts.*

Before the early 1960s, concrete structures were designed with lower stresses, thicker sections, and shorter spans than are common today. Deflections were rarely a problem then. Today, many structures are balanced between the competing demands of highly stressed members using less labor and material and, hence, less cost and deflection behavior that is satisfactory.

The objectives of deflection calculations are as follows:

1. For preliminary design, establish concrete outlines that will result in acceptable deflections most of the time.

2. For final design, assure satisfactory deflection response by checking the deflection of critical members under all design conditions.

3. If deflection is excessive, select corrective measures.

8.2 DEFLECTION BEHAVIOR

Perhaps the most important step in computing deflections is to sketch the deflected shape of the structure, especially if its geometry or loading is somewhat complicated (see Fig. 13.4). Computation of deflection magnitude will be meaningless if the

---

*Laypersons are frequently bosses, owners, building officials, tenants, and other persons in a position to command attention.

engineer has the wrong concept of deflection response. Horizontal members can deflect upward as well as downward and vertical members can deflect in either direction. Sometimes a member is in double curvature and deflects in both directions. One load may cause a member to deflect in one direction, and another load may cause the same member to deflect in the opposite direction. If an engineer has difficulty visualizing the direction of deflection, experimentation with a simple model of heavy paper, balsa wood, plastic, or other flexible material should clarify the deflection response.

There are several situations common to reinforced concrete structures that tend to increase the complexity of deflection calculations over that required for other structural materials. The increased complexity can be reduced or even eliminated by a proper approach to design. These situations are as follows:

1. Degree of continuity. Most concrete structures have some fixity at supported ends of members, whereas most structures of other materials have almost no fixity at supports so that members are treated as simple spans. The matter of continuity is considered in Section 8.3 of this text.

2. Moment of inertia. Reinforced concrete is not a homogeneous, isotropic material for which section properties can be computed using concrete outlines and ordinary, simple concepts from strength of materials. At low load levels, before the concrete has been cracked, the moment of inertia I_g can be approximated by using the gross concrete outlines. Transformed area* of reinforcing steel could also be included but is usually not, because it normally has little effect on the initial elastic deflection Δ_i.

   After concrete has been heavily loaded once, it is cracked extensively and the moment of inertia I_{cr} depends on the concrete in compression and the transformed area of steel in tension. Concrete in tension is disregarded. Transformed area of reinforcement in compression should also be included but is usually not, because it normally has little effect on initial elastic deflections.

   At intermediate load levels, concrete is cracked only slightly, and the moment of inertia I_e is interpolated between I_g and I_{cr}.
   Calculation of the moment of inertia is considered in Section 8.4.

3. Long-time deflection. Concrete is a highly viscous material as it creeps over a period of time. Also, concrete shrinkage is resisted by embedded reinforcement which tends to cause warping in the same general direction as deflection due to loads. Thus, the initial elastic deflection Δ_i will be increased significantly over a long time. Long-time deflection Δ_{LT} is considered in Section 8.5.

*The concept of transformed area is discussed in most texts on structural mechanics. Simply stated, transformed area is nA_s, where $n = E_s/E_c$.

Deflection Limits

ACI Committee 435 has examined in detail the criteria for limiting deflection [1]. Their report lists criteria under four categories: sensory acceptability, service-ability of structure, effect on nonstructural elements, and effect on structural elements. For each criteria, examples, deflection limitation values, and the portion of total deflection on which the deflection limitation is based are given. The most important criteria for limiting deflection can be summarized as follows.

Structures that sag or droop noticeably are unacceptable to most people. This situation can be avoided by cambering as well as by stiff members.

Building occupants may become annoyed or alarmed if the structural frame vibrates under conditions of ordinary use. Acceptable deflection response is still based on subjective opinion. Stiffer members reduce objectionable vibrations, but other important factors affecting the magnitude of vibrations or human sensitivity to them are damping in the floor system, natural frequency of the structure, frequency of the exciting force, and how often it occurs.

Roofs, outdoor decks, and other slabs that might have water on the surface should slope to a drain or at least remain flat so as to not pond water noticeably. Bowling alleys, gymnasiums, and members supporting certain sensitive equipment must also remain level or deflect less than the tolerance of the equipment.

Masonry, plaster, and other brittle materials supported by the structure can be damaged by excessive deflection after construction. Windows, partitions, and other nonstructural building elements can be damaged if a structural member above them deflects and comes to bear on the nonstructural element.

Some deflection limitations apply to only a portion of the total deflection. For example, masonry walls are stressed and cracked only by that portion of the total deflection that occurs after the walls are constructed. This portion is called the incremental deflection and is frequently only a small portion of the total deflection.

Deflection limitations are usually stated in terms of a span-to-deflection ratio. However, absolute values of deflection are appropriate in some instances, especially for spans exceeding about 30 ft.

Table A8.1 lists ACI Code limitations on deflection.

8.3 CONTINUITY

Using any of several methods, the maximum initial deflection Δ_i in an elastic member is given by

$$\Delta_i = \frac{\beta_w W \ell^3}{EI} \tag{8.1}$$

where β_w = a coefficient that depends on the degree of fixity at supports, the variation in moment of inertia along the span, and the distribution of loading.

W = total service load on the member.

EI = flexural rigidity of the member, or the product of modulus of elasticity E and moment of inertia I.

The maximum deflection may also be expressed as

$$\Delta_i = \frac{\beta_m M \ell^2}{EI} \tag{8.2}$$

where β_m = a coefficient that depends on the same variables as β_w but is applied to moments instead of loads.

M = maximum positive moment for beams on two supports and maximum negative moment at the support for cantilevers.

Table A8.2 lists deflection coefficients β_w and β_m for common loading and support conditions. Note that the spread in values for β_m is much less than the spread for β_w. This is significant for simplified procedures discussed in Section 8.5.

Support conditions for most concrete members are neither free (zero negative moment) nor fixed. A convenient approach to such conditions is to consider the downward deflection due to loads separate from the upward deflection due to end moments. The two deflection calculations can then be added algebraically to obtain the actual total deflection. The maximum deflections do not necessarily occur at the same point for uniform and concentrated loads and for end moments nor do they necessarily occur at mid-span or under a concentrated load. However, the error in total deflection obtained by summing individually calculated deflections will be much less than the range of error discussed in Section 8.5.

Using the conjugate-beam method, deflection at a given point in a beam from a straight line connecting two other points A and B equals the bending moment at that point for a beam supported at A and B and loaded with the elastic M/EI diagram. Referring to Fig. 8.1(c) and considering points A and B to be at the supports, it is

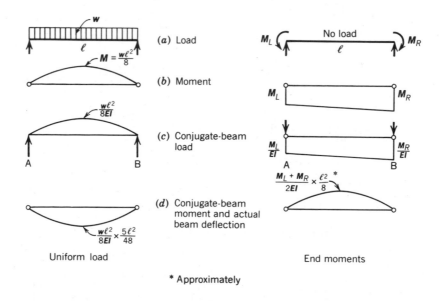

FIGURE 8.1   Deflections by conjugate-beam method.

sufficiently accurate to calculate deflection due to end moments as

$$\Delta_i = \left(\frac{M_L}{2} + \frac{M_R}{2}\right)\frac{\ell^2}{8EI} \tag{8.3}$$

where M_L and M_R = moments at the left and at the right supports, respectively.

Example 8.1   What is the deflection of a 36-in.-wide by 20-in.-deep beam under concentrated and uniform loads, as shown in Fig. 8.2. Assume flexural rigidity $EI = 4.1 \times 10^7$ k-in^2 (see Ex. 8.6). Loads are service loads, and end moments are computed for the case of maximum mid-span moments and not maximum end moments used in design.

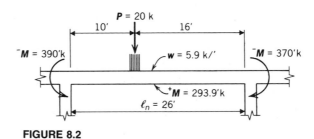

FIGURE 8.2

Solution   Compute deflections separately for uniform loads, concentrated load, and end moments.

Uniform load deflection $= \dfrac{5w\ell^4}{384EI}$

$$\frac{5}{384} \times \frac{5.9 \times (26)^4 \times (12)^3}{4.1 \times 10^7} = 1.48'' \ \downarrow$$

Concentrated load deflection $\simeq \dfrac{Pa^2b^2}{3EI\ell}$

[Equation taken from AISC Manual (2).]

$$\frac{20 \times (10 \times 16)^2(12)^3}{3 \times 4.1 \times 10^7 \times 26} = 0.28'' \ \downarrow$$

End moment deflection (Use Eq. 8.3)

$$\left(\frac{390 + 370}{2}\right) \times \frac{26^2 \times 12^3}{8 \times 4.1 \times 10^7} = \underline{1.35''} \ \uparrow$$

$$\text{Net } \Delta_i = 0.41'' \ \downarrow$$

Cantilevers

Equations for computing the deflection of a cantilever under various loading conditions are well known and given in many textbooks and handbooks. Most such equations give the deflection from a line tangent to the beam at the support, but the support joint of most cantilevers will rotate under load. The rotation may increase or decrease the deflection of a cantilever from its original position, depending on the span and loading conditions. The component of deflection at the top of cantilevered retaining walls due to rotation or tilting of the footing is frequently more than the component due to flexure in the wall itself (see Section 12.5).

Using the conjugate-beam method, the rotation at point B in a beam from a straight line connecting two points A and B equals the shear at point B on a conjugate beam loaded with the elastic **M/EI** diagram between points A and B. For example, considering the simple cantilever beam with a hold-down span in Fig. 8.3 in which points A and B are at the supports, rotation θ of the beam at support B is

$$\theta = \frac{P\ell_2}{EI} \times \frac{\ell_1}{3}$$

Deflection of the cantilever due to rotation of its support is simply the rotation times the length of the cantilever.

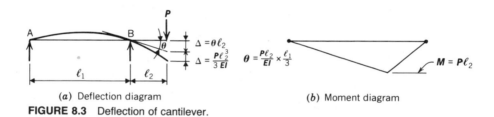

(a) Deflection diagram   (b) Moment diagram

FIGURE 8.3   Deflection of cantilever.

Example 8.2   What is the deflection of a 36-in.-wide by 20-in.-deep cantilever beam under concentrated and uniform loads, as shown in Fig. 8.4? Assume that the flex-

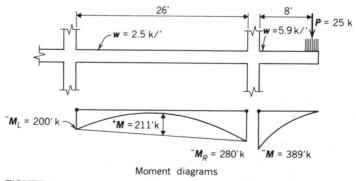

Moment diagrams

FIGURE 8.4

ural rigidity of the cantilever $EI = 3.0 \times 10^7$ ksi and of the beam span $EI = 4.1 \times 10^7$ ksi. Loads and moments are service loads and moments, and load on the hold-down span is the minimum load expected during the service life of the structure.

Solution   First calculate rotation of the beam at the support; then calculate deflection due to the rotation, the uniform load, and the concentrated loads separately, adding them algebraically for the total deflection.

Rotation of beam at support due to end moments

$$\theta = \left(\frac{^-M_L}{6} + \frac{^-M_R}{3} \right) \times \frac{\ell}{EI}$$

$$\theta = \left(\frac{200}{6} + \frac{280}{3} \right) \times \frac{26 \times (12)^2}{4.1 \times 10^7} = 11.57 \times 10^{-3} \qquad \circlearrowright$$

Rotation of beam at support due to load $= \dfrac{^+M\ell}{3EI}$

$$\theta = \frac{211 \times 26 \times (12)^2}{3 \times 4.1 \times 10^7} = \underline{6.42 \times 10^{-3}} \qquad \circlearrowleft$$

Net rotation $= 5.15 \times 10^{-3}$ $\circlearrowright$

Deflection due to rotation at support $= \theta\ell$

$$\Delta_i = \theta\ell = 5.15 \times 10^{-3} \times 8 \times 12 = 0.49'' \;\downarrow$$

Deflection due to concentrated load $= \dfrac{P\ell^3}{3EI}$

$$\Delta_i = \frac{25 \times (8)^3 \times (12)^3}{3 \times 3.0 \times 10^7} = 0.25'' \;\downarrow$$

Deflection due to uniform load $= \dfrac{w\ell^4}{8EI}$

$$\Delta_i = \frac{5.9 \times (8)^4 \times (12)^3}{8 \times 3.0 \times 10^7} = \underline{0.17''} \;\downarrow$$

Total $\Delta_1 = 0.91'' \;\downarrow$

In this example, deflection at the end of the cantilever was increased more than 100% by rotation of the support.

8.4 MOMENT OF INERTIA

Moment of inertia I of a concrete member depends not only on its dimensions but also on the amount of cracking in the concrete. Before cracking, I is based on the gross transformed section and designated I_g. After *extensive* cracking, I is based on the transformed cracked section, assuming no tensile strength in the concrete and designated I_{cr}. With only a little cracking, tensile strength of the concrete still contributes to stiffness, and the effective moment of inertia I_e is between I_g and I_{cr}. Flexural cracking does not necessarily result in cracks that are easily visible to the naked eye, especially in members with well-distributed reinforcement.

Deflection calculations usually assume that the moment of inertia can be based on conditions under total service load. The alternative is to assume that, on first loading, deflection calculations for intermediate load levels less than the total are based on the condition of cracking at those intermediate load levels. That is, dead load deflection might be based on I_g if stresses under dead load do not crack the member, and total service load deflection might be based on I_{cr} or I_e if stresses under total service load do crack the member. This alternative assumption is not normally used because it adds to the deflection calculation complexity and because, during construction, a concrete structure frequently supports stockpiled construction materials and shoring to support slabs being cast above. These construction loads often equal the total service loads used in design. Thus, conditions under total service load more nearly represent conditions under partial load also. In any case, once a member has been cracked for any reason, its stiffness will be best represented by I_{cr} or I_e.

Gross Section (Uncracked I_g)

Before a member is cracked, I_g can be calculated using the usual concepts from strength of materials. Flanges contribute significantly to stiffness and should be included up to the limit permitted by the ACI Code in design for strength (see Fig. 2.3). Engineers can refer to Fig. A8.1 to reduce the computational effort or simply estimate the increase in stiffness due to flanges.

The contribution of reinforcement to stiffness can be estimated by including the area of steel transformed to an equivalent area of concrete, $A_s(n - 1)$. The modular ratio n ($n = E_s/E_c$) is reduced by 1.0 to account for concrete replaced by steel. The transformed area of reinforcement is most effective in stems of T-beams where it is farthest from the neutral axis. This refinement is usually disregarded as it normally adds little to the total I_g.

From structural mechanics, the moment M_{cr} causing flexural cracking in a concrete member can be calculated using

$$M_{cr} = \frac{f_r I_g}{y_t} \tag{8.4}$$

where f_r = modulus of rupture (The ACI Code specifies a value $f_r = 7.5\sqrt{f_c'}$ for normal weight concrete);

and y_t = distance from the neutral axis of the gross section, neglecting reinforcement, to extreme fiber in tension.

The value of the modulus of rupture f_r given above should be used when computing deflections to determine whether the structure meets deflection limits in the ACI Code (Table A8.1). For other situations, the engineer may wish to avoid overestimating deflection. In such cases, a slightly higher but more realistic value for the average modulus of rupture can be used, $f_r = 2.3f_c'^{2/3}$ [3]. In addition, the actual average concrete strength could be used instead of the specified minimum strength f_c'. Usually the average strength is 10% to 20% higher than the specified minimum strength. If concrete used in the structure is tested for the modulus of elasticity E_c and the modulus of rupture f_r, accuracy of deflection computations might be improved by using the test values.

Cracked Section (I_{cr})

The moment of inertia of a cracked section is calculated from structural mechanics, disregarding concrete in tension and transforming reinforcement into an equivalent area of concrete by multiplying A_s by the modular ratio n. Load levels and concrete stresses are in the elastic range of the concrete stress-strain curve (see Fig. 1.3) so that inelastic concepts used in developing the ultimate strength of a section in Chapter 2 are inappropriate.

Referring to Fig. 8.5, the location of the neutral axis is computed by taking moments of the transformed area about the neutral axis.

$$\frac{b(kd)^2}{2} = nA_s(d - kd)$$

Solving the quadratic equation for kd yields

$$kd = \sqrt{\left(\frac{nA_s}{b}\right)^2 + \frac{2nA_s d}{b}} - \frac{nA_s}{b}$$

Letting $A_s = \rho bd$

$$k = \sqrt{(\rho n)^2 + 2\rho n} - \rho n \tag{8.5}$$

The moment of inertia is computed about the neutral axis

$$I_{cr} = \frac{b(kd)^3}{3} + nA_s(d - kd)^2$$

Simplifying,

$$I_{cr} = nA_s(1 - k)jd^2 \tag{8.6}$$

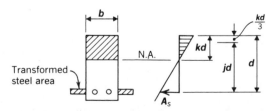

FIGURE 8.5   Elastic analysis of rectangular beam.

where jd = internal lever arm by elastic theory

$$(1 - k/3)d$$

Flexural rigidity is the product of the modulus of elasticity and the moment of inertia

$$E_c I_{cr} = E_s A_s (1 - k) jd^2 \qquad (8.7)$$

Values for the cracked flexural rigidity can be computed using Eqs. 8.5 and 8.7 or, more simply, by using Fig. A8.2 for concrete strengths ranging from 3 to 5 ksi.

Effective Moment of Inertia (I_e)

On first cracking, the stiffness of a concrete member does not suddenly drop from the gross moment of inertia I_g to the cracked moment of inertia I_{cr} because a single crack that may not even extend to the neutral axis affects stiffness only slightly. As the moment is increased, more cracks form over a larger portion of the span and gradually reduce the stiffness to the cracked section stiffness. Studies of available test data have resulted in an empirical equation adopted by the ACI Code giving a smooth transition between I_g and I_{cr}. This equation can be restated as

$$I_e = I_{cr} + \left(\frac{M_{cr}}{M_a}\right)^3 (I_g - I_{cr}) \qquad (8.8)$$

where M_{cr} = cracking moment, as computed by Eq. 8.4,
and M_a = maximum service load moment in the member.

Note that when $M_a/M_{cr} = 2$, the increase in stiffness from a cracked section is only 12½% of the difference between I_g and I_{cr}. For this reason, cracked section stiffness I_{cr} is usually used when $M_a/M_{cr} > 2$ unless a higher level of accuracy is required.

A rectangular beam and a T-beam of typical proportions are shown in Fig. 8.6 to illustrate the variation in moment of inertia as the amount of tensile flexural reinforcement is varied. Note that the flange has a large effect on gross moment of inertia I_g but a smaller effect on cracked moment of inertia I_{cr}.

Variation in Moment of Inertia

Even though concrete outlines may be uniform over the full length of a concrete member, the amount and location of reinforcement is frequently not, and the degree of cracking also usually varies. To avoid lengthy, tedious calculations that might even take considerable time on a small computer, it is convenient to have a single value for flexural rigidity EI for a span. Several methods have been proposed to arrive at that single value [4].

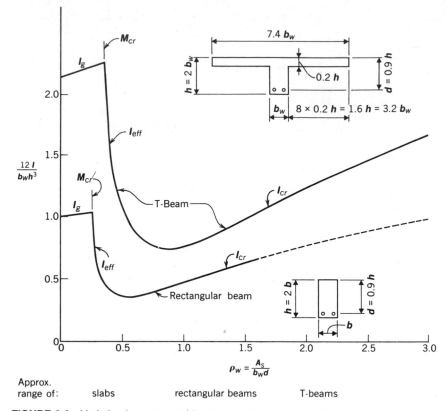

FIGURE 8.6 Variation in moment of inertia as a function of ρ_w. $f'_c = 3$ ksi.

1. Use the mid-span value for flexural rigidity alone for simple span and continuous beams and the value at the support for cantilever beams. The empirical method for determining I_e takes into account the variation in the amount of cracking along the beam. This is the simplest method to use.*

   Methods 2 and 3 are mentioned here because they have been given some attention in the past. They are rarely used today because they require more calculation labor without increasing the accuracy of the result.

2. Use a simple average of values for flexural rigidity at mid-span and at the ends for continuous beams. This method is also relatively simple to use but is not as accurate as method 1 when there is a large difference between flexural rigidities at the ends and at mid-span.

3. Use an average of values at mid-span and at the ends for continuous beams weighted according to the moments at each section. This method is more accurate than method 2 and about as accurate as method 1.

---

*To belabor a much made point, the engineer is paid to produce designs of economical, constructible, and workable structures and not elegant, supposedly rigorous, mathematical computations.

The reason for the accuracy of method 1, as well as its simplicity, can be appreciated by referring to Fig. 8.7. **M/EI** diagrams are drawn for a continuous, uniformly loaded beam using I_{cr}, I_g, and I_e. In Fig. 8.7(a), moments near the ends crack the beam, but moments at mid-span do not. In Fig. 8.7(b), the moment at mid-span cracks the beam, but moments at the ends do not. Figure 8.7(c) illustrates the situation just after the first cracking near mid-span. Remembering that deflection equals the moment in a conjugate beam loaded with the elastic **M/EI** diagram, it is obvious that deflection is most affected by the elastic moment diagram near mid-span. Hence, a single mid-span value gives reasonably accurate results for both simple span and continuous beams. Deflection is not greatly affected by magnitude of the elastic **M/EI** diagram near the ends of the beam. In Fig. 8.7(c), note that the first crack does not cause a large immediate deviation from the **M/EI**$_g$ curve.

To account for variations in the moment of inertia, another method that approaches the "theoretically correct" method outlined in Ref. 4 is to compute the flexural rigidities continuously along the beam, taking into account the cracking at all points, and then use any of the classical techniques such as moment area or conjugate beam. The increase in accuracy using this method is marginal and probably does not justify the large increase in computational labor.

Torsional Rotation and Deflection

Sometimes it is necessary to calculate the rotation of a member with torsional moment or the deflection of an element supported by the member but eccentric to its center of gravity. In such cases, the twist θ over a length ℓ_t of a member can be calculated as

$$\theta = \frac{\Sigma T}{GK} \tag{8.9}$$

where ΣT = summation of the service torsional moment along length ℓ_t of the member (see Fig. 8.8).

GK = Torsional rigidity similar to the flexural rigidity EI.

For an uncracked member it is sufficiently accurate to take

$$G = 0.3E_c$$

$$K_g = \frac{\Sigma x^3 y}{3}$$

$$GK_g = 0.1E_c \Sigma x^3 y \tag{8.10}$$

FIGURE 8.7 *M/EI* diagrams for computing deflection by conjugate-beam method.

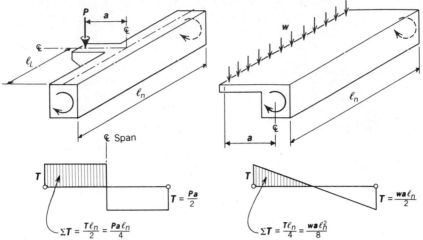

FIGURE 8.8   Twist on an isolated beam due to eccentric loads.

where x = shorter overall dimension of the rectangular part of the cross section and
y = longer overall dimension of the rectangular part of the cross section.

Component rectangles of flanged sections should be selected to yield the largest value of $\Sigma x^2 y$ as in design for torsion (see Section 6.7).

For a torsionally cracked member, it is sufficiently accurate to use an equation developed by Lampert [5].

$$GK_{cr} = \frac{E_s(x_o y_o)^2 A_n}{2(x_o + y_o)s} \times (1 + m) \qquad (8.11)$$

where E_s = modulus of elasticity of steel
x_o and y_o = shorter and longer dimension, respectively, between longitudinal corner bars,
s = stirrup spacing,
A_n = Area of one stirrup leg,

$$m = \frac{(A_s + A'_s)s}{2A_n(x_o + y_o)}$$

A_s = Area of longitudinal tension steel
A'_s = Area of longitudinal compression steel

Deflection of an element supported by a twisting member can be calculated by multiplying the angle of twist in the member at the element by the distance from the center of gravity of the member (see Fig. 8.9).

For most cases in which twist or deflection due to twisting members is significant, the member will be at least partially torsionally cracked, and torsional rigidity given by Eq. 8.11 should be used as it will give an upper bound to the deflection. Torsional rigidity of an uncracked member is usually much larger than that of the same

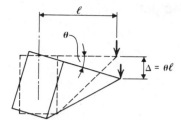

FIGURE 8.9 Deflection due to beam rotation.

member torsionally cracked, but there is little research available to indicate the proper transition from uncracked to cracked rigidities.

Procedures for calculating torsional rotation are not intended for application to moment distribution procedures.

Example 8.3 What is the twist in an 11-ft length of rectangular beam 16 in. wide by 24 in. deep with a uniform service torsional moment of 50 ft-kips? The beam is reinforced with 4 #9 bottom and 2 #9 top bars, as shown in Fig. 8.10. Shear reinforcement is #4 bars at 5 in. c/c. Assume $f'_c = 3$ ksi.

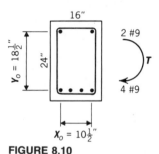

FIGURE 8.10

Solution

Using Eq. 8.11, $2(x_o + y_o) = 2(10.5 + 18.5) = 58$, and cracked torsional rigidity is

$$GK_{cr} = \frac{29 \times 10^3 (10.5 \times 18.5)^2 \times 0.20}{58 \times 5} \times \left(1 + \frac{6.00 \times 5}{0.2 \times 58}\right)$$

$$= 2.7 \times 10^6 \text{ ksi}$$

Using Eq. 8.9, torsional rotation or twist is

$$\theta = \frac{50 \times 12 \times (11 \times 12)}{2.7 \times 10^6} = 2.9 \times 10^{-2} \text{ rad.} \approx 1.7°$$

Actual twist may be less than that calculated if the beam remains partially uncracked.

8.5 CALCULATION PROCEDURES

The labor in preparing deflection calculations can be considerably reduced by the judicious selection of a few, most critical members in a structure for which deflection calculations will be made and disregarding all other members. The success of this approach depends on the skill of the engineer in selecting critical members.

Labor in preparing deflection calculations can also be reduced by first deciding the reason for limiting the deflection and directing the calculations to that end.

Finally, deflection calculations can be minimized by determining the deflection limit and span and then selecting an appropriate calculation method by referring to Fig. 8.11. Note that in Fig. 8.11 there are no precise lines of demarcation between methods. Experienced engineers will consider computation time available, the importance of the member and its deflection response, and the importance that owners and users of the structure will assign to proper deflection behavior before selecting a calculation method. Some details of the normal or extended calculation methods can be used in a simpler method as the situation warrants. For these reasons, an engineer may want to start with the simplest calculation method and extend it if results of the first calculation indicate a potential deflection problem.

Example 8.4 What calculation method should be used for a solid slab spanning 15 ft when the total deflection is limited to $\ell/240$ and the incremental deflection for a masonry wall supported on the slab is limited to $\ell/480$?

Solution Referring to Fig. 8.11, the first limitation indicates that method 1, Indirect Calculation, would be satisfactory, but the second limitation indicates that method #2, Simple Calculation, would be required. If slab stiffness is well in excess of indirect calculation requirements, simple calculations might not be necessary.

Example 8.5 What calculation method should be used for a 40-ft beam if the incremental deflection limit is $\ell/360$ and the total deflection limit is 1 in.?

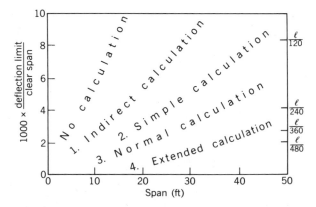

FIGURE 8.11 Recommended calculation procedures.

Solution   One inch in a 40' span is $\ell/480$. Referring to Fig. 8.11, method 4, Extended Calculation, should be used. Experience indicates long spans in reinforced concrete frequently have excessive deflection.

Most deflection calculations are directed toward the response under service loads and not factored or ultimate loads. Under short-time service loads, concrete is nearly elastic, as shown in Fig. 1.2. Deflection under loads exceeding service loads by a substantial amount must be calculated by methods that take into account the inelastic behavior of concrete under short-time loading. Such calculations are not normally performed in a design office.

In the design for strength, appropriate factors of safety are needed. In contrast, deflection calculations should not use a factor of safety if an accurate assessment of the expected deflection is desired. For example, the actual average modulus of elasticity and modulus of rupture should be used when calculating deflections. Indeed, liberal assumptions may be desirable if an overestimate of deflections would be troublesome as, for example, in determining the camber in a floor that must deflect at least to a level position.

Because concrete is a material with properties that have considerable scatter in test data, engineers should not expect to be able to calculate the precise amount of deflection of any given member. ACI Committee 435 has concluded that there is approximately a 90% chance that the deflections of a particular beam will be within the range of 20% less to 30% more than the calculated value [6]. This conclusion is based on a meager amount of data, much of which was used to develop deflection calculation procedures. Engineers certainly should not expect to improve on this accuracy when field conditions are unknown and actual material properties have not been measured. ACI Committee 435 conclusions should be considered the best obtainable accuracy in deflection computations.

To meet ACI Code requirements, it is necessary to calculate deflection by ACI Code procedures and compare the result to certain ACI Code limits (Table A8.1). The ACI Code procedures are noted in the discussion below. Engineers should recognize that ACI Code procedures may not always give realistic values of deflection, but their use is necessary to meet ACI Code requirements. Four general approaches to calculating deflection corresponding to those shown in Fig. 8.11 are discussed below.

Indirect Calculation

Deflection can be limited indirectly by limiting the span-to-depth ratio of a member or by limiting the stress level.

A.   Span-to-depth ratios have been adopted partially by the ACI Code to limit deflections. For beams and one-way slabs, minimum thickness is given by Table A8.3 when members do not support or are not attached to partitions or other construction likely to be damaged by large deflections. This is an important qualification as many members *do* support or are attached to fragile building elements. Members within the limits of Table A8.3 will meet ACI Code requirements.

The ACI Code requires that flat slabs and flat plates be limited to a span-to-depth ratio ℓ_n/h of 36 when design $f_y = 40$ ksi is used and 32.7 when design $f_y = 60$ ksi is used (Code 9.5.3). In rectangular panels and panels with stiff beams on the sides, the ACI Code permits somewhat thinner slabs. These Code provisions are not discussed here because there is a high risk of unacceptable deflection when slabs are thinner than the limits given above. The ACI Code also requires that the thickness be increased 10% in an edge panel when the edge (spandrel) beam has a stiffness ratio α less than 0.8, where α is the ratio of flexural stiffness of the beam section to flexural stiffness of a width of slab bounded laterally by the centerline of the adjacent panel. Flat slabs may not be less than 4 in. thick, and flat plates may not be less than 5 in. thick. The ACI Code permits the use of these limits for slabs supporting fragile building elements, but experienced engineers will use caution in doing so.* Slabs with a high ratio of live load to dead load where a high percentage of the live load is permanent, as in libraries and warehouses, will experience more than the normal deflection.

Two-way slabs supported on all four sides by stiff beams, where α exceeds 0.8, may be as thin as $\ell_n/40$ when design $f_y = 40$ ksi is used and $\ell_n/36.4$ when design $f_y = 60$ ksi is used. Even thinner slabs may be used in some circumstances. The ACI Code permits the use of these limits for slabs supporting fragile building elements, but caution is advised here also.

B. Experience indicates that flexural members remaining essentially uncracked at service loads will generally not have excessive deflection. This condition can easily be checked using structural mechanics. For rectangular members, the smallest moment causing flexural cracking M_{cr} is

$$M_{cr} = f_r S = \frac{7.5\sqrt{f_c'}bh^2}{6}$$

where f_r = modulus of rupture given by the ACI Code as $7.5\sqrt{f_c'}$ for normal weight concrete.

The section modulus of T-beams can be approximated by increasing the section modulus half as much as the moment of inertia for tensile stress at the bottom of the stem $[1 + (\beta_g - 1)/2]$ and increasing the section modulus twice as much as the moment of inertia for tensile stress at the top of the flange $[1 + (\beta_g - 1)2]$. The factor β_g can be taken from Fig. A8.1. The error introduced in this approximation is less than 10% for $b/b_w \le 15$ and $h_f/h \le 0.4$, except that it errs on the low side by up to 30% for tensile stress at the top of the flange and high ratios of b/b_w.

Simple Calculation

For a quick estimate of deflection, use the mid-span moment at service loads as calculated for strength design (or maximum moment in cantilevers). If only factored

---

*Partitions near columns are most likely to be damaged. In some cases, deflection causes rotation of a partition abutting a column and opens a large crack between wall and column near the ceiling, whereas there is no visible crack in the same joint near the floor.

moment is available, divide it by the estimated average load factor to obtain service moment. If concentrated loads or variable uniform loads are present, or if end moments are not equal to zero or fixed end moments, use an appropriate coefficient from Table A8.2 and compute deflection from the mid-span moment and the equation for a simple span uniformly loaded beam (Case I in Table A8.2). Use the clear span for beams and one-way slabs.

Estimate whether or not the beam is cracked by using Eq. 8.4, and calculate either I_g or I_{cr}. Use Fig. A8.1 to assist in calculating I_g and Fig. A8.2 to assist in calculating I_{cr}. Do not include the effects of compression reinforcement. Estimate the effect of creep and shrinkage of long-term deflection by using Fig. A8.3. Do not calculate incremental deflections.

Normal Calculation

For a more careful estimate of deflection, use actual service loads and moments and calculate deflection using normal methods of structural mechanics.

Compute the cracking moment for the mid-span section. If service moment is less than the cracking moment, calculate I_g and use it in the deflection calculations. If service moment is more than cracking moment, calculate I_{cr} and I_e and use the latter in the calculations. Computation of I_g, I_{cr}, and I_e by methods in Section 8.4, neglecting compression reinforcement (and tension reinforcement in the gross section), will meet ACI Code requirements.

Deflection of concrete increases with time because of creep and shrinkage of concrete in the compression zone of flexural members. If compression reinforcement is present, long-time deflection is reduced because steel does not creep or shrink and the steel tends to take a larger share of the compression force than at initial loading.

The *additional* (*not* total) long-time deflection, including the effects of compression reinforcement, may be estimated by multiplying immediate deflection by a factor λ. Factor λ is given by an empirically derived equation (Code 9.5.2.5) [7].

$$\lambda = \frac{\xi}{1 + 50\rho'} \tag{8.12}$$

where ξ = an empirical time-dependent factor for sustained loads equal to
        2.0 for loads lasting 5 years or more
        1.4 for loads lasting 12 months
        1.2 for loads lasting 6 months
        1.0 for loads lasting 3 months.
ρ' = reinforcement ratio for compression reinforcement, which equals A_s'/bd

The effect of compression reinforcement on additional long-time deflection can be seen in Fig. A8.3, which plots the values of λ for sustained load. Total long-time deflection Δ_{LT} can be computed using

$$\Delta_{LT} = (1 + \lambda)\Delta_i \tag{8.13}$$

When appropriate for the deflection limitation under consideration, calculate incremental deflections using Fig. A8.3 to estimate additional deflection for intermediate time periods. Computation of long-time deflection by the method just outlined will meet ACI Code requirements.

Extended Calculation

For a discussion of more precise calculation methods, see Refs. 8 and 9. Factors considered in more precise calculations can be categorized under loading, flexural rigidity, creep and shrinkage, and fixity.

Although all of the factors listed below can have a significant influence on deflection in some circumstances, those factors that are most likely to affect deflection adversely are noted by a double asterisk (**). Extended calculations may result in better estimates of deflection, but they do *not* meet ACI Code requirements.

Factors Affecting Loading
- Loading history, including the age at loading and unloading each increment and the proportion of long-time versus transient loading.**
- Actual loads rather than those assumed in strength design.

Factors Affecting Flexural Rigidity
- Modulus of elasticity of concrete E_c.**
- Modulus of rupture of concrete f_r.**
- Actual location of reinforcement as built (for investigation of completed structures).**
- Amount and location of compression reinforcement in computing I_g and I_{cr}.
- Amount and location of tension reinforcement in computing I_g.
- Variation in flexural rigidity EI along the span due to change in cross section, change in amount of reinforcement, and degree of cracking.

Factors Affecting Long-time Deflection
- Volume-to-surface ratio (higher ratios reduce shrinkage and creep).**
- Creep coefficient for concrete used in the structure.
- Shrinkage coefficient for concrete used in the structure.
- Ambient relative humidity (higher humidity reduces shrinkage and creep).
- Ambient temperature (higher temperatures increase shrinkage and creep).

Factors Affecting Fixity
- Rigidity of joints (weak joints or joints with inadequately anchored reinforcement may allow some rotation and larger deflections).
- Support transverse to span (as in slabs nominally designed as spanning one way but actually receiving some support from parallel side members).

- Moment distribution based on actual conditions of loading and stiffness of members.

Two-Way Slab Systems

ACI Committee 435 has prepared a comprehensive survey of calculation methods for two-way slab systems [10]. The simplest and most straightforward method for flat plates without the stiffening effects of column-line beams, drop panels, and column capitals is the Wide Beam Method. It lacks the rigor of classical methods but gives results that are well within the range of accuracy for slabs of normal spans. For slabs with beams, column capitals, or drop panels, the Equivalent Frame Method (EFM) [11], or finite element method, would be appropriate. The ACI Code is silent on procedures for computing deflection of two-way slab systems, except for indirect calculations noted above.

The Wide Beam Method considers the deflection of a slab panel in one direction separately from that in the other direction, later adding the two contributions to obtain the total deflection at any point of interest. Full service loading must be used in the calculations for each direction.

Referring to Fig. 8.12(a), the slab in any panel is considered to act as a broad, shallow beam equal in width to the panel width ℓ_2 and having a span ℓ_1. At this stage, the slab is considered to be resting on unyielding support lines at two opposite edges.

The average deflection is obtained, but the difference in deflection of middle and column strips should be accounted for. Because the greater part of the span is subjected to positive bending, it is sufficiently accurate to use the lateral distribution factors for positive bending moment to compute deflection in column and middle strips. For example, for an interior slab panel without beams, column strips are assigned 60% of the moment in the positive bending section. Accordingly, deflection on the column line Δ_c can be estimated as $(0.60/0.50 =)$ 1.2 times the average deflection. The deflection thus obtained would be appropriate to assess the risk of damage to a partition parallel to the span on the column centerline. Similarly, deflection of the middle strip Δ_m from a line tangent to the slab at the supporting column lines (e.g., to assess the potential damage to a partition parallel to and in the center of the middle strip) can be estimated as $(0.40/0.50 =)$ 0.8 times the average deflection.

A similar calculation is performed in the perpendicular direction [Fig. 8.12(b)]. Mid-panel deflection, for example, is obtained as the sum of the average mid-span deflection of the slab panel in two directions or, perhaps, more precisely, the deflection of the column strip in one direction plus the deflection of the middle strip in the other direction [see Fig. 8.12(c)]. Note that the span center-to-center of supports is used rather than the clear span. This introduces a small error on the conservative side (i.e., overstating computed deflection).

For the special case of a two-way slab supported on four sides by rigid beams, moment and deflection are reduced by torsion in the slab. An approximate but conservative solution for secondary slabs of moderate span is to compute mid-panel deflection as the deflection of a one-way slab spanning the short direction of the panel. If the ratio of long side to short side of the panel is two or less, the mid-panel

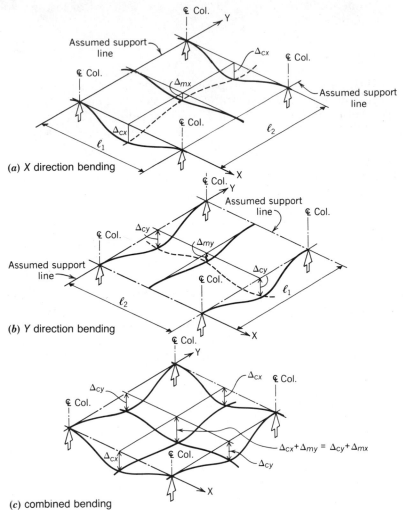

(a) X direction bending

(b) Y direction bending

(c) combined bending

FIGURE 8.12 Basis of wide beam method. (Courtesy of the American Concrete Institute.)

deflection can be reduced to 0.8 of the equivalent one-way slab deflection. Deflection calculations need only be made in the short direction and not in two directions as for flat plates because the rigid supports on all four sides carry part of the load and thus reduce deflection from that of a one-way slab system.

8.6 REDUCING DEFLECTION

When calculated deflections are greater than that permitted by the Code or greater than the engineer believes is tolerable, several options are available to make deflection and deflection limits compatible.

1. Increase concrete dimensions. An increased depth is more efficient than an increased width but poses other problems, such as interference with mechanical systems (piping and ventilating ducts) or the ceiling of room space below. After preliminary design of a structure has been established, architects and owners will be especially resistant to increasing structural depth. Increasing the width of a member increases its gross moment of inertia I_g proportionally with a proportional reduction in deflection, but increasing the width of the stem of a T-beam may increase its cracked moment of inertia I_{cr} only marginally. Of course, slabs cannot be increased in width.

2. Increase flexural tension reinforcement. This option is especially effective for highly cracked members ($M_a/M_{cr} > 2$) with a low percentage of flexural tension reinforcement. It will have little effect on uncracked members.

3. Provide compression reinforcement or increase the amount (see Fig. A8.3). It is especially effective in reducing long-time deflections.

4. Camber slabs and beams. Camber is effective only for reducing or eliminating visual problems, ponding water, and maintaining alignment.* It is completely ineffective for reducing potential damage to walls and other fragile construction. Walls are damaged by deflection from initial position and not by the curve of the initial position itself. The initial position is the position of the structural element when the walls are first constructed and may be straight, humped up, or sagged downward. The structural member may have been constructed with a hump or a sag, or these may have been the result of movements after construction of the structural member but before construction of the wall.

   Experience shows that contractors have difficulty building camber properly. Therefore, it should only be used for special situations and where repetition permits training of construction crews or where qualified instructors are present to help construction workers build cambered formwork and finish concrete with a camber properly. Only camber in excess of about 1 in. should be specified in order to avoid trivial measurements that are difficult to control in the field. Some engineers require contractors to employ surveyors to measure camber.

5. Provide slip joints in partitions, windows, and other fragile construction below a concrete structure. Properly designed and constructed, such joints can eliminate concern that deflection might damage nonstructural elements below or beside concrete slabs and beams.

6. Reassess the deflection limits. They might be too conservative.

7. Delay construction of walls and equipment sensitive to excessive deflection until the remaining incremental deflection is acceptable. This option must be used judiciously to avoid delaying construction unnecessarily. Construction

*The human eye can detect very small deviations from a straight line, and some people are bothered by deviations that they can detect. Furthermore, laypersons easily understand camber or deflection measured long after construction is complete, whether it is relevant or not.

delays may cost far more than other options and may be hard to enforce. On the other hand, some delay is a normal part of the construction process.

8. Provide shorter spans, lower loads, or additional supporting beams and columns. This option is severely limited by architectural considerations, especially after preliminary design has been established and approved.

9. Use partial or full prestressing to balance dead load deflection.

Improper construction procedures more often result in high deflection than in low strength. Therefore, engineers should take special precautions in specifying construction of marginal members that are vulnerable to large deflection from improper construction. Following are some items to be considered:

1. Do not permit loading of the structure in excess of design loads under any condition. Members designed to be uncracked may become cracked with a consequent reduction in moment of inertia. Furthermore, compressive stress in concrete may enter the inelastic range with higher than expected deflection. Flat plates designed for superimposed loading of 60 to 80 psf (apartment live load of 40 psf plus partitions) are especially vulnerable to overloading because the weight of the slab above may be more than the design load. Some studies of shoring and reshoring in multistoried buildings under construction indicate that shoring loads can impose a load on just one of the floors below of two to three times the weight of the fresh concrete plus formwork and workers.

2. Do not permit loading of a structure before concrete reaches its design strength. Early loading can crack concrete that would otherwise remain uncracked and load the concrete when its modulus of elasticity is lower than expected. Failure to observe precautions 1 and 2 are the most common cause of excessive deflection created by improper construction procedures.

3. Ensure that top steel is not displaced, especially in thin members.* Moment of inertia is proportional to the square of effective depth for cracked sections so that a reduction of 20% in d results in an increase of 36% in deflection.

4. Ensure proper shoring and construction to prevent a built-in sag or negative camber. Heavy concrete floors supported on shores bearing on soft, muddy soil foundations are especially critical. Negative camber does not affect later deflection behavior but does present an unsightly appearance and allows ponding of water on the slab.

5. Require thorough, long-term curing for thin members such as joists with 2- to 4½-in.-thick top slabs. Thin slabs dry out faster than thicker members and, hence, shrink more. In a joist system, shrinkage of the top slab can cause more shrinkage warping than in a solid slab, rectangular beam, or T-beam of normal proportions. The top slab is far from the tension reinforcement that resists shrinkage, thus causing a large warping moment.

*Construction workers are notoriously unaware of the engineer's careful calculations and laborious layouts of delicate bar placements. Workers run roughshod over slab steel like vintners stomping grapes. After all, where else do they have to walk?

6. Avoid anything that would reduce the modulus of rupture f_r or the modulus of elasticity E_c or increase shrinkage or creep even though a reduction in compressive strength might be acceptable. In addition to lower compressive strength, a high water content, high water-to-cement w/c ratios, certain admixtures, and some aggregates can reduce f_r and E_c or adversely affect shrinkage and creep.

Example 8.6   Calculate flexural rigidity EI of the beam in Ex. 8.1 using normal calculation procedures. Assume a 4-in.-thick flange of maximum width permitted by the ACI Code. The beam cross section is shown in Fig. 8.13. Assume 6.4 sq. in. of positive flexural reinforcement and $f_c' = 3$ ksi.

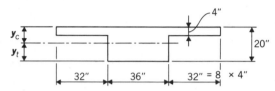

FIGURE 8.13

Solution

Find the location of the neutral axis by taking moments about the top edge:

$$36 \times 20 = 720 \times 10 = 7200 \text{ in.}^3$$

$$64 \times 4 = \underline{256} \times 2 = \underline{512}$$
$$976 7712 \text{ in.}^3$$

$$y_c = \frac{7712}{976} = 7.9''$$

$$y_t = 20 - 7.9 = 12.1''$$

Moment of inertia:

$$\frac{36 \times (20)^3}{12} = 24{,}000 \text{ in.}^4$$

$$720 \times (10 - 7.9)^2 = 3{,}175$$

$$64 \times (4)^3/12 = 340$$

$$256 (7.9 - 2)^2 = \underline{8{,}911}$$

$$I_g = 36{,}400 \text{ in.}^4$$

$$\text{Modulus of rupture } f_r = \frac{7.5\sqrt{3000}}{1000} = 0.411 \text{ ksi}$$

$$M_{cr} = \frac{0.411 \times 36{,}400}{12.1 \times 12} = 103.1'k$$

From Ex. 8.1, $^{+}M = 293.9'k = 2.8M_{cr}$; therefore, use the cracked moment of inertia.

For a simpler procedure, use Fig. A8.1,

$$\text{Flange width} = 2 \times 32'' + 36'' = 100''$$

$$\frac{b}{b_w} = \frac{100}{36} = 2.8, \qquad \frac{h_f}{h} = \frac{4}{20} = 0.20$$

$$\beta_g = 1.5$$

Assume that an increase in section modulus for tension in the stem of the T-beam is about one-half the increase in moment of inertia, or about 1.25.

$$S = 1.25 \times \frac{36 \times (20)^2}{6} = 3000 \text{ in.}^3$$

and
$$M_{cr} = \frac{0.411 \times 3000}{12} = 102.7'k$$

Calculate cracked moment of inertia from basic principles (refer to Fig. 8.14).

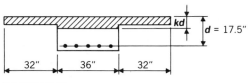

FIGURE 8.14

Location of neutral axis of elastic section:

$$E_c = 57\sqrt{3000} = 3.12 \times 10^3 \text{ ksi}$$

$$n = \frac{E_s}{E_c} = \frac{29 \times 10^3}{3.12 \times 10^3} = 9.3$$

$$64 \times 4(kd - 2) + 36kd\left(\frac{kd}{2}\right) = 6.4 \times 9.3(17.5 - kd)$$

$$18(kd)^2 + (256 + 59.5)kd - 512 - 1042 = 0$$

$$(kd)^2 + 17.53kd - 86.3 = 0$$

$$kd = \sqrt{\left(\frac{17.53}{2}\right)^2 + 86.3} - \frac{17.53}{2} = 4.01''$$

Moment of inertia, I_{cr}

$$\frac{100 \times (4)^3}{3} = 2{,}133$$

$$6.4 \times 9.3(17.5 - 4.0)^2 = \underline{10{,}847}$$

$$I_{cr} = 13{,}000 \text{ in.}^4$$

$$EI_{cr} = 3.12 \times 10^3 \times 13{,}000 = 4.05 \times 10^7 \text{ ksi}$$

For a simpler computation, calculate moment of inertia using Fig. A8.2.

$$\rho = \frac{6.4}{100 \times 17.5} = 0.37\%, \qquad \beta_{cr} = 0.77$$

$$EI_{cr} = 0.77 \times 10^2 \times 100 \times (17.5)^3 = 4.13 \times 10^7 \text{ ksi}$$

Note that the error in reading Fig. A8.2 is well within the range of accuracy of deflection calculations.

Example 8.7 Compute deflection of the beam in Ex. 2.4 (Figs. 2.9 and 2.11) and assess the acceptability of the calculated deflections.

Solution From Fig. 8.11, for a span of 25', simple or normal calculation methods can be used, depending on sensitivity to deflections of the structure and the construction supported by it.

As a rectangular beam, $S = \dfrac{24(17)^2}{6} = 1156 \text{ in.}^3$

As a T-beam, assume $S \simeq 1156 \times 1.25 = 1440 \text{ in.}^3$

$$M_{cr} = f_y S = \frac{7.5\sqrt{4000} \times 1440}{12{,}000} = 57.1'\text{k}$$

From Ex. 2.4, total load moment $M_a = 190$

Because $M_a > 3M_{cr}$, use I_{cr}

Calculate I_{cr} using Fig. A8.2

Flange width $= 16 \times 3 + 24 = 72''$

$${}^+A_s = 4.68^{\square''}, \qquad \rho = \frac{4.68}{72 \times 14.5} = 0.45\%$$

$$\beta_{cr} = 0.98$$

$$EI_{cr} = 0.98 \times 10^2 \times 72(14.5)^3 = 2.2 \times 10^7 \text{ ksi}$$

Note that if M_{cr} had been closer to M_a, more accurate methods for calculating M_{cr} might have been required and calculation of I_g and I_e might have been necessary.

Calculate deflections using mid-span moment and the equation from Table A8.2 for a continuous, uniformly loaded span.

$$\Delta_1 = \frac{1}{16} \times \frac{190 \times 12(25 \times 12)^2}{2.2 \times 10^7} = 0.58''$$

From Ex. 2.4, live load moment = 86′k

Dead load moment = 104″k

$$LL\ \Delta = \frac{0.58 \times 86}{190} = 0.26''$$

$$DL\ \Delta_1 = \frac{0.58 \times 104}{190} = 0.32''$$

With no compression reinforcement, $\lambda = 2$

$$LL\ \Delta \qquad\qquad = 0.26$$

$$DL\ \Delta_{LT} = 0.32(1 + 2) = \underline{0.96}$$

$$\text{Total}\ \Delta_{LT} = 1.22''$$

Deflection limits:

1. Total deflection may be excessive for appearance and ponding water and may require cambering, increased concrete outlines, or addition of compression reinforcement.

2. Live load $\dfrac{\Delta}{\ell} = \dfrac{0.26}{25 \times 12} = \dfrac{1}{1150}$ OK

3. Incremental deflection after partitions are built:

$$1.22 - 0.32 = 0.90'' = \frac{\ell}{330}$$ NA

Recalculate incremental deflection if construction of partitions is delayed 2 months and flexural reinforcement is added to the beam so that $A_s = 6.00^{□''}$ and $A'_s = 4.00^{□''}$. For a more accurate analysis, assume that concrete is 80% of the dead load and partitions are 20% of the dead load.

$$\rho = \frac{6.00}{72 \times 14.5} = 0.57\% \qquad \rho' = \frac{4.00}{72 \times 14.5} = 0.38\%$$

From Fig. A8.2, $\beta_{cr} = 1.1$

$$EI_{cr} = 1.1 \times 10^2 \times 72(14.5)^3 = 2.4 \times 10^7 \text{ ksi}$$

$$\Delta_1 = \frac{1}{16} \times \frac{190 \times 12(25 \times 12)^2}{2.4 \times 10^7} = \underline{0.53''}$$

$$\text{LL } \Delta = 0.53 \times \frac{86}{190} = 0.24''$$

$$\text{Concrete DL } \Delta_i = 0.8 \times 0.53 \times \frac{104}{190} = 0.23''$$

$$\text{Partition DL } \Delta_i = 0.2 \times 0.53 \times \frac{104}{190} = \underline{0.06''}$$

$$\text{Total initial deflection} = 0.53''$$

Deflection after 5 years,

Using Eq. 8.12, $\lambda = \dfrac{2}{1 + 50 \times 0.0038} = 1.68$

or read $\lambda = 1.7$ from Fig. A8.3

$$\text{LL } \Delta = = 0.24''$$

$$\text{DL} \Delta_{LT} = (0.23 + 0.06)(1 + 1.68) = \underline{0.78''}$$

$$\text{Total } \Delta_{LT} = 1.02''$$

From Fig. A8.3, read $\lambda = 0.7$ for $\rho' = 0.38\%$
and load duration = 2 months.

$$\text{At 2 months, concrete DL } \Delta = 0.23(1 + 0.7) = \underline{0.39}$$

$$\text{Net incremental } \Delta = 0.63''$$

$$0.63'' = \ell/480$$

Because $\dfrac{\ell}{480} = \dfrac{\ell}{480}$, deflection is $\underline{\text{OK}}$

If the partitions are especially sensitive to cracking and the building is an important one subject to much public scrutiny and critical review, further precautions may be desirable, such as enlarging the concrete outlines or adding even more flexural and compression reinforcement.

Example 8.8

A. Estimate the minimum slab thickness required for the flat slab in Ex. 4.1 (shown in Fig. 4.17), using indirect calculation methods.

B. Compute short- and long-term deflections at the center of panel 2-3, B-C, using simple calculation methods.

C. Assess the acceptability of the deflections. If deflection exceeds limits, suggest corrective measures.

Solution A   Because $f_y = 60$ ksi is used,

$$\text{Minimum thickness} = \frac{\ell_n}{32.7} = \frac{(15-1)12}{32.7} = 5.1''$$

$$= \frac{(16-1.33)12}{32.7} = 5.4''$$

$5\frac{1}{2}''$ slab is OK

From Ex. 4.1, $\alpha = 1.17 > 0.8$; therefore, a 10% increase in slab depth is not required in end panels.

Using a $5\frac{1}{2}''$ slab, $\mathbf{S} = \dfrac{12 \times (5.5)^2}{6} = 60.5\,\text{in.}^3/'$

From Ex. 4.1, the average load factor

$$= \frac{207\ \text{psf}}{139\ \text{psf}} = 1.49$$

In N-S direction,

Col. strip width = 7'-6"

$$\text{Col. strip }^+\mathbf{M} = \frac{17.5'\text{k}}{7.5 \times 1.49} = 1.57'\text{k}/'$$

In E-W direction,

Col. strip width = 7'-11"

$$\text{Col. strip }^+\mathbf{M} = \frac{16.9'\text{k}}{7.92 \times 1.49} = 1.43'\text{k}/'$$

Maximum bending stress in the plain concrete slab at service loads,

$$\frac{\mathbf{M}}{\mathbf{S}} = \frac{1.57 \times 12}{60.5}$$

$$= 0.312\ \text{ksi} < \frac{7.5\sqrt{3000}}{1000} = 0.411\ \text{ksi} \qquad\qquad \underline{\text{OK}}$$

Because the slab is essentially uncracked in positive moment regions, deflection behavior should be satisfactory.

Solution B   From Solution A to Ex. 8.8 above, service moments are less than the cracking moment so that I_g may be used.

$$I_g = \frac{12(5.5)^3}{12} = 166.4\ \text{in.}^4/'$$

For $f'_c = 3$ ksi, $E_c = 3.12 \times 10^3$ ksi

Using coefficient $\beta_m = 1/16$ from Table A8.2 for a continuous, uniformly loaded span and taking panel moments from Ex. 4.1,

N-S: $^+M = 29.2'$k, $\ell_1 = 16'\text{-}0''$, $\ell_2 = 15'\text{-}0''$

$$\Delta_i = \frac{1}{16} \times \frac{29.2 \times 12(16 \times 12)^2}{3.12 \times 10^3 \times 166.4 \times 15} = 0.10''$$

E-W: $^+M = 28.2'$k, $\ell_1 = 15'\text{-}0''$, $\ell_2 = 16'\text{-}0''$

$$\Delta_i = \frac{1}{16} \times \frac{28.2 \times 12(15 \times 12)^2}{3.12 \times 10^3 \times 166.4 \times 16} = \underline{0.08''}$$

Total live and dead load
initial deflection at midpanel, $\Delta_i = 0.18''$

Because $A'_s = 0$, $\lambda = 2$ for load duration over 5 years.

Live load (40 psf)

$$\Delta = \frac{0.18 \times 40}{139} = 0.05''$$

Initial slab dead load (69 psf)

$$\Delta_i = \frac{0.18 \times 69}{139} = 0.09$$

Initial other dead load (30 psf)

$$\Delta_i = \frac{0.18 \times 30}{139} = 0.04$$

Added long-term dead load

$$\Delta_{LT} = (0.09 + 0.04) \times 2 = \underline{0.26}$$

Total live and dead load long-term
deflection at midpanel,

$$\Delta = 0.44''$$

Solution C

For live load:

$$\ell = 16 \times 12 = 192''$$

$$0.05 = \frac{\ell}{3840} < \frac{\ell}{360} \qquad\qquad \underline{\text{OK}}$$

Incremental deflection for partitions (assume 30 psf "other" dead load is for partitions):

$$0.44 - 0.09 = \frac{\ell}{550} < \frac{\ell}{480} \qquad\qquad \underline{\text{OK}}$$

If construction of partitions is delayed 3 months, $\lambda = 1.0$ from Fig. A8.3, and incremental deflection $= 0.44 - (0.09 \times 2) = 0.26''$

$$0.26 = \frac{\ell}{740} < \frac{\ell}{480} \qquad\qquad \underline{\text{Better}}$$

PROBLEMS

1. Using indirect calculations, what thickness is required for a flat slab in an exterior bay with 18-in. square columns spaced 22×26 ft on centers with no spandrel beam?
2. Using indirect calculations, what minimum depth of a beam with one end continuous is required to span 31 ft to carry a roof with masonry partitions built tight under the beam?
3. What size beam is required to resist a service moment of 675 ft-kips and remain essentially uncracked?
4. Compute short- and long-time deflections at the center of flat slab panel 1-2, A-B in Ex. 4.1 (Fig. 4.17), assuming that early or heavy loading has cracked the slab.
5. Compute the effective moment of inertia for the beam in Ex. 8.7 if maximum service moment is 75 ft-kips.
6. In Exs. 8.1 and 8.6, assume that maximum mid-span service moment is 130 ft-kips. What value for the flexural rigidity should be used in deflection calculations?
7. For a $12 + 4$ pan slab with 6-in. joists at 36 in. c/c (similar to Ex. 3.7) with bottom flexural reinforcement of 1 #6 plus 1 #5 bar with ¾ in. cover and $f_c' = 3$ ksi:
   A. What is the gross moment of inertia of the slab?
   B. What is the cracking moment at mid-span?
   C. What is the cracked moment of inertia in the mid-span region?
   D. What moment of inertia should be used if service moment is 48.1 ft-kips?
8. How much will the slab in Problem 7 deflect at 5 years under a service dead load of 336 plf and transient service live load of 300 plf per joist, if the clear span is 28 ft in an interior bay of approximately equal spans on each end?
9. Referring to Fig. 8.15, how much vertical movement should the joint between window and beam be designed to accommodate? Assume a 31-ft clear end span with a service moment at mid-span of 168 ft-kips of which 20% is live load moment.

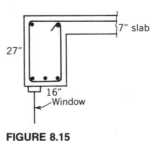

27″

7″ slab

16″
Window

FIGURE 8.15

10. Estimate the long-time deflection at the tip of the cantilever slab in Ex. 6.6 (Fig. 6.28) assuming no movement in the support at each end of the beam.

Selected References

1. ACI Committee 435. "Allowable Deflections," (ACI 435.3R-68) (Reaffirmed 1984), *ACI Journal*, V.65, No.6, pp 433–444, June 1968. See also *ACI Manual of Concrete Practice. Part 3*, American Concrete Institute, Detroit, 1985.
2. *Manual of Steel Construction.* American Institute of Steel Construction, New York, 1980.
3. Jerome M. Raphael. "Tensile Strength of Concrete," *ACI Journal*, V.81, No. 2, pp 158–165, March-April 1984.
4. ACI Committee 435. "Deflections of Continuous Concrete Beams," (ACI 435.5R-73) (Reaffirmed 1984), *ACI Journal*, V.70, No.12, pp 781–787, December 1973. See also *ACI Manual of Concrete Practice. Part 4*, American Concrete Institute, Detroit, Michigan, 1985.
5. Paul Lampert. "Postcracking Stiffness of Reinforced Concrete Beams in Torsion and Bending," *Analysis of Structural Systems for Torsion*, Sp-35; pp 385–433, American Concrete Institute, Detroit, 1973.
6. ACI Committee 435. "Variability of Deflections of Simply Supported Reinforced Concrete Beams," (ACI 435.4R-72) (Reaffirmed 1984), *ACI Journal*, V.69, No.1. pp 29–35, (January 1972). See also *ACI Manual of Concrete Practice. Part 4*, American Concrete Institute, Detroit, 1985.
7. Dan E. Branson. "Compression Steel Effect on Long-Time Deflections," *ACI Journal*, V.68, No.8, pp 555–559, August 1971.
8. ACI Committee 435. "Deflections of Reinforced Concrete Flexural Members," (ACI 435.2R-66) (Reaffirmed 1984), *ACI Journal*, V.63, No.6, pp 637–674, June 1966. See also *ACI Manual of Concrete Practice. Part 4*, American Concrete Institute, Detroit, 1985.
9. ACI Committee 209, Subcommittee 2. "Predictions of Creep, Shrinkage, and Temperature Effects in Concrete Structures," *Designing for Effects of Creep, Shrinkage, Temperature in Concrete Structures*, SP-27, pp 51–94, American Concrete Institute, Detroit, 1971.
10. ACI Committee 435. "Deflection of Two-Way Reinforced Concrete Floor Systems: State-of-the-Art Report," (ACI 435.6R-74) (Reaffirmed 1984), *Deflections of Concrete Structures*, SP-43, pp 55–82, American Concrete Institute, Detroit, 1974. See also *ACI Manual of Concrete Practice. Part 3*, American Concrete Institute, Detroit, 1985.
11. Arthur H. Nilson and Donald B. Walters, Jr. "Deflection of Two-Way Floor Systems by the Equivalent Frame Method," *ACI Journal*, V.72, No.5, pp 210–218, May 1975.

CHAPTER 9

COLUMNS

9.1 OBJECTIVES

All members that carry an axial compression load are defined as compression members regardless of the presence or magnitude of bending. The most common compression members are columns and bearing walls, but arch ribs and horizontal members in rigid frames are also compression members even though bending may predominate. In this chapter, compression members are called columns, but the design principles apply to other members in compression as well.

There are three types of columns:

1. Concrete members reinforced with longitudinal bars restrained laterally with ties, called tied columns.

2. Concrete members reinforced with longitudinal bars restrained laterally with spirals, called spiral columns.

3. Concrete members with embedded longitudinal structural steel shapes or structural steel pipes filled with concrete, with or without added longitudinal reinforcing bars, called composite columns.

Of these three types, illustrated in Fig. 9.1, the first is by far the most common.

Assuming concrete outlines have been established in preliminary design, the objectives in design of columns are the following:

1. Select concrete strengths to be used and the locations where each concrete strength class will be used.

2. Select the number, size, and arrangement of longitudinal bars and, in composite columns, the structural steel shape.

3. Select the number, size, type, and spacing of ties or spirals.

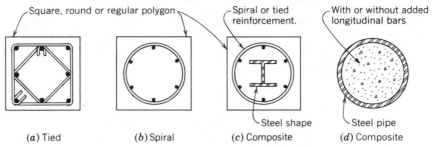

FIGURE 9.1 Types of concrete columns. *(a)* Tied; *(b)* Spiral; *(c)* composite.

Splices of longitudinal bars were discussed in Chapter 5, Anchorage and Bar Development. Special problems associated with transmitting axial loads and moments through floor framing are discussed in Chapter 10, Joints.

9.2 COLUMN BEHAVIOR

The ACI Code requires a strength reduction factor $\phi = 0.70$ for tied columns and $\phi = 0.75$ for spiral columns (Code 9.3.2.2). For composite columns with spiral reinforcement meeting requirements of the ACI Code, $\phi = 0.75$. For other composite columns $\phi = 0.70$. The larger value for spiral columns recognizes the greater toughness or ductility of such columns. The face shell on a tied column will spall when the maximum load is reached, and failure follows immediately. When the maximum load is first reached in a spiral column, the face shell spalls but the column continues to support load while large deformations take place, as shown in Fig. 9.2. Spirals confine concrete in the core so that it can continue to support load.

The strength of all columns depends more on the strength of concrete and other uncertainties than does the strength of flexural members; hence, the capacity reduction factors for all types of columns are lower than those for flexural members.

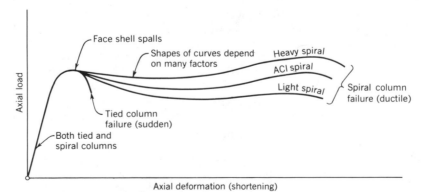

FIGURE 9.2 Idealized load-deformation curves of columns.

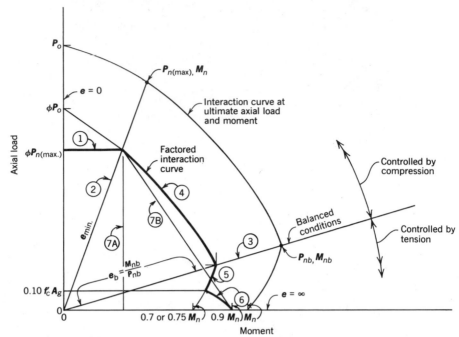

FIGURE 9.3   Typical column interaction diagram. (Circled numbers refer to discussion in Section 9.2.)

All columns in structures have some moment coincident with the axial load. By examining a typical axial load-moment interaction diagram in Fig. 9.3, it is evident that the presence of moment on a column usually does not reduce its axial load capacity proportionally. Numbered features of the diagram are explained below.

1. The point P_o is the maximum axial load capacity for a concentrically loaded, very short column. In practical situations, the maximum usable axial load capacity $\phi P_{n(max)}$ is somewhat less to allow for a minimum amount of moment. Factored service load P_u must never exceed axial load capacity $\phi P_{n(max)}$. Some sources of moment in a column not normally computed are the following:

   - Loads may be applied eccentrically as from a column above intended to be concentric but constructed off center from the column below. (Such eccentricity can be within specified construction tolerances.)
     Eccentricities can also be introduced by foundations constructed off center or by uneven settlement of foundation soils.

   - Centerline of the column may not coincide with centerline of the beam framing into it.

   - The frame may introduce moments in the column not anticipated by calculation. These moments may be the result of creep or shrinkage of concrete.

- Columns may be constructed so that the centroid of reinforcement does not coincide with the geometric center of the column. Such eccentricity could be within specified construction tolerances.
- Variations in materials in the column could introduce additional eccentricities. For example, understrength bars could be accidentally used, or concrete on one face of the column could be damaged by fire, by inadequate curing, or by improper placing procedure.

2. At $\phi P_{n(\max)}$ the column will resist some moment without any reduction of usable axial load capacity. It is convenient to think of this moment-axial load relationship as an eccentricity of axial load, that is, $e = M/P$. The maximum moment capacity when axial load is at a maximum $\phi P_{n(\max)}$ is $e_{\min} \phi P_{n(\max)}$, where $e_{\min}$ is about 5 to 10% of the column thickness, depending on the amount and placement of reinforcement within the column and the concrete and steel strengths. At the point of maximum nominal axial load strength $P_{n(\max)}$ at minimum eccentricity $e_{\min}$ and the corresponding maximum nominal moment strength M_n, all reinforcement will be at yield and concrete will be at its crushing strength even though strain across the section will vary.

3. As the moment is increased from $e_{\min}$ and axial load is reduced, compressive strain on one face of the column is reduced to zero. As the eccentricity is further increased toward balanced conditions e_b, tensile strain on the tension face of the column is increased until reinforcement in that face reaches yield. At this point, balanced strain conditions exist at a cross section, that is, tension reinforcement reaches strain corresponding to its yield strength f_y just as concrete in compression reaches its assumed ultimate strain of 0.003. Reinforcement on the compressive face of the column may or may not be at compressive yield, depending on column dimensions and strengths of concrete and steel. At balanced conditions, eccentricity is e_b, nominal axial load strength is P_{nb}, and nominal moment strength M_{nb} is at a maximum except for members with very low or no axial load.

4. Between $e_{\min}$ and e_b, failure is initiated by crushing of the concrete. Near $e_{\min}$, reinforcement in the face of the column with the largest compressive strain will be at compressive yield with a stress of f_y. As conditions approach e_b, strain in the face of the column with the smallest compressive strain will reduce to zero and then reverse, reaching tensile yield strain at e_b as noted in the previous paragraph.

5. For large moments when $e > e_b$, failure is initiated by yielding of the tension steel. As the steel continues to elongate, cracks become larger, the neutral

axis shifts toward the compression side, and a secondary failure occurs by crushing of the concrete (as described in Chapter 2, Beam Flexure). The interaction diagram approaches a point of pure moment with no axial load with strength reduction factor $\phi = 0.75$ for spiral columns and $\phi = 0.70$ for tied columns.

6. Because strength reduction factors ϕ for columns are smaller than for flexural members, the ACI Code provides for a smooth transition toward $\phi = 0.9$ for pure flexure when axial load is very small.

7A, 7B. For manual design when graphs, tables, or computers are not available, it is convenient to have simple but conservative interaction curves. The curves shown here are straight lines.

Due to the many variations in dimensions, concrete and steel strengths, and bar arrangements, the interaction curve of a particular column can take one of many different shapes, as shown in Fig. 9.12. The shape of interaction curves for all columns are related, however, to the curve in Fig. 9.3, and all interaction curves have the same features as described above.

Concrete columns are far less likely to buckle than are columns of other materials. Nevertheless, slenderness effects must be considered under some circumstances. In addition to high ratios of length to thickness (or ℓ/r), columns in unbraced frames subject to sidesway require consideration of slenderness. Concrete columns do not buckle in the traditional sense, but stress in slender columns is higher than in short, stout columns. Slenderness of concrete columns* is considered in Section 9.4.

In axially loaded compression members with small or no eccentricity, as loads approach the ultimate, there is a nonelastic redistribution of stresses. Concrete reaches its ultimate stress and continues to deform until it reaches its ultimate strain. If concrete and steel reinforcement are perfectly bonded, strain or deformation of steel is the same as in concrete, and stress in the steel is proportional to strain until the steel reaches its yield point. At this point, the column has reached its ultimate load-carrying capacity. Application of a small additional load will result in further deformation or straining of both concrete and steel and, ultimately, a crushing of concrete and buckling of the bars (see Fig. 9.4(d)).

Crushing of the face shell of concrete outside longitudinal bars usually occurs first. With tied columns, buckling of the bars and crushing of concrete in the core follow almost immediately. In spiral columns, both longitudinal bars and core concrete are heavily restrained laterally. As a result, spiral columns can continue to deform while supporting the same load, that is, they are ductile. Ductility is a desirable structural feature to give warning of impending failure as the maximum load is approached.

---

*"Slender concrete column" may seem a contradiction in terms to anyone who has seen laboratory tests and countless examples from everyday life of skinny metal objects that buckle under load and snap back to their original position when the load is removed. Engineers can best visualize a slender concrete column as one that has increased stresses due solely to its slenderness.

9.3 ANALYSIS OF CROSS SECTION

In a rectangular section, as in Fig. 9.5(a), for small eccentricities of load, the neutral axis lies far outside the section, and concrete reaches ultimate strain throughout the cross section. Summing the axial forces on the section gives the following:

$$P_n = C_c + C_s' + C_s$$
$$P_n = A_{gc}0.85f_c' + A_s'f_s' + A_sf_s \tag{9.1}$$

where A_{gc} = area of the concrete, which equals $A_g - A_{st}$
and A_g = gross area of the section
A_{st} = total area of longitudinal reinforcement.

Taking moments about A_s,

$$P_ne' = C_c\left(d - \frac{h}{2}\right) + C_s'(d - d')$$

$$P_ne' = A_{gc}0.85f_c'\left(d - \frac{h}{2}\right) + A_s'f_s'(d - d') \tag{9.2}$$

Maximum Axial Load

For a section where eccentricity of load equals zero, $e = 0$, strain is the same throughout the cross section and $f_s' = f_s = f_y$. Equation 9.1 then becomes

$$P_o = A_{gc}0.85f_c' + A_{st}f_y \tag{9.3}$$

P_o is shown on the y-axis of Fig. 9.3.

When eccentricity of load is zero, the load passes through the centroid of the section. The centroid coincides with the geometric center if reinforcement is placed symmetrically. For asymmetric reinforcement, where $A_s \neq A_s'$, the load must pass through the *plastic centroid* if zero eccentricity is to be achieved. The plastic centroid is the combined center of gravity of the concrete area times the ultimate concrete stress $(0.85f_c')$ and the center of gravity of the steel area times the steel yield stress. An eccentricity e' from the center of the steel on one face A_s was used above and will continue to be used in derivation for simplicity. Eccentricity e' should not be confused with eccentricity e from the centroid of the section.

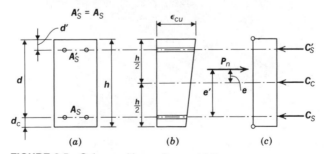

FIGURE 9.5 Column with small eccentricity.

FIGURE 9.4 Test columns after failure under concentric axial load. *(a)* Spiral column with $\ell_u/h = 6.6$; *(b)* spiral column with $\ell_u/h = 9.8$; *(c)* spiral column with thick face shell.

FIGURE 9.4 *(d)* round tied columns with vertical steel $= 0.04\,\boldsymbol{A_g}$. Note spalling of face shell on all columns, confinement of core concrete in spiral columns and, in tied columns, buckling of bars and splitting of core concrete. (Photographs provided by the University of Illinois Engineering Experiment Station.)

Columns are important members of a structural frame because failure of a column is usually more serious than failure of a beam, slab, or other member. Therefore, a larger factor of safety is assigned to columns than to other members. In addition, design procedures must allow for accidental eccentricities, such as those described in Section 9.2. The ACI Code, Section 10.3.5, requires that the design axial load strength $\phi P_{n(\max)}$ of columns must not be taken greater than the following:

$$\phi P_{n(\max)} = 0.85\phi P_o \text{ for spiral and composite columns}$$
$$\phi P_{n(\max)} = 0.80\phi P_o \text{ for tied columns}$$

When these identities and the value of $A_{gc} = A_g - A_{st}$ are substituted into Eq. 9.3, the maximum axial load on spiral and composite columns is given by

$$\phi P_{n(\max)} = 0.85\phi[0.85f'_c(A_g - A_{st}) + f_y A_{st}] \tag{9.4}$$

where $\phi = 0.75$

The maximum load on tied columns is given by

$$\phi P_{n(\max)} = 0.80\phi[0.85f'_c(A_g - A_{st}) + f_y A_{st}] \tag{9.5}$$

where $\phi = 0.70$.
Refer to note 1 in Fig. 9.3.

Minimum Eccentricity

The magnitude of moment a column can resist depends on the amount and placement of reinforcement in the column. Columns with the maximum amount of reinforcement in the tension face generally resist the largest moment. Usually it is not obvious to construction workers which face is the tension face.* Furthermore, columns in double curvature might have tension in two opposite faces, and columns in biaxial bending might have tension in all four faces of the column. For these reasons, longitudinal reinforcement in columns is almost always placed symmetrically about two axes or at least one axis. Common patterns of reinforcement are shown in Fig. 9.6. In general, spiral columns and columns with circular ties [Fig.

*Nor do workers always understand the concept of tension versus compression.

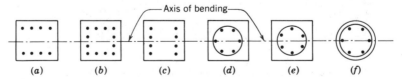

FIGURE 9.6 Typical patterns of column longitudinal reinforcement.

9.6(*d*), 9.6(*e*), 9.6(*f*)] do not resist moment as efficiently as columns with rectangular ties. Furthermore, it should be obvious that columns with reinforcement in faces parallel to the axis of bending [Fig. 9.6(*a*)] will resist moment more efficiently than columns with other bar arrangements. Normally, only those members subjected to relatively large moments and small axial loads are reinforced asymmetrically. For such members, engineers must take care to clearly specify the location of reinforcement.

Referring to Fig. 9.3, note 2, it is clear that a column will resist some moment when the axial load is limited to $\phi P_{n(\max)}$. Under these conditions, the moment divided by the maximum axial load is called the minimum eccentricity

$$\frac{\phi M_n}{\phi P_{n(\max)}} = e_{\min}$$

The section will always resist the maximum axial load eccentric by as much as $e_{\min}$ or less under all conditions permitted by the Code.

For tied rectangular columns with all longitudinal reinforcement placed symmetrically in the two faces parallel to the axis of bending (including four bar columns with one bar in each corner) as in Fig. 9.6(*a*), the minimum eccentricity can be determined approximately by multiplying both sides of Eq. 9.5 by e'/ϕ and setting it equal to Eq. 9.2 for the case of $P_n = P_{n(\max)}$.

$$P_{n(\max)}e' = 0.80[0.85f'_c(A_g - A_{st}) + f_y A_{st}]e'$$

$$= 0.85f'_c(A_g - A_{st})\left(d - \frac{h}{2}\right) + A'_s f'_s(d - d')$$

For a symmetrical section,

$$A'_s = 0.5A_{st}$$

$$\frac{d - d'}{2} = d - \frac{h}{2}$$

and

$$e' = \left(d - \frac{h}{2}\right) + e_{\min}$$

Assuming $f'_s = f_y$, dividing through by $0.85f'_c$, and letting $K = \dfrac{f_y}{0.85f'_c}$

$$0.80[A_{gc} + KA_{st}]\left[\left(d - \frac{h}{2}\right) + e_{\min}\right] = A_{gc}\left(d - \frac{h}{2}\right) + KA_{st}\left(d - \frac{h}{2}\right)$$

dividing through by $(d - h/2)$

$$0.80[A_{gc} + KA_{st}]\left[\frac{1 + e_{\min}}{d - h/2}\right] = A_{gc} + KA_{st}$$

dividing by $[A_{gc} + KA_{st}]$, substituting $(h/2 - d_c)$ for $(d - h/2)$, and solving for e_{min},

$$e_{min} = 0.25 \left(\frac{h}{2} - d_c \right) \tag{9.6}$$

where d_c = concrete cover plus one-half bar diameter (see Fig. 9.5).

At minimum eccentricity e_{min} (note 2 in Fig. 9.3), axial force is not at the center of the cross section, and reinforcement on the face of the column with the smallest compressive strain has not necessarily reached yield. Also, the assumption of $f'_s = f_y$ is not necessarily valid at e_{min} for all cases; thus Eq. 9.6 is not exact. It is conservative except for $d/h \geqslant 0.9$. For example, Eq. 9.6 is not conservative by about 4% when $f'_c = 4$ ksi, $\rho_g = 1\%$, and $d/h = 0.95$. (This latter condition prevails in a 54-in. column with 2 in. cover on #11 bars.)

For rectangular columns with the longitudinal reinforcement uniformly distributed along the sides, as in Fig. 9.6(c), steel near the neutral axis does not reach yield, and the minimum eccentricity can be conservatively approximated by

$$e_{min} = 0.20 \left(\frac{h}{2} - d_c \right) \tag{9.7}$$

For columns with longitudinal reinforcement uniformly distributed in a circular pattern, as in Fig. 9.6(d), 9.6(e), and 9.6(f), most steel is not fully effective in resisting moment, and the minimum eccentricity can be conservatively approximated by

$$e_{min} = 0.14 \left(\frac{h}{2} - d_c \right) \tag{9.8}$$

Table A9.1 gives the minimum eccentricity e_{min}/h as a function of depth for symmetrical rectangular columns with symmetric reinforcement using Eqs. 9.6, 9.7, and 9.8. All three equations are conservative by as much as 50% for normal conditions. The ratio of e_{min} by accurate analysis to e_{min} by approximate analysis is given in Table 9.1.

TABLE 9.1 Ratio of e_{min} by Accurate Analysis to e_{min} by Eqs. 9.6, 9.7, and 9.8

		ρ_g	
γ	Example	1%	8%
0.60	12″ col., #6 bars 14″ col., #11 bars	1.30 to 1.50	1.00 to 1.15
0.70	16″ col., #7 bars 18″ col., #11 bars	1.15 to 1.35	0.97 to 1.10
0.80	27″ col., #11 bars	1.00 to 1.20	0.96 to 1.05
0.90	54″ col., #11 bars	0.94 to 1.08	0.94 to 1.03

Example 9.1 Referring to Example 6.4,

   A. Is the compressive stress of 1.28 ksi in the diagonal strut acceptable? (See comment 2 in Ex. 6.4.)

   B. Is anchorage of 8 #11 bars in 20 × 20-in. compression block of the floor construction satisfactory? (See comment 2 in Ex. 6.4.)

   C. Design the 24-in. square columns at the ends of the transfer girder.

Solution A If floors are 20″ thick, unsupported height of the girder is

$$14 - (2 \times 1.67) = 10.67' \text{ and } \frac{\ell_u}{r} = \frac{10.67}{0.3 \times 1.67} = 21.3$$

Assume that the girder is laterally braced and that slenderness effects do not require a reduction in strength (slenderness is discussed in Section 9.4).

Using Eq. 9.5, the stress permitted on concrete in a tied column is

$$\frac{\phi P_{n(\max)}}{A_{gc}} = 0.80\phi(0.85f_c')$$

$$= 0.8 \times 0.7 \times 0.85 \times 4 = 1.9 \text{ ksi}$$

Actual stress from Ex. 6.4 is 1.28 ksi.   <u>OK</u>

Solution B Eight tension bars are anchored by extending beyond the outer face of the end column (see Fig. 6.17). If the full stress in these bars is to be developed, the resulting force is transmitted by anchorage to the concrete, which then abuts the end of the girder. The resulting compression force in the concrete is resisted by the horizontal component of the diagonal compression strut, as shown by the diagram of forces in Fig. 9.7.

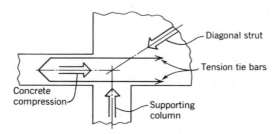

FIGURE 9.7 Diagram of forces in girder at the supporting column.

The force in 8 #11 tension bars is $\phi A_s f_y$

$$0.9 \times 8 \times 1.56 \times 60 = 674 \text{ k}$$

If the bars are anchored in a section of concrete 20″ deep by 20″ wide, the average stress on concrete will be

$$\frac{674}{20 \times 20 - 8 \times 1.56} = 1.74 \text{ ksi}$$

As shown above, stress permitted on concrete is 1.9 ksi, and anchorage of tension bars is assured.

Tension bars occupy $(8 \times 1.56)/(20 \times 20) = 0.031$ of the cross section of the anchor block. In this example, it is apparent that only a few additional tension tie bars could be anchored without enlarging the anchor block, increasing the concrete strength, or taking other precautions.

Solution C Computations and discussion are keyed to the notes in Fig. 9.8. Conservatively assume that the girder web does not contribute to the axial load capacity of the 24″ square column built integrally with the girder web at each end.

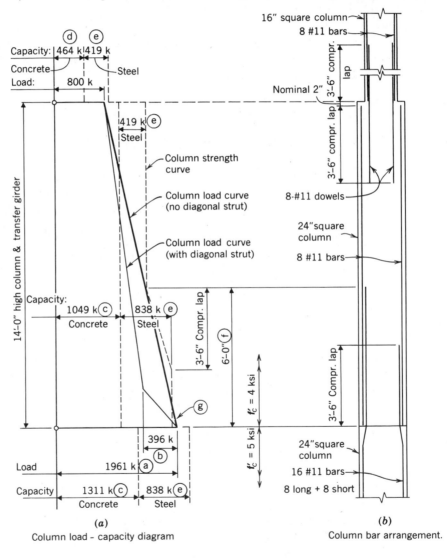

(a)

Column load - capacity diagram

(b)

Column bar arrangement.

FIGURE 9.8 Column at end of transfer girder in Example 9.1.

a. Axial load on column supporting the girder,

Supported columns: 2×800 = 1600 k

Floor load: $28 \times 2 \times 4$ = 224

Girder load: $28 \times 14 \times 0.25 \times 1.4 = \underline{\ \ 137}$

$$\boldsymbol{P_u} = 1961 \text{ k}$$

b. Using the same assumptions as in comment 2 in Ex. 6.4, the vertical reaction of the diagonal strut is

$$800 \left(\frac{214 - 108}{214} \right) = 396 \text{ k}$$

c. Assuming a braced column with capacity unreduced by slenderness effects or by large eccentricity, strength of concrete in the column using Eq. 9.5 is

$$0.80 \times 0.70 \times 0.85 \times 4(24 \times 24 - 16 \times 1.56) = 1049 \text{ k for } \boldsymbol{f_c'} = 4 \text{ ksi}$$

$$= 1311 \text{ k for } \boldsymbol{f_c'} = 5 \text{ ksi}$$

d. Strength of concrete in the 16″ square column is

$$0.80 \times 0.70 \times 0.85 \times 4(16 \times 16 - 8 \times 1.56) = 464 \text{ k}$$

e. Strength of the reinforcing steel in the columns is

$$0.80 \times 0.70 \times 60 \times 8 \times 1.56 = 419 \text{ k for } 8 \text{ \#11 bars}$$

$$= 838 \text{ k for } 16 \text{ \#11 bars.}$$

f. The height to which 16 #11 bars are required can be measured from a sketch, similar to Fig. 9.8, drawn to scale or computed.

$$\text{Height} = \frac{(1961 - 1049 - 419)}{1961 - 800} \times 14 = 5'\text{-}11'' \text{ (say } 6'\text{-}0'')$$

but not less than compression lap length = 3′-6″ for #11 bars.

g. For a short distance (about 1 ft) above the bottom of the girder, the column load is greater than the capacity by $(1961 - 1049 - 838) = 74$ k. If this excess is carried by concrete alone, the maximum stress increase will be about 7% $(74 \times 100/1049)$. As will be seen in Chapter 10, a stress slightly in excess of that allowed on the column is acceptable at the joint in a column. In this example, the concrete floor surrounds and is larger than the column, thus adding to the safety.

Required bar selection and arrangement is shown in Fig. 9.8.

Balanced Strain Conditions

Balanced strain conditions (see note 3 in Fig. 9.3) occur when tension reinforcement just reaches tensile yield strain at the same time that concrete in the compression face of the column reaches ultimate compressive strain. Referring to Fig. 9.9, the depth to neutral axis at balanced strain conditions c_b can be determined by geometry

$$c_b = \frac{\epsilon_{cu}}{\epsilon_y + \epsilon_{cu}} d$$

Substituting 0.003 for ϵ_{cu} and f_y/E_s for ϵ_y and inserting 29,000 ksi for f_y yields

$$c_b = \frac{87}{f_y + 87} d \tag{9.9}$$

By summing axial forces in Fig. 9.9(c),

$$P_{nb} = C_c + C'_s - T$$

Disregarding the reduction in C_c caused by the displacement of concrete in compression by compression reinforcement,

$$P_{nb} = a_b b 0.85 f'_c + A'_s f'_s - A_s f_s \tag{9.10}$$

Taking moments about the tension reinforcement in Fig. 9.9(c)

$$P_{nb}e' = C_c \left(d - \frac{a_b}{2} \right) + C'_s(d - d')$$

$$P_{nb}e' = a_b b 0.85 f'_c \left(d - \frac{a_b}{2} \right) + A'_s f'_s(d - d') \tag{9.11}$$

Assuming a symmetrical section so that $A_s = A'_s$ and assuming that compression steel also reaches yield, solving Eq. 9.10 for a_b gives

$$a_b = \frac{P_{nb}}{0.85 f'_c b} \tag{9.12}$$

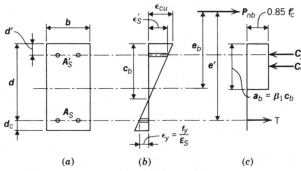

FIGURE 9.9 Column with large eccentricity.

Setting Eq. 9.12 equal to $\beta_1 c_b$ and using the value of c_b from Eq. 9.9,

$$\beta_1 \frac{87}{f_y + 87} d = \frac{P_{nb}}{0.85 f'_c b}$$

Solving for P_{nb},

$$P_{nb} = 0.85 \beta_1 f'_c bd \frac{87}{f_y + 87} \tag{9.13}$$

To obtain the eccentricity at balanced conditions e_b and the moment at balanced conditions M_{nb}, substitute the value of P_{nb} from Eq. 9.10 and the value of a_b from Eq. 9.12 into Eq. 9.11 and solve for e'_b then e_b and M_{nb}. Manual solutions are unwieldy, but exact solutions by computer are available in the form of graphs [1] or tables [2]. To meet Code requirements for safety, the maximum factored axial load is

$$P_u = \phi P_n$$
$$P_u = 0.70 P_n \text{ for tied columns.}$$

Capacity Controlled by Compression

For eccentricities smaller than balanced eccentricity ($e < e_b$), failure is by crushing of concrete, and tension reinforcement has not reached yield point so that $f_s < f_y$. Referring to Fig. 9.9, depth to neutral axis is

$$c = \frac{\epsilon_{cu}}{f_s/E_s + \epsilon_{cu}} d \tag{9.14}$$

Substituting 0.003 for ϵ_{cu}, $c = a/\beta_1$ and $E_s = 29,000$ ksi, and solving for f_s,

$$f_s = 87 \left(\frac{\beta_1}{a/d} - 1 \right) \tag{9.15}$$

If this value of f_s is substituted in Eq. 9.10, and the value of a is eliminated by using Eq. 9.11, a cubic equation for the value of P_n is the result. The value of P_n may be obtained directly by computer or manually by successive approximations. To meet Code requirements for safety, the maximum factored axial load $P_u = \phi P_n$

$$P_u = 0.70 P_n \text{ for tied columns.}$$

Values of P_u for a wide range of variables have been tabulated in the form of graphs or tables as mentioned earlier. In Fig. 9.3, the portion of the curve marked by note 4 indicates the range of eccentricities in which column capacity is controlled by compression.

Capacity Controlled by Tension

For eccentricities larger than balanced eccentricity ($e > e_b$), failure is initiated by yielding of tension steel (see note 5 in Fig. 9.3). As the steel continues to elongate beyond yield, cracks widen and become longer. The neutral axis shifts toward the compression face until a secondary failure occurs by crushing of the concrete.

When tension steel yields, $f_s = f_y$. Assuming compression steel also yields ($f_s' = f_y$), ignoring concrete area displaced by compression steel and assuming a symmetrical cross section and reinforcement, Eqs. 9.10 and 9.11 become

$$P_n = 0.85f_c'ab \qquad (9.16)$$

$$P_n e' = P_n \left(d - \frac{a}{2} \right) + A_s'f_y(d - d') \qquad (9.17)$$

Solving Eq. 9.16 to give the depth of the rectangular stress block

$$a = \frac{P_n}{0.85f_c'b}$$

and then substituting this value into Eq. 9.17 gives a quadratic equation

$$\frac{P_n^2}{2 \times 0.85f_c'b} + (e' - d)P_n - A_s'f_y(d - d') = 0$$

Solving the quadratic equation for P_n gives the ultimate axial load for eccentricities exceeding balanced conditions e_b.

$$P_n = 0.85f_c'bd \left[1 - \frac{e'}{d} + \sqrt{\left(1 - \frac{e'}{d}\right)^2 + 2\rho K_f \left(1 - \frac{d'}{d}\right)} \right] \qquad (9.18)$$

where $\rho = \dfrac{A_s}{bd} = \dfrac{A_s'}{bd}$ and

$$K_f = \frac{f_y}{0.85f_c'}$$

To account for the concrete displaced by compression steel, replace f_y with ($f_y - 0.85f_c'$) in computing K_f or replace K_f with ($K_f - 1$) in Eq. 9.18.

For shallow sections with high-strength steel, it may be necessary to verify that the compression steel has reached yield when the concrete reaches its ultimate strain of $\epsilon_c = 0.003$, as assumed in developing Eq. 9.18. As a guide, compression steel will reach yield unless the section is somewhat shallower than that given by Eq. 2.18, which was developed for the case of bending alone. By making a strain compatibility analysis using a diagram similar to Fig. 9.9(b), determine the strain in compression steel ϵ_s'. If $\epsilon_s'E_s < f_y$ then $f_s' = \epsilon_s'E_s$ should be used in Eqs. 9.10 and 9.11 and the ultimate axial load computed in a manner similar to the development of Eq. 9.18.

To meet Code requirements for safety, the maximum factored axial load $P_u = \phi P_n$

$P_u = 0.70P_n$ for tied columns.

When computing the area of required reinforcement (tension and compression) in beams and other flexural members with small compression axial load

($<0.10f'_cA_g$) it is conservative to neglect the effect of axial load if reinforcement is symmetrical, that is, if $A_s = A'_s$. For asymetrical reinforcement (tension reinforcement with no compression reinforcement or a small amount of compression reinforcement), it is conservative to neglect the effect of axial load when computing the required tension reinforcement. Compression reinforcement is required when axial load exceeds the difference between an axial force of 75% of balanced tension reinforcement and the axial force of reinforcement required for flexure.

$$A'_s f_y = P_u - (0.75A_{sb}f_y - A_s f_y)$$

where $A_{sb} = \rho_b bd$, and ρ_b is given by Eq. 2.13. Table A2.1 tabulates values of $0.75\rho_b$ for common conditions.

Dividing by f_y,

$$A'_s = P_u/f_y + A_s - 0.75A_{sb} \tag{9.19}$$

In beams and other flexural members with tension axial load, reinforcement required to resist the tension force should be added to reinforcement required to resist flexure if one face of the member is always in compression. If the entire cross section of a member is in tension under some loading conditions, tension reinforcement must be distributed throughout the tension zone of the section.

Example 9.2 Given the T-beam shown in Fig. 9.10, what additional reinforcement is required for an axial load $P_u = 85$ k when the tension reinforcement required for flexure only at mid-span is $^+A_s = 4.32^{\square''}$ and at the ends of the span, is $^-A_s = 5.45^{\square''}$? (Negative moment bars are not shown in Fig. 9.10.) Assume $f'_c = 3$ ksi and $f_y = 60$ ksi.

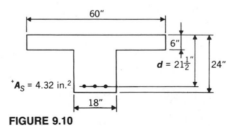

FIGURE 9.10

Solution   Neglecting the contribution of the beam stem below the flange, the maximum compressive force available for positive moment is

$$0.85f'_c h_f b\beta_1 = 0.85 \times 3 \times 6 \times 60 \times 0.85 = 780 \text{ k}$$

and 75% of balanced tension reinforcement is

$$0.75\,^+A_{sb} \geq 0.75 \times \frac{780}{60} = 9.75^{\square''}$$

$$A'_s = \frac{85}{60} + 4.32 - 9.75 < 0$$

Therefore, no compression reinforcement is needed at mid-span. At the ends of the span,

$$0.75^-\boldsymbol{A}_{sb} = 0.016 \times 18 \times 21.5 = 6.19^{\square''}$$

Compression reinforcement required at the ends of the span:

$$\boldsymbol{A}'_s = \frac{85}{60} + 5.45 - 6.19 = 0.68^{\square''}$$

Transition in Strength Reduction Factor

Both nominal axial load capacity and nominal flexural strength of compression members are reduced by an appropriate single value of ϕ, that is, $\phi = 0.75$ for spiral columns and 0.70 for tied columns and other reinforced members. When a member resists bending only, the nominal flexural strength is reduced by $\phi = 0.90$. When a member resists flexure primarily and axial compression is very small, the ACI Code provides a smooth transition between $\phi = 0.90$ and 0.75 or 0.70 (Code 9.3.2.2).

For members in which $\boldsymbol{f}_y$ does not exceed 60 ksi, with symmetric reinforcement, and with $(\boldsymbol{h} - 2\boldsymbol{d}')/\boldsymbol{h}$ not less than 0.70 [e.g., an 18-in.-deep member (or deeper) with 2 in. cover to #11 bars], ϕ may be increased linearly to 0.90 as $\phi\boldsymbol{P}_n$ decreases from $0.10\boldsymbol{f}'_c\boldsymbol{A}_g$ to zero. A large majority of flexural members with small compression will be within these limitations (see Fig. 9.11).

For members outside the limitations of the previous paragraph, ϕ may be increased linearly to 0.90 as $\phi\boldsymbol{P}_n$ decreases from $0.10\boldsymbol{f}'_c\boldsymbol{A}_g$ or $\phi\boldsymbol{P}_{nb}$, whichever is smaller, to zero.

The effect of these provisions is shown by note 6 in Fig. 9.3.

The strength of sections other than symmetrical rectangular sections can be developed in a manner similar to that above, but the calculations are longer and more involved. Such calculations are normally performed only in the development of design aids.

Approximate Interaction Diagrams

When columns are designed manually without the use of graphs, tables, or computers, it is convenient to use an approximate but conservative procedure that is

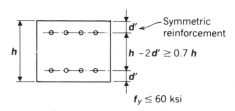

FIGURE 9.11   Conditions permitting linear interpolation of ϕ from $\boldsymbol{P}_u = 0$ to $\boldsymbol{P}_u = 0.10\,\boldsymbol{f}'_c\boldsymbol{A}_g$.

simpler, shorter, and easier to use than the exact procedures discussed above. Two procedures are possible.

The first procedure is to use the maximum axial load capacity $\phi P_{n(\text{max})}$ and the associated moment capacity at minimum eccentricity as the axial load and moment capacity for all eccentricities (note 7A in Fig. 9.3). When used in conjunction with approximate procedures for computing moment capacity (Eqs. 9.6, 9.7, and 9.8) and the ratio of reinforcement to gross area of the column $\rho_g = 0.01$, results are conservative by understating the moment capacity at balanced conditions by about one-third. As the area of reinforcement increases, the procedure becomes more conservative. When the eccentricity is larger than balanced conditions (an uncommon condition for columns), the understatement of moment capacity becomes less. When $\rho_g < 0.015$, $f_c' > 4$ ksi, and $P_u/A_{gc} < 500$ psi, this procedure may be slightly unconservative under some patterns of reinforcement.

The second procedure is to use a straight line interaction diagram from the point of the maximum axial load and the associated moment capacity at minimum eccentricity ($\phi P_{n(\text{max})}$, ϕM_n) to the point of zero axial load and maximum moment capacity ($\phi P_n = 0$, ϕM_n) (note 7B in Fig. 9.3). The procedure is conservative for all conditions if the capacity reduction factor ϕ applied to the moment is the one used for columns ($\phi = 0.75$ or 0.70). If $\phi = 0.90$ is applied to moment, the procedure will be conservative when $\rho_g f_y/f_c' < 0.50$. When $\rho_g f_y/f_c' > 0.50$, the procedure will be conservative, except when ϕP_n is less than about $0.10 f_c' A_g$. This procedure is conservative by about one-third under the most extreme conditions.

Figure 9.12 shows interaction curves for a range of typical conditions encountered in practice. By plotting the approximate interaction curves on Fig. 9.12, engineers can judge the degree by which each procedure is conservative.

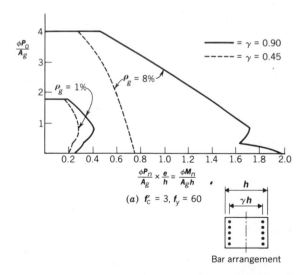

FIGURE 9.12 Typical column interaction diagrams. (———) $\gamma = 0.90$; (– – –) $\gamma = 0.45$. (Courtesy of the American Concrete Institute.)

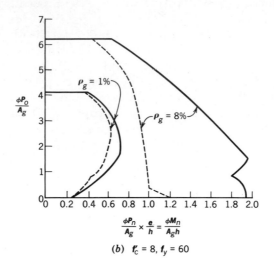

$$\frac{\phi P_n}{A_g} \times \frac{e}{h} = \frac{\phi M_n}{A_g h}$$

(b) $f'_c = 8$, $f_y = 60$

FIGURE 9.12   *(Continued)*

Example 9.3   For a 20-in. square tied column with 8 #11 longitudinal bars arranged symmetrically on all 4 sides, (A) What is the maximum axial load capacity? (B) Using approximate procedures, what moment can be sustained when P_u is maximum? (C) If the axial load $P_u = 1100$ k, what moment is permitted on the column, using approximate procedures? Assume $f'_c = 5$ ksi and $f_y = 60$ ksi.

Solution A $A_g = 400^{□''}$, $A_{st} = 8 \times 1.56 = 12.48^{□''}$
Using Eq. 9.5,

$P_u = 0.80 \times 0.70[0.85 \times 5(400 - 12.48) + 60 \times 12.48] = 1342$ k

Solution B   Using Eq. 9.7,

$$d = 20 - 2 - 0.7 = 17.3'', \qquad h = 20''$$

$$e_{min} = 0.20\left(\frac{20}{2} - 2.7\right) = 1.46''$$

$$M_u = P_u e_{min} = 1.46 \times \frac{1342}{12} \geq 163'k$$

Solution C   For bending alone, assume 3 #11 bars in tension face, $A_s = 4.68^{\square''}$
Using Eq. 2.21, assuming $(\phi f_y j) = 4$, and using $\phi = 0.90$ for flexure only,

$$M_u = 4.68 \times 4 \times 17.3 = 324'\text{k}$$

By straight line interpolation referring to Fig. 9.13,

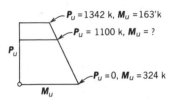

- $P_u = 1342$ k, $M_u = 163'$k
- $P_u = 1100$ k, $M_u = ?$
- $P_u = 0$, $M_u = 324$ k

P_u

M_u

FIGURE 9.13   Interaction diagram.

$$M_u \geqslant 163 + (324 - 163) \times \frac{(1342 - 1100)}{1342} = 192'\text{k}$$

$$\text{Eccentricity } e = \frac{(192 \times 12)}{1100} = 2.1'' \quad \text{or} \quad e/h = 0.10$$

(By a longer and more precise procedure using design aids, $M_u = 315'$k. Engineers must decide for each case whether the extra calculation effort is worth the savings in construction cost.)

Biaxial Bending

The discussion up to this point has concerned columns with bending about one axis only. However, columns with bending about the two principal axes are common, an example being corner columns in a structural frame. Interior columns in frames with an irregular column pattern or with highly variable loading could also experience biaxial bending.

Based on the same assumptions as were used to develop equations for the strength of columns with uniaxial bending (strain compatibility, ultimate concrete strain equal to 0.003, rectangular stress block, etc.), it is possible to calculate the strength of columns in biaxial bending [3]. Because the computations are lengthy and tedious, graphic design aids have been prepared using a computer [1]. The strength of a column in biaxial bending can be visualized as the surface of a solid, somewhat resembling a cone, that passes through the interaction diagram for bending about the X-axis and the interaction diagram for bending about the Y-axis, as shown in Fig. 9.14.

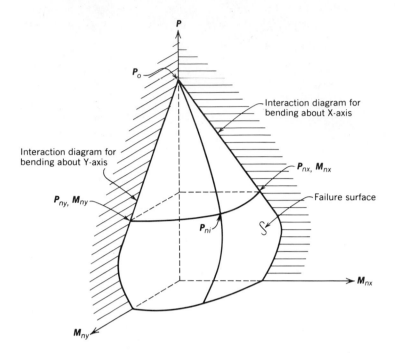

FIGURE 9.14   Interaction diagram for biaxial bending. (Courtesy of the American Concrete Institute.)

Bresler [4] has suggested a relatively simple approximation of biaxial bending in the form

$$\frac{1}{P_{ni}} = \frac{1}{P_{nx}} + \frac{1}{P_{ny}} - \frac{1}{P_o}$$

(9.20)

where P_{ni} = Nominal axial load strength for biaxial bending with eccentricities e_x and e_y

P_{nx} = Nominal axial load strength considering eccentricity about the X-axis e_x only. (e_y = 0),

P_{ny} = Nominal axial load strength considering eccentricity about the Y-axis e_y only. (e_x = 0), and

P_o = Nominal axial load strength at zero eccentricity (see Eq. 9.3).

All terms in Eq. 9.20 may be divided by ϕ to use the design strengths resisting factored loads. Based on his computer studies, Bresler reports that Eq. 9.20 "is in excellent agreement with calculated theoretical values and with test results, the maximum deviation being 9.4%, and average deviation being 3.3%." Conservative engineers may wish to add up to 9% to the axial load capacity, especially for columns with heavy biaxial bending, to account for potentially unconservative results when using Eq. 9.20.

The limiting condition for Eq. 9.20 is bending about one axis only. For example, if $e_y = 0$, then $P_{ny} = P_o$ and Eq. 9.20 reduces to $P_{ni} = P_{nx}$, as it should. Fig. A9.1 presents a graphic solution to Eq. 9.20. A study of this figure will show that small eccentricities in one direction do not result in a proportional reduction in moment strength in the other direction. When the numerical sum of eccentricities in the two directions does not exceed the minimum eccentricity in either direction, no reduction in axial load strength is necessary.

Equation 9.20 is probably not reliable when $P_n \leq 0.20 P_o$. In such cases dominated by bending, if the reinforcement is placed symmetrically, it is conservative and satisfactory to ignore the axial load and to design the section for biaxial bending only, because the axial load has the effect of reducing the tensile force in the steel. Where moment predominates, most sections will be reinforced on the tension faces only. In such cases, engineers should compute compressive stresses in concrete and add compression reinforcement in the corner with the highest compressive stress if the sum of the moment resisting capacities R_c for the two directions approaches or exceeds that permitted on concrete (See Eq. 2.27 and the discussion on uniaxial bending in Section 9.3, under Capacity Controlled by Tension.)

Example 9.4 Consider a 12×16-in. tied column with 6 #11 longitudinal bars, which resists a moment about the strong axis of $M_{ux} = 150'k$ and a moment about the weak axis of $M_{uy} = 50'k$. What is the axial load capacity? Assume $f'_c = 4$ ksi and $f_y = 60$ ksi.

Solution $A_g = 12 \times 16 = 192^{□''}$

$$A_{st} = 6 \times 1.56 = 9.36^{□''}$$

Using Eq. 9.3,

$$\phi P_o = 0.7[(192 - 9.36)0.85 \times 4 + 9.36 \times 60] = 828 \text{ k}$$

About the strong axis of the column,

$$\frac{\phi M_{nx}}{A_g h} = \frac{150 \times 12}{192 \times 16} = 0.586$$

From the graphs in Fig. 9.15, when $\rho = \dfrac{9.36}{192} = 0.049$

$$\frac{\phi P_{nx}}{A_g} = 2.35 \text{ ksi when } \gamma = 0.60$$

$$= 2.65 \text{ ksi when } \gamma = 0.75$$

where γ is the ratio of the distance between reinforcing bars and total thickness of the column.

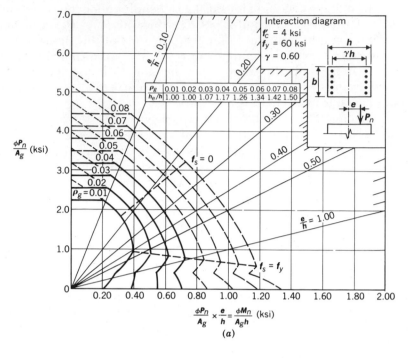

(a)

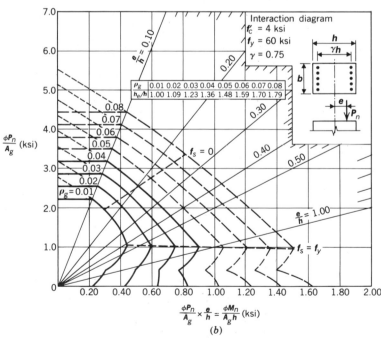

(b)

FIGURE 9.15 Column interaction diagrams from Ref. 1. (Courtesy of the American Concrete Institute.)

By interpolation when $\gamma = \dfrac{16 - 2.7 \times 2}{16} = 0.66$

$$\frac{\phi P_{nx}}{A_g} = 2.47$$

$\phi P_{nx} = 2.47 \times 192 = 474$ k

About the weak axis, when $e < e_{min}$,

$$\phi P_{ny} = 0.8\phi P_o = 0.8 \times 828 = 662 \text{ k}$$

From Fig. 9.15(*a*), $\quad \dfrac{\phi M_n}{A_g h} = 0.28$

from which $\phi M_n = \dfrac{0.28 \times 192 \times 12}{12} = 53'\text{k} > 50'\text{k}$

For biaxial bending, using Eq. 9.20,

$$\frac{1}{\phi P_{ni}} = \frac{1}{474} + \frac{1}{662} - \frac{1}{828} = 0.002412$$

$\phi P_{ni} = 414$ k

9.4 SLENDERNESS EFFECTS

A slender column is a long column that has a small cross section with respect to its length so that its slenderness reduces its strength. A slender column fails by buckling at an axial load less than the axial load capacity of a short column. (The latter was discussed in Section 9.3.) Design of a concrete column for slenderness effects is similar to design of columns constructed of other materials. The measure of slenderness is ℓ_u/r, where ℓ_u is the unbraced length of the column and r is the radius of gyration $= \sqrt{I/A}$. For simplicity, r may be taken as $0.30h$ for square and rectangular concrete columns and $0.25h$ for round concrete columns, where h is the overall column thickness. The presence of reinforcing steel will usually increase the radius of gyration a little but can be conservatively disregarded. For rectangular columns, the largest ℓ_u/r of the two axes should be used.

In addition to slenderness, the ultimate buckling load depends on the amount of rotational restraint at each end of the column and the amount of lateral movement of one end with respect to the other end that is permitted. This can best be visualized by reference to Fig. 9.16 for typical idealized and practical conditions.

In Fig. 9.16, Cases I, II, V, and VI are idealized conditions that never occur in actual structures. Ends of columns are never completely free to rotate nor are they ever fully fixed so that no rotation can occur. These cases are useful theoretical conditions against which an engineer can compare actual column conditions. A braced column is one in which lateral movement of the top with respect to the bottom

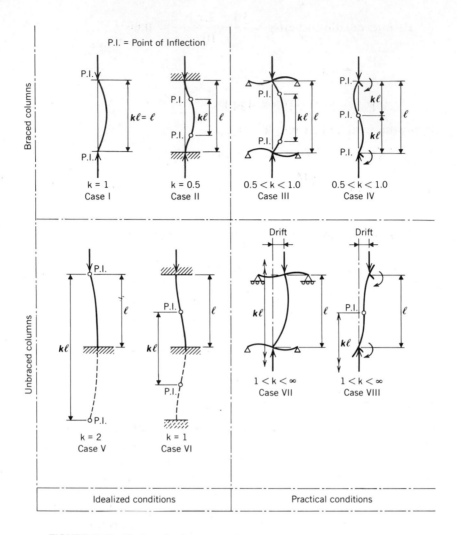

FIGURE 9.16   Modes of column curvature.

is restrained. Likewise, an unbraced column is one in which lateral movement is permitted.

Case I is an idealized braced column with pinned ends so that no restraining moment is introduced. All other column cases are referenced to this one.

Case II is an idealized braced column with fully fixed ends. A slender column may buckle in either direction but will be in double curvature so that two points of inflection will occur within its length. The slenderness $k\ell/r$ is exactly one-half that of Case I.

Case III is a practical braced column. Framing members at each end of the column do not introduce moments into the column but do provide restraint against rotation of the ends of the column. Such conditions exist with uniformly loaded symmetrical floor framing around a column and with a footing concentric with a column. Slenderness $k\ell/r$ will be between that of Cases I and II, depending on the amount of rotational restraint at the ends of the column.

Case IV is a practical braced column where floor framing at both ends of the column introduces moment into the column such that it is bent in double curvature and one point of inflection occurs between the ends. Slenderness $k\ell/r$ will be between that of Cases I and II, depending on the amount of rotation permitted at the ends of the column.

Case V is an idealized unbraced column. The top has no rotational or lateral restraint, but the bottom is fully fixed. Slenderness $k\ell/r$ is just double that of Case I. Case V is similar to a flagpole, but slenderness of a real "flagpole" column should be taken as somewhat more than twice its length due to imperfect fixity of the bottom end.

Case VI is an idealized unbraced column with both ends fully fixed against rotation but with no lateral restraint at the top. Slenderness $k\ell/r$ is equal to that of Case I.

Case VII is a practical unbraced column similar to Case III, except for lack of bracing. No point of inflection occurs within the column, which is bent in single curvature. Slenderness $k\ell/r$ is greater than for Case I, depending on the amount of rotation permitted at the ends of the column. If no rotational restraint is provided, $k\ell/r$ = infinity and the column is unstable. It would fall even under no load.

Case VIII is a practical unbraced column similar to Case IV, except for lack of bracing. This is the common case of an edge column in an unbraced frame. Slenderness is always greater than for Case I.

The ACI Code recognizes three methods for taking slenderness into account in the design of columns. A brief description of each method and the conditions under which each should be used follows.* More detailed descriptions are given later.

1. Strength Reduction or R Method   When the strength of a column is reduced by its slenderness, both axial load and moment capacity are reduced, depending on ℓ_u/r, amount of eccentricity, and curvature in braced columns. In addition, for unbraced columns, the reduction depends on duration of loading and relative stiffnesses of members framing into the column at each end. Design length ℓ_u' may be longer than the real unsupported length, depending on the stiffnesses of columns and other framing members. The R Method is approximate, empirical, relatively easy to use, and should be considered wherever applicable. It is limited to $\ell_u/r \le 100$ for braced columns and $\ell_u'/r \le 40$ for unbraced columns. It should not be used for columns hinged at both ends, that is, columns approaching Case I in Fig. 9.16. The R Method may be overly conservative if the column is relatively slender, has large eccentricities, has loads of long duration, has high concrete and steel strengths, has a low steel percentage, and is a component of an unbraced frame.

---

*The choice between methods can be viewed as the choice between a straightforward, unsophisticated, "quick and dirty" method versus an enlightened, refined, longer method versus an elegant, erudite but painfully slow method. Sometimes, all three methods yield essentially the same results. The engineer's skill is in deciding when to be sophisticated and when to be artless.

2. Moment Magnifier Method   This method is permitted as an approximate method under ACI Code Section 10.11 and is described in detail in the ACI Code Commentary [5]. The moments computed by an ordinary frame analysis are multiplied by a "moment magnifier," which is a function of the factored axial load P_u and the critical buckling load P_c for the column. This design method embodies some of the design provisions of the working stress procedure for steel beam-columns included in the AISC specifications for structural steel for buildings [6]. The Moment Magnifier Method is more rational and more accurate than the R Method but still uses some simplifying assumptions. It should be used if the R Method is not applicable or if the R Method results in large reductions in capacity and there is a financial incentive to reduce construction costs because many columns will be built from the same design. The Moment Magnifier Method may not be used when ℓ_u/r or ℓ_u'/r exceeds 100.

3. Second-Order Analysis Method or P-Δ Method   Design of columns is based on forces and moments determined from analysis of the structure. Such analysis must take into account the influence of axial loads and variable moment of inertia on member stiffness and fixed-end moments, the effect of deflections on moments and forces, and the effects of duration of loads. Generally, the moments from a **P-Δ** analysis are a better approximation to the real moments than those of the Moment Magnifier Method. For sway frames or lightly braced frames, economies may be achieved by the use of **P-Δ** analyses. A **P-Δ** analysis is required when ℓ_u/r or ℓ_u'/r exceeds 100. Such conditions rarely occur in the practice of most engineers.

Strength Reduction Method

The first step in application of the strength reduction or R Method is to determine whether the column is braced. (See ACI Code Commentary, Section 10.11.) A column will be braced against sidesway if the horizontal displacements within a story do not significantly affect the moments in the structure.

The ACI Code Committee states in the Commentary to the Code that "A column may be assumed braced if located in a story in which the bracing elements (shearwalls, shear trusses, or other types of lateral bracing) have a total stiffness, resisting lateral movement of the story, at least six times the sum of the stiffnesses of all the columns within the story. With this amount of lateral stiffness, lateral deflections of the story will not be large enough to affect the column strength significantly."

When shearwalls provide the bracing for columns,

$$\Sigma(EI/\ell)_{\text{wall}} \geqq 6\Sigma(EI/\ell)_{\text{col}} \tag{9.21}$$

Implicit in the Code Commentary statement and Eq. 9.21 are the following assumptions.

1. Shearwalls are designed to resist nearly all lateral loads applied to the structure. Because lateral loads will be resisted by the various members in proportion to their relative stiffnesses, shearwalls will resist at least six-sevenths of the lateral load.

In some cases, this requirement may be more severe than the requirement to meet Eq. 9.21.

2. Foundations for shearwalls are at least as stiff as the minimum required stiffness for the shearwalls. That is, the lateral drift at the top of the shearwalls due to rotation of their foundations is no more than the lateral drift due to flexure, as in Fig. 9.17. This limit is equivalent to a portal frame in which the lateral drift due to rotation of joints is no more than the lateral drift due to flexure in columns. Joint rotation is a function of relative stiffnesses of beam members in the portal frame. Foundation stiffness may be assured by anchoring the shearwall to a basement wall with a stiffness at least equal to shearwall stiffness or by using rigid foundations on stiff foundation soils. Foundation stiffness may also be provided by a moment couple between a first floor slab at grade and a basement floor slab or foundation.

3. Actual stiffnesses **EI** are used in Eq. 9.21. For simplicity, uncracked flexural stiffnesses EI_g may be used in Eq. 9.21 if shearwalls remain uncracked under factored loads. Alternatively, it is conservative to use the cracked flexural stiffness EI_{cr} for shearwalls and the uncracked flexural stiffness EI_g for columns.

4. Shearwalls are located and designed to resist twisting of the structure about a vertical axis, as shown in Fig. 9.18. Normally this will require at least three shearwalls, two of which should be equal in size and symmetrically placed.

5. Overall height of the structure and shearwalls is not more than 75 ft. In taller structures, interaction between shearwall and frame will become more pronounced, and significant moments in columns will occur even though stiffnesses of shearwalls meet Eq. 9.21.

The above discussion on shearwalls applies equally well to vertical trusses and other forms of bracing. Only the method of computing stiffness will differ. For example, shear deformations should be considered in trusses but may be neglected in walls.

Columns that are essentially unrestrained at the ends, as in Case I in Fig. 9.16, cannot be designed using the R Method as the results are not conservative for this case.

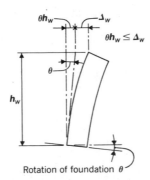

FIGURE 9.17   Lateral drift of shear wall.

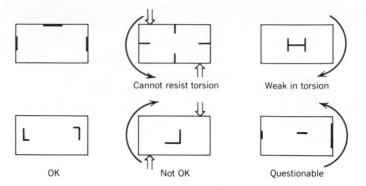

FIGURE 9.18   Plan of buildings braced by shear walls.

Example 9.5   In a building with a floor framing plan as shown in Fig. 9.19, what size shearwalls are required to assure bracing of the columns so that a braced frame may be assumed? Interior columns are 18 in. square and exterior columns are 12 × 18 in. Clear story height is 10 ft.

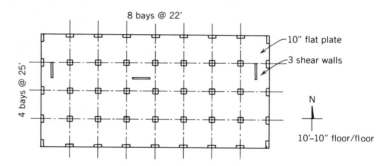

FIGURE 9.19

Solution

In Eq. 9.21, assume E_c and ℓ are the same for both columns and walls, so that $\Sigma I_{wall} \geqslant 6\Sigma I_{col}$.

In the North-South direction,

Interior columns $I = \left(\dfrac{18 \times 18^3}{12}\right) \times 21 =$ 183,700 in.^{4}

End columns $I = \left(\dfrac{12 \times 18^3}{12}\right) \times 10 =$ 58,300

Side columns $I = \left(\dfrac{18 \times 12^3}{12}\right) \times 14 =$ $\underline{36,300}$

$\Sigma I_{col} =$ 278,300 in.^{4}

Minimum $\Sigma I_{wall} = 278{,}300 \times 6 = 1{,}670{,}000$ in.^{4}

Try 2 walls, 8″ thick: $I = \dfrac{\Sigma h \ell_w^3}{12}$ or $\ell_w = \sqrt[3]{\dfrac{12 I}{\Sigma h}}$

Required length of walls $= \sqrt[3]{\dfrac{12 \times 1{,}670{,}000}{8 \times 2}} = 107.8''$

say 9′-0″ long, minimum.

In the East-West Direction,

Interior columns, same as N-S direction $=$ 183,700 in.^{4}

End columns $I = \left(\dfrac{18 \times 12^3}{12}\right) \times 10 =$ 25,900 in.^{4}

Side columns $I = \left(\dfrac{12 \times 18^3}{12}\right) \times 14 = $ $\underline{81{,}600}$

$\Sigma I_{col} =$ 291,200 in.^{4}

$\Sigma I_{wall} = 6 \times 291{,}200 = 1{,}750{,}000$ in.^{4}

Try one wall, 8″ thick:

Required length of wall $= \sqrt[3]{\dfrac{12 \times 1{,}750{,}000}{8}} = 138''$

say 11′-6″ long, minimum.

Foundations for shearwalls must also be checked for stiffness and the walls checked for flexural cracking. In this building, assume that the walls are cracked and that the cracked stiffness is 1/4 of the uncracked stiffness.

In the N-S direction,

$\Sigma I_{wall} = \dfrac{1}{4} \times \dfrac{\Sigma h \ell_w^3}{12}$ or $\ell_w = \sqrt[3]{\dfrac{48 I}{\Sigma h}}$

Required length of 8″ walls $= \sqrt[3]{\dfrac{48 \times 1{,}670{,}000}{8 \times 2}} = 171''$

say 14′-3″ long.

Required length of 12″ walls $= \sqrt[3]{\dfrac{48 \times 1{,}670{,}000}{12 \times 2}} = 150''$

say 12′-6″ long

Required length of walls in E-W direction is computed in the same manner.

If necessary, stresses in the shearwalls and their actual flexural stiffness can be computed. The minimum length might then fall between the lengths computed above for uncracked and cracked flexural stiffnesses (9'-0" and 14'-3" for two 8" walls). This would require an iterative procedure.

Braced Columns If a column is braced and the end conditions are such that the column is bent in double curvature (one or two points of inflection between ends of the column, as in Cases II and IV of Fig. 9.16), the capacity reduction factor **R**, applied to both axial load and moment, is

$$R = 1.32 - 0.006 \frac{\ell_u}{r} \leq 1.0 \tag{9.22}$$

When $\ell_u/r < 54$, no reduction in capacity need be made. When floor members do not introduce moment into the column, it will be bent in buckling, as shown in Case III of Fig. 9.16. If floor members are sufficiently stiff so that the column does not approach unrestrained buckling, as in Case I, Eq. 9.22 may be used. As a guide, the ratio of the sum of stiffnesses of columns to the sum of stiffnesses of floor members at each end of the column should not exceed unity. (See the discussion for unbraced columns below.)

If a column is braced and the end conditions are such that the column is bent in single curvature (no inflection points between ends of the column), and if the eccentricity at either end does not exceed 0.10**h** (where **h** is the overall thickness of the column), the capacity reduction factor **R**, applied to both axial load and moment capacity, is

$$R = 1.23 - 0.008 \frac{\ell_u}{r} \leq 1.0 \tag{9.23}$$

When $\ell_u/r < 29$, no reduction in capacity need be made. Under the same conditions but with an eccentricity exceeding 0.10**h**, the capacity reduction factor **R**, applied to both axial load and moment capacity, is

$$R = 1.07 - 0.008 \frac{\ell_u}{r} \leq 1.0 \tag{9.24}$$

A braced column bent in single curvature would look something like Case III in Fig. 9.16 but without inflection points. This condition would occur when members framing into the column at one end impart moment to the column in opposite rotation from the moment imparted by members framing into the column at the other end.

For example, it is conservative to assume a one-story portal frame, as in Fig. 9.20(*a*), would be bent in single curvature even though the foundation would offer some restraint and theoretically force the column into double curvature. As a guide, double curvature should only be considered if the inflection point is within the middle half of the column length. Marginal cases should be designed by the Moment Magnifier Method if the assumption of single curvature is too conservative.

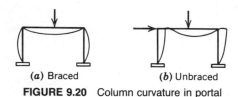

(a) Braced   *(b)* Unbraced

FIGURE 9.20 Column curvature in portal frames.

For a braced column, when axial load is less than $0.10f'_cA_g$, the capacity reduction factor may be increased linearly to **R** = 1 for flexure without axial load in the same manner as described in Section 9.3 for the axial load-moment interaction diagram (see description of note 6 in Fig. 9.3).

Unbraced Columns If a column is unbraced, it is necessary to first compute the stiffnesses of all columns and flexural members framing into the joint at each end of the column. Stiffness **K** is computed in the conventional manner where $K = EI/\ell$ and ℓ is the center-to-center span or length. In the direction for which moments are being considered, the ratio of the sum of stiffnesses of all columns framing into a joint to the sum of all stiffness of flexural members framing into a joint is computed for each end of the column and the average is taken to obtain $\alpha_m = \Sigma K_{col}/\Sigma K_{flex}$. Using α_m, compute the effective unsupported length ℓ'_u from

$$\ell'_u = (0.78 + 0.22\alpha_m)\ell_u \geq \ell_u \qquad (9.25)$$

Equation 9.25 is a method for increasing the effective length of columns based on the restraint at the ends of the columns, where ℓ'_u is the equivalent of $k\ell$ in Fig. 9.16. For unbraced columns, the effective unsupported length ℓ'_u should be used for all subsequent calculations on slenderness of the column.

When $\ell'_u/r > 40$, the column cannot be designed using the R Method. The engineer should then use the Moment Magnifier Method or a Second-Order Analysis.

The negative moment reinforcement ratio must be at least $^-\rho = 0.01$ in all flexural members bracing the column at each end. This will assure flexural members sufficiently stiff to brace the column as assumed by the R Method. (Referring to Chapter 8, Deflection, members lightly reinforced to resist moment are more likely to crack and will have a lower flexural stiffness, once cracked.) In most cases, this minimum reinforcement will be met. If $^-\rho < 0.01$ is provided, columns cannot be designed by the R Method. Alternatively, the R Method can be used and the negative reinforcement in bracing members increased to the minimum $^-\rho = 0.01$. In the latter case, negative moment reinforcement should be extended into bracing members as it would be if required by analysis.

Within the above limitations, where the design of an unbraced column is governed by lateral loads of short duration, such as wind and earthquake, the capacity reduction factor is given by Eq. 9.24, substituting ℓ'_c for ℓ_u. Single curvature [Fig. 9.20(*b*)] versus double curvature (Fig. 9.16, Case VIII) is not considered for unbraced frames because sidesway increases moments significantly in either case.

Where the design of an unbraced column is governed by lateral loads, gravity loads, or other loads of long duration, the capacity reduction factor **R**, applied to both axial load and moment capacity, is

$$\mathbf{R} = 0.97 - 0.008 \, \frac{\ell'_u}{r} \leq 1.0 \tag{9.26}$$

Design for loads of long duration is more conservative because creep increases lateral drift and therefore the secondary moments.

In borderline cases, it may be necessary to check slender unbraced columns for both conditions, that is, for lateral loads of short duration using Eq. 9.24 and gravity loads of long duration using Eq. 9.26.

A flowchart for using the R Method is presented in Fig. 9.21 to help visualize the procedure. Equations 9.22, 9.23, 9.24, and 9.26 are shown graphically in Fig. A9.2.

For design, both the axial load and moment obtained by conventional elastic analysis are increased by dividing them by the appropriate **R** factor. The increased load and moment are then used to design the column, following the procedures discussed in Section 9.3.

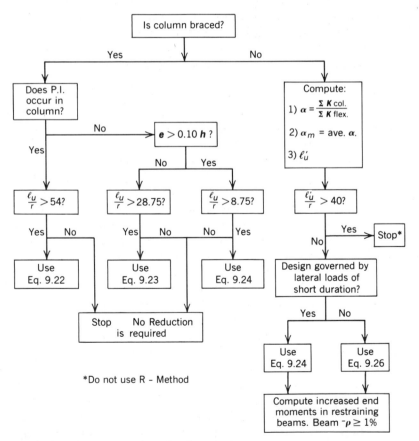

FIGURE 9.21 Flowchart for using R method.

When using the R Method with unbraced columns, it is necessary to increase moments in members bracing the column to account for the secondary moments caused by sidesway. These secondary moments also occur in the column but are accounted for indirectly by the capacity reduction factor. In Fig. 9.22, subscripts ℓ and s refer to capacities of long and short columns, respectively. By similar triangles, the intercept on the line of eccentricity for the short column,

$$RM_s = P_\ell e_s$$

and

$$M_\ell = M_s \times \frac{P_o - P_\ell}{P_o - P_s}$$

Substituting $M_s = P_\ell e_s/R$ and remember that $R = P_\ell/P_s$ as previously defined, and simplifying, gives

$$M_\ell = P_\ell e_s \left(\frac{1 - P_\ell/P_o}{R - P_\ell/P_o} \right) \tag{9.27}$$

Equation 9.27 is shown graphically in Fig. A9.3.

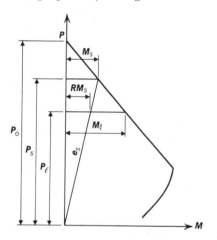

FIGURE 9.22 Approximation for increased moments in members bracing a column against side sway. (Courtesy of the American Concrete Institute.)

Example 9.6 If the structure in Ex. 9.5 is modified by removing shearwalls, changing all columns to 18 in. square and adding 20 × 20-in. spandrel beams on all four sides, what reduction in strength of the columns is necessary to allow for slenderness effects? What moment in North-South direction should be used in design of an interior column-slab connection? Factored axial load on an interior column is 940 k, and factored moment to be transferred between an interior column and a slab due to wind in the North-South direction is 58′k and due to unbalanced live load is 72′k. Select vertical reinforcement for interior columns. Assume $f'_c = 5$ ksi and $f_y = 60$ ksi.

Solution

All columns must deflect together because they are tied together with rigid floor slabs. Therefore, assume slenderness can be evaluated by taking the sum of stiffnesses of all columns divided by the sum of stiffnesses of all flexural members framing into all columns in the direction for which slenderness is being considered.

$$\Sigma I_{col} = \left(\frac{18 \times 18^3}{12}\right) \times 45 = 394{,}000 \text{ in.}^4$$

Because concrete strength is the same for both columns and slabs, the modulus of elasticity will also be the same and E_c can be eliminated from the computations; thus, the sum of column stiffnesses above and below a floor is

$$\Sigma K_{col} = \frac{2\Sigma I_{cols}}{\ell} = \frac{2 \times 394{,}000}{12 \times 10.83} = 6060 \text{ in.}^3$$

The entire width of the slab is not effective in bracing the column because most of the slab does not frame directly into the column. Ratios of effective width to total width of 25% to 50% are commonly assumed. Conservatively assume that 25% of the width of the slab is effective.

$$\Sigma K_{slabs} = \frac{\Sigma I}{\ell} = \frac{0.25(22 \times 12) \times 10^3(8 \times 4 \times 2)}{12 \times (25 \times 12)} = 1170 \text{ in.}^3$$

where $(8 \times 4 \times 2) = 8$ bays wide by 4 bays long by 2 ends per bay.
For spandrel beams at ends,

$$\Sigma K_{beams} = \frac{\Sigma I}{\ell} = \frac{(20 \times 20^3/12) \times (2 \times 4 \times 2)}{12 \times 10.83} = \underline{1640 \text{ in.}^3}$$

$$\Sigma K_{floor} = 2810 \text{ in.}^3$$

Assuming the same stiffness conditions at each end of each column, the average ratio of ΣK_{col} to ΣK_{slab} at the two ends of the columns, α_m, is

$$\alpha_m = \frac{\Sigma K_{col}}{\Sigma K_{slab}} = \frac{6060}{2810} = 2.16 > 1.0$$

Using Eq. 9.25,

$$\ell'_u = (0.78 + 0.22 \times 2.16)10 \times 12 = 150''$$

and $\dfrac{\ell'_u}{r} = \dfrac{150}{0.3 \times 18} = 27.9$

Because ℓ'_u/r does not exceed 40, use of the R Method is acceptable.

Lateral loads due to wind are of short duration only. Assume that gravity live loads are placed symmetrically and therefore do not cause significant sidesway.

Using Eq. 9.24,

$$R = 1.07 - 0.008 \times 27.9 = 0.84$$

For an interior column, neglecting slenderness effects, design axial load $P_u = 940$ k and design moment $M_u = 58 + 72 = 130'$k from given data.

As a long column,

$$\text{Design axial load } P_\ell = \frac{P_u}{R} = \frac{940}{0.84} = 1119 \text{ k}$$

$$\text{and Design moment } M_\ell = \frac{M_u}{R} = \frac{130}{0.84} = 155'\text{k}$$

$$\text{Eccentricity } e_s = \frac{130 \times 12}{940} = 1.66''$$

From Eq. 9.3, assuming 12 #10 longitudinal bars ($A_s = 15.24^{□''}$),

$$P_o = 0.85 \times 5 \times (18^2 - 15.24) + 15.24 \times 60 = 1312 + 914 = 2226 \text{ k}$$

Column capacity $\phi P_n = 0.8 \times 0.7 P_o = 1246 \text{ k} > 1119 \text{ k}$ <u>OK</u>

$$\text{min } e \approx 0.25 \left(\frac{h}{2} - d_c \right) = 0.25 \left(\frac{18}{2} - 2.6 \right) = 1.6''$$

Because actual eccentricity $e = 1.66''$ is less than 4% more than min $e = 1.6''$ computed by simplified and slightly conservative procedures, column moment is acceptable.

The minimum negative moment reinforcement in the slab strip centered on the column with a width of 25% of the panel width is

$$0.01(22 \times 12 \times 0.25)8.5 = 5.61^{□''}$$

The slab must be designed for a moment that is greater than that produced by factored dead, live, and wind loads. To evaluate this moment, use Eq. 9.27 to compute the moment to be transferred between column and slab M_ℓ

$$\text{Slab } M_\ell = \frac{1119 \times 1.66}{12} \times \frac{1 - 1119/2226}{0.84 - 1119/2226} = 155 \times 1.47 = 229'\text{k}$$

The factor 1.47 can be checked by referring to Fig. A9.3.

The increased moment M_ℓ should be used in computing shear stress in the slab around the column or column capital.

Because the reduction in capacity is 0.16 $(1 - 0.84)$ for interior columns, the potential for construction cost savings is significant.

Assume that a more refined analysis results in a reduction in column strength of only 8% (one half of 16%).

$$\text{Design axial load } P_\ell = \frac{940}{0.92} = 1022 \text{ k}$$

$$\text{Concrete capacity} = 0.8 \times 0.7 \times 1312 = \underline{735}$$

$$\text{Required capacity of reinforcment} = 287 \text{ k}$$

$$\mathbf{A}_s\text{reqd} = \frac{287}{0.8 \times 0.7 \times 60} = 8.55^{\square\prime\prime}$$

Potentially use 6 #11 bars, $\mathbf{A}_s$prov. $= 6 \times 1.56 = 9.36^{\square\prime\prime}$

Bar area of $(15.24 - 9.36)$ $5.88^{\square\prime\prime}$ per column would be saved with a length $= (12'\text{-}10'') + (4'\text{-}8'') = 17'\text{-}6''$.

Total savings in one story of interior columns

$$= 21 \text{ columns} \times 5.88^{\square\prime\prime} \times 3.4 \text{ plf} \times 17.5' \times \$0.25/\text{lb.} = \$1836$$

Using Eq. 1.7, the maximum time an engineer should spend to accomplish this amount of savings is

$$\mathbf{T} = \frac{1836 \times 0.5}{0.50 \times 2 \times 3} = 306 \text{ minutes} \simeq 5 \text{ hours.}$$

Furthermore, the arbitrary assumption that slenderness of all columns may be analyzed as a unit is not necessarily conservative. If interior columns had been analyzed separately,

$$\alpha_m = \frac{\Sigma \mathbf{K}_{\text{col}}}{\Sigma \mathbf{K}_{\text{slab}}} = \frac{6060/45}{1170/32} = 3.68$$

$$\ell'_u = (0.78 + 0.22 \times 3.68) \ 120 = 191''$$

$$\frac{\ell'_u}{\mathbf{r}} = \frac{191}{0.3 \times 18} = 35.3$$

$$\mathbf{R} = 1.07 - 0.008 \times 35.3 = 0.79$$

$$\text{and } \mathbf{P}_\ell = \frac{940}{0.79} = 1194 \text{ k}$$

$$\mathbf{M}_\ell = \frac{130}{0.79} = 165' \text{ k}$$

Twelve #10 longitudinal bars would still be satisfactory.

For the above reasons, analysis of column slenderness by the Moment Magnifier Method (see Ex. 9.7) or by Second-Order Analysis might be desirable.

Moment Magnifier Method

In actual practice, many engineers will use the R Method more often than the Moment Magnifier Method because the first method requires fewer calculations with less chance for error, as well as saving the engineer's valuable time. Furthermore, the majority of columns require little if any increase in strength to allow for slenderness effects.

In the Moment Magnifier Method, the moments in all frame members, columns and beams, computed by a conventional frame analysis (first-order analysis), are increased or magnified. The amount of increase approximates the moments caused by axial load on the columns times the maximum deflections at either end or mid-height of the columns, as shown in Fig. 9.23. Because creep will increase the deflection caused by loads of long duration, the amount of increase is more for moments caused by loads of long duration than for moments of short duration. The amount of increase becomes larger as the load on the columns approaches Euler's critical buckling load for an elastic column.

Use of the Moment Magnifier Method is explained by the following step-by-step procedure. For a detailed explanation, see Ref. 7.

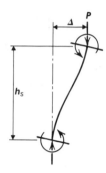

FIGURE 9.23   *P-Δ* column moments.

Step 1   Compute frame moments and axial loads by conventional elastic frame analysis. Considering a column, at the end with the larger moment, divide the moment into moments M_{2s} due to loads that result in appreciable sidesway and moments M_{2b} due to loads that result in no appreciable sidesway. M_{2s} will usually be caused by lateral loads, and M_{2b} will usually be caused by gravity loads, but extremely asymmetrical frames or gravity loads can also cause appreciable sidesway in some circumstances. Similarly, for each column, at the end with smaller moment, divide the moment into M_{1s} and M_{1b}.

When computed moments are small, the ACI Code requires that a minimum moment be used to evaluate column slenderness. When the column is braced, M_{2b} must be at least equal to the moment computed using an eccentricity of $(0.6 + 0.03h)$ in. The value of h in inches is the thickness of the column in the direction for which moments are being determined. When the column is unbraced, the value of M_{2b} (for loads not causing sidesway) may be the computed value, but M_{2s} (for loads that do cause sidesway) must be at least equal to the moment computed using an eccentricity of $(0.6 + 0.03h)$ in. For both braced and unbraced columns, the minimum eccentricity must be used for each of the principal axes separately.

Step 2 Decide whether or not the frame is braced against sidesway. The approximate procedure given in the discussion on the R Method is satisfactory within the limits given. Other, more accurate procedures are also available (see Ref. 8).

Step 3 Evaluate the effective unsupported length of the column by computing the stiffness of all members (in the plane being considered) framing into the joints at the top and bottom of the column. Thus, the ratio of stiffnesses at one end of the column

$$\Psi_A = \frac{\Sigma(EI/\ell) \text{ of compression members}}{\Sigma(EI/\ell) \text{ of flexural members}}$$

where EI is computed in the usual manner for the gross uncracked section, and ℓ is the center-to-center length of each member.

The ratio of stiffnesses at the other end of the column, Ψ_B, is computed in the same manner.

With the two ratios Ψ_A and Ψ_B, select the effective length k from the Jackson and Moreland Alignment Charts in Fig. A9.4[5], or compute k from the following equations. Factor k is the same as that illustrated in Fig. 9.16.

For braced columns, the effective length factor k may be taken as the smaller of Eq. 9.28 or 9.29.

$$k = 0.7 + 0.05(\Psi_A + \Psi_B) \leq 1.0 \tag{9.28}$$

$$k = 0.85 + 0.05\Psi_{\min} \leq 1.0 \tag{9.29}$$

Ψ_A and Ψ_B = values of Ψ at ends A and B of a column.
and $\Psi_{\min}$ = the smaller of Ψ_A or Ψ_B.

For unbraced columns restrained against rotation at both ends, the effective length factor k may be taken as follows:

For $\Psi_m < 2$,

$$k = \frac{20 - \Psi_m}{20}\sqrt{1 + \Psi_m} \tag{9.30}$$

For $\Psi_m \geq 2$

$$k = 0.9\sqrt{1 + \Psi_m} \tag{9.31}$$

where Ψ_m is the average of the Ψ values at the two ends of the column.

For unbraced columns hinged at one end, the effective length factor k may be taken as follows:

$$k = 2.0 + 0.3\Psi \tag{9.32}$$

where Ψ is the value at the restrained end.

A quick evaluation of k without much computation may be desirable for use in Step 4, especially if the engineer suspects that slenderness effects might not need to be considered. If slenderness effects are to be considered, then k must be evaluated accurately.

Step 4 Evaluate whether or not slenderness must be considered. For a braced column, the effects of slenderness may be neglected when

$$\frac{k\ell_u}{r} < 34 - \frac{12M_{1b}}{M_{2b}} \tag{9.33}$$

where the ratio M_{1b}/M_{2b} is positive when the column is bent in single curvature, as in Case I in Fig. 9.16, and negative when the column is bent in double curvature, as in Cases II and IV in Fig. 9.16. Use the actual value of the end moments M_{1b} and M_{2b} in Eq. 9.33 rather than the minimum moment discussed in Step 1. If there is essentially no moment at both ends of the column, use a value of 1 for the ratio M_{1b}/M_{2b}.

For an unbraced column, the effects of slenderness may be neglected when $k\ell_u/r < 22$.

Step 5 Assuming that the effects of slenderness must be considered, compute Euler's critical buckling load P_c. In computing the critical load, the selection of stiffness EI must consider the variations due to cracking, creep, and the nonlinearity of the concrete stress-strain curve. The Code accepts precise calculations that agree with test data.

In lieu of a more accurate calculation, the flexural stiffness EI may be taken as either

$$EI = \frac{(E_c I_g/5) + E_s I_{se}}{1 + \beta_d} \tag{9.34}$$

or, conservatively,

$$EI = \frac{E_c I_g/2.5}{1 + \beta_d} \tag{9.35}$$

where β_d = absolute value of the ratio of maximum factored dead load moment to maximum factored total load moment and is always positive.

I_{se} = moment of inertia of reinforcement about centroidal axis of member cross section.

The critical buckling load P_c may then be computed as

$$P_c = \frac{\pi^2 EI}{(k\ell_u)^2} \tag{9.36}$$

Step 6 Compute moment magnification factor δ_b to reflect the effects of member curvature between ends of the column. Also compute moment magnification factor δ_s for unbraced frames to reflect lateral drift resulting from lateral and gravity loads. In a braced frame, $\delta_s = 1.0$.

$$\delta_b = \frac{C_m}{1 - \frac{P_u}{\phi P_c}} \geq 1.0 \tag{9.37}$$

$$\delta_s = \frac{1}{1 - \dfrac{\Sigma P_u}{\phi \Sigma P_c}} \geq 1.0 \tag{9.38}$$

where P_u and P_c are the factored axial load and the critical load, respectively, for an individual column,

ΣP_u and ΣP_c are the sum of the factored axial loads and the critical load, respectively, for all columns in the story under consideration.

P_c is computed for each column for the direction being considered.

For columns braced against sidesway and without transverse loads between supports,

$$C_m = 0.6 + 0.4 \frac{M_{1b}}{M_{2b}} \geq 0.4 \tag{9.39}$$

For all other cases, $C_m = 1.0$.

The ratio M_{1b}/M_{2b} is positive if the column is bent in single curvature and negative if the column is bent in double curvature. Use the actual value of end moments M_{1b} and M_{2b} in Eq. 9.39 rather than the minimum moment discussed in Step 1. If there is essentially no moment at both ends of the column, use a value of 1 for the ratio M_{1b}/M_{2b}.

By comparing Cases IV and VIII in Fig. 9.16, it is obvious that the column in Case VIII must resist a larger moment than the frame moment computed in the normal manner. The additional moment is equal to the axial load times the lateral drift, or $P\Delta$.

Step 7   Compute the magnified factored moment M_c.

$$M_c = \delta_b M_{2b} + \delta_s M_{2s} \tag{9.40}$$

In frames effectively braced against sidesway, the second term of Eq. 9.40 becomes zero. When shear walls or diagonal bracing is tall and slender, that is, when it is flexible, the second term in Eq. 9.40 must be considered because the bracing may not be entirely effective.

Step 8   Columns are then designed using the axial load computed in Step 1 and the magnified moment computed in Step 7. The axial load need not be magnified.

Example 9.7   Given the structure in Ex. 9.6 (Fig. 9.19) without shear walls, for what moments should the interior columns be designed using the Moment Magnifier Method? Select longitudinal reinforcement for interior columns. The factored axial load on corner columns is 265 k and the factored axial load on other edge columns is 500 k. Assume that the factored moment to be transferred between an interior column and the slab due to wind is 58′k at the top of the column and 45′k at the bottom of the column.

Solution: Use the step-by-step procedure.

Step 1 $M_{2b} = 72'k$ $M_{1b} = 72'k$ $P_u = 940\ k$

$M_{2s} = 58'k$ $M_{1s} = 45'k.$

min $e = (0.6 + 0.03 \times 18) = 1.14''$

min $M_{2s} = \dfrac{1.14 \times 940}{12} = 89.3'k$

Step 2   Frame is not braced.

Step 3   For interior column, $\Sigma K_{col} = \dfrac{2(18 \times 18^3/12)}{12 \times 10.83} = 134.6$

Assume that 25% of the width of the slab is effective.

$$\Sigma K_{slab} = \dfrac{2 \times 0.25 \times 22 \times 12 \times 10^3}{12(12 \times 25)} = 36.7$$

$$\Psi_A = \Psi_B = \dfrac{\Sigma K_{col}}{\Sigma K_{slab}} = \dfrac{134.6}{36.7} = 3.67$$

From Eq. 9.31,

$k = 0.9\ \sqrt{1 + 3.67} = 1.94$

Step 4   By inspection $k\ell_u r > 22$, and slenderness must be considered.

Step 5   Assume moments due to gravity loads are all due to unbalanced live loads because all spans are equal. Thus, $\beta_d = 0$. Because the amount of reinforcement in the column is unknown at this stage of design, use Eq. 9.35 to determine the flexural stiffness of interior columns.

$$EI = \dfrac{3.1 \times 10^3 \times 18^4/12}{2.5} = 1.08 \times 10^7$$

Using Eq. 9.36 to obtain the critical buckling load,

$$P_c = \dfrac{\pi^2 \times 1.08 \times 10^7}{(1.94 \times 12 \times 10.00)^2} = 1967\ k$$

Step 6   Because $M_{1b} = M_{2b}$ and the column would normally be bent in double curvature, $C_m = 0.6 - 0.4 \geqslant 0.4.$

The moment magnification factor δ_b can now be computed using Eq. 9.37.

From Ex. 9.6, $P_u = 940\ k$

$$\delta_b = \dfrac{0.4}{1 - \dfrac{940}{0.7 \times 1967}} = 1.26$$

The factored axial load and the critical load of all columns in a story must be computed for use in Eq. 9.38.

Total factored axial load:

Interior columns $940 \times 21 = 19{,}740$ k

Side columns $500 \times 20 = 10{,}000$

Corner columns $265 \times 4 = \underline{1{,}060}$

$\Sigma P_u = 30{,}800$ k

For corner columns (see calculation for Example 9.6):

$$K_{\text{slab}} = \frac{0.25(11 \times 12)10^3/12}{12 \times 25} = 9.2$$

$$K_{\text{beam}} = \frac{20 \times 20^3/12}{12 \times 25} = \underline{44.4}$$

$$K_{\text{floor}} = 53.6$$

$$K_{\text{col}} = \frac{2(18 \times 18^3/12)}{12 \times 10.83} = 134.6$$

$$\Psi_A = \Psi_B = \frac{\Sigma K_{\text{col}}}{\Sigma K_{\text{floor}}} = \frac{134.6}{53.6} = 2.51$$

$$k = 0.9\sqrt{1 + 2.51} = 1.69$$

$$P_c = \frac{\pi^2 \times 1.08 \times 10^7}{(1.69 \times 12 \times 9.17)^2} = 3098 \text{ k}$$

For other columns along East and West sides,

$$K_{\text{slab}} = 9.2 \times 2 = 18.4$$

$$K_{\text{beam}} = 44.4 \times 2 = \underline{88.8}$$

$$K_{\text{floor}} = 107.2$$

$$K_{\text{col}} = 134.6$$

$$\Psi_A = \Psi_B = \frac{\Sigma K_{\text{col}}}{\Sigma K_{\text{floor}}} = \frac{134.6}{107.2} = 1.25$$

$$k = \frac{20 - 1.25}{20}\sqrt{2.25} = 1.41$$

$$P_c = \frac{\pi^2 \times 1.08 \times 10^7}{(1.41 \times 12 \times 9.17)^2} = 4455 \text{ k}$$

For other columns along the North and South sides,

$$K_{slab} = K_{floor} = 9.2 \times 2 = 18.4$$

$$K_{col} = 134.6$$

$$\Psi_A = \Psi_B = \frac{\Sigma K_{col}}{\Sigma K_{floor}} = \frac{134.6}{18.4} = 7.31$$

$$k = 0.9\sqrt{1 + 7.31} = 2.59$$

$$P_c = \frac{\pi^2 \times 1.08 \times 10^7}{(2.59 \times 10 \times 12)^2} = 1100 \text{ k}$$

Total critical load:

Interior columns $1677 \times 21 = 35{,}200$ k

Corner columns $3098 \times 4 = 12{,}400$

E-W side columns $4455 \times 6 = 26{,}700$

N-S side columns $1100 \times 14 = \underline{15{,}400}$

$$\Sigma P_c = 89{,}700 \text{ k}$$

Using Eq. 9.38,

$$\delta_s = \frac{1}{1 - \dfrac{30{,}800}{0.7 \times 89{,}700}} = 1.96$$

Step 7   The magnified factored moment M_c is computed using Eq. 9.40.

$$M_c = 2.00 \times 72 + 1.96 \times 89.3 = 319'k$$

Step 8   Design interior columns for $P_u = 940$ k and $M_c = 319'$k,

Eccentricity $e = \dfrac{319 \times 12}{940} = 4.08'' > e_{min}$ by inspection.

Using the interaction diagrams in Fig. 9.24,

$$\phi M_n \geqslant M_u \quad \text{and} \quad \phi P_n \geqslant P_u$$

$$\frac{\phi M_n}{A_g h} \geqslant \frac{319 \times 12}{324 \times 18} = 0.66$$

$$\frac{\phi P_n}{A_g} \geqslant \frac{940}{324} = 2.90$$

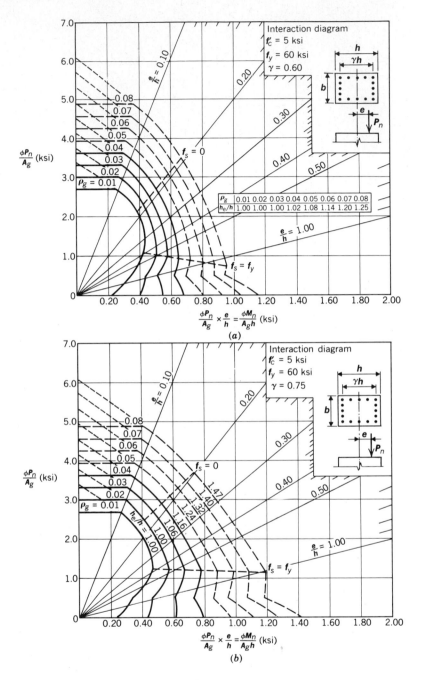

FIGURE 9.24 Column interaction diagrams (from Ref. 9.1). (Courtesy of the American Concrete Institute.)

Required $\rho = 5.5\%$ for $\gamma = 0.75$

$\rho = 6.7\%$ for $\gamma = 0.60$

Use $\rho = 5.9\%$ for $\gamma = 0.7$ by interpolation.

$A_s = 324 \times 0.059 = 19.1^{\square''}$

Use 12 #11 bars, $A_s$prov. $= 18.72^{\square''}$

Compare this result with the 12 #10 bars selected by the R Method in Ex. 9.6.

Second-Order Analysis

A second-order analysis, or $P\Delta$ analysis, includes the effects of sway deflections on the axial loads and moments in a frame. The analysis will give a better approximation of the real loads and moment than will either the R Method or the Moment Magnifier Method. It is especially appropriate for tall, lightly braced frames where columns smaller than those resulting from use of the other two methods might be justified. It is also appropriate for structures of sufficient magnitude or extent to generate construction cost savings larger than the cost of additional calculation effort.

As a minimum, the following items must be considered in a $P\Delta$ analysis.

1. Realistic moment-curvature or moment-end rotation relationships based on conditions at the ultimate limit state.
2. The effect of foundation rotations on lateral drift.
3. The effect of axial load on the stiffness and carry-over factors, especially for slender columns.
4. Differential shortening due to creep and its effect on lateral drift.
5. Moment in the column determined by the amount of lateral drift of the frame and deflection of the column itself.

Guidance in performing a $P\Delta$ analysis is given in Refs. 8 and 9.

9.5 COMPOSITE COLUMNS

Sometimes it is advantageous to use both structural steel and reinforced concrete in a column. Such members are called composite columns. When properly designed, composite columns are stronger than columns composed of either material alone. There may be other benefits, such as fireproofing the structural steel portion of the member. Figure 9.1 illustrates two typical composite columns. One type consists of a structural steel shape, usually an H section, enclosed by a square, rectangular, or round concrete column. Lateral reinforcement may be either spirals or ties. The other type consists of a round pipe or square or rectangular tubing filled with concrete, either with or without longitudinal bars.

Design of composite columns is governed by ACI Code Section 10.14. Requirements can be summarized as follows.

Axial load carried by concrete or longitudinal bars must be transferred to concrete in direct bearing on the composite column concrete. This can be accomplished by lugs or brackets on structural steel bearing on the concrete or by concrete beams or slabs as in normal concrete construction. Concrete-filled pipes and tubes present practical difficulties in meeting this requirement. Merely filling a steel pipe with concrete does not necessarily increase its axial load strength.

A composite column can be evaluated for the effects of slenderness in the same manner as a reinforced concrete column. Rules in the Code for estimating the radius of gyration **r** for concrete columns (see Section 9.4) are overly conservative for concrete-filled tubing and do not apply to other composite columns. For composite members, the radius of gyration **r** must be computed as

$$r = \sqrt{\frac{(E_c I_g/5) + E_s I_t}{(E_c A_g/5) + E_s A_t}} \tag{9.41}$$

where I_t = moment of inertia of structural steel shape, pipe, or tubing about the centroidal axis of the composite member cross section.

A_t = area of structural steel shape, pipe, or tubing in a composite section.

In Eqs. 9.34 and 9.35, both **EI** for concrete and for steel have been reduced because creep tends to increase the stress in steel. The steel might reach yield and thus reduce the effective combined **EI**. In composite columns, steel is a large percentage of the cross section, and load transfer to steel is not significant. A reduction in **EI** for steel is not necessary. Thus, **EI** may be computed as either the value given by Eq. 9.35 or by

$$EI = \frac{(E_c I_g/5) + E_s I_t}{1 + \beta_d} \tag{9.42}$$

The wall of a structural steel pipe or tube enclosing a concrete core must be thick enough to ensure that the steel will reach longitudinal yield before buckling outward. For rectangular tubes, the thickness of the steel tube must not be less than

$$b \sqrt{\frac{f_y}{3E_s}}$$

where f_y = the yield point of the steel tube
and b = width of the compression face of the tube.

For round tubes or pipes, the thickness must not be less than

$$h \sqrt{\frac{f_y}{8E_s}}$$

where h = overall diameter of the tube.

Concrete-filled pipes and tubes may (but need not) also enclose longitudinal bars. If used, longitudinal bars contribute to the strength of the member. Ties or spirals are not required, but a few ties should be used to hold the bars in position.

Where concrete surrounds a structural steel core, longitudinal bars at least 1% but not more than 8% of the net area of concrete section (gross area less the area of the structural steel core) must be used. Longitudinal bars can be used in computing column strength. They can be used in computing slenderness effects only if enclosed in a spiral. The minimum concrete strength is $f'_c = 2.5$ ksi, and the maximum structural steel strength is $f_y = 50$ ksi. The purpose of these requirements is to ensure that the steel core reaches yield at strains that can be sustained in the concrete without spalling the face shell.

In columns with a steel core, longitudinal bars must be enclosed with either ties or a spiral. Spirals must conform to the same requirements as for reinforced concrete columns. Tied columns with an H section core may have a thin section of concrete along two faces. To maintain the concrete around the structural steel core and to prevent separation at high strains between the steel core and the concrete in tied columns, a longitudinal bar shall be located at every corner of a rectangular cross section, with other longitudinal bars spaced not farther apart than one-half the least side dimension of the composite member. Additional requirements are placed on size and spacing of lateral ties (see the next section).

9.6 DESIGN CONSIDERATIONS

This section deals with several matters that must be considered in the design of most columns.

Minimum and Maximum Dimensions

The ACI Code does not limit the size of a concrete column. However, minimum concrete cover and clearance between bars, as well as practical considerations of concrete placement and accidental eccentricities, will limit tied columns to a minimum thickness of about 8 in. The clear distance between longitudinal bars and between a contact lap splice and adjacent splices or bars must be not less than 1.5 bar diameters nor 1½ in. (Code 7.6.3 and 7.6.4). Concrete cover must be at least 1½ in. for ties and 2 in. for longitudinal bars, except where concrete is cast against and permanently exposed to earth, in which case the cover must be at least 3 in. for both ties and longitudinal bars (Code 7.7).

The minimum size of spiral columns is about 12 in.; however, the high cost of spirals will generally dictate larger minimum sizes.

The area of longitudinal reinforcement in noncomposite columns must be at least 1% but not more than 8% of the gross area of the column (Code 10.9). If a column is made larger than required to carry load, not less than one-half the total area may be used to determine minimum reinforcement and design strength. This is equivalent to permitting the area of longitudinal reinforcement to be as little as 0.5% of the gross area of the column. The total area of concrete is used in structural analysis, design for effects of slenderness, and design of other members framing to the column, but the column reinforcement must be selected by using, for design only, that portion of the concrete area that results in a minimum steel area of 1% (Code 10.8.4).

The minimum number of longitudinal bars is 4 for bars within rectangular or circular ties, 3 for bars within triangular ties, and 6 for bars within spirals (Code 10.9.2). The maximum number of bars is limited only by the clearance between bars and between contact lap splices and adjacent splices or bars. Groups of 2, 3, or 4 longitudinal bars may be bundled in contact to meet spacing requirements and to facilitate the placement of concrete (Code 7.6.6).

Irregular Columns

The cross section of columns is most often in the shape of a square, rectangle, or circle. Other regular polygons, such as hexagons and octagons, can also be used. To simplify it, design may be based on the largest circular section that can be enclosed within the column outline (Code 10.8.3). When irregular polygons are used, the placement of ties and longitudinal bars must follow the concrete outlines closely. Large areas of the cross section must not be left unreinforced. "L"-shaped columns, for example, are reinforced as two intersecting rectangular columns (see Fig. 9.25). In design, they may be treated as a unit for the purpose of determining slenderness. When a column is built monolithically with a wall or other irregular cross section, the outer limit of the effective concrete cross section for design of the column is 1½ in. outside the spiral or tie reinforcement (Code 10.8.2).

7 longitudinal bars, minimum.

FIGURE 9.25 Bar arrangement in L-shaped column.

Selection of Concrete Strength

The selection of concrete strength to be used in a particular structure depends on a number of considerations. In most areas of the United States, concrete strength $f'_c = 3$ ksi is readily available. Most engineers would not consider using a lower strength of concrete. Higher strength concrete carries proportionally more load but only costs slightly more. For large structures with heavily loaded columns, consider concrete with strength as high as is available. The reliability of production of high-strength concrete is a limiting factor. Small projects may not warrant the effort and expense of quality assurance to produce limited quantities of high-strength concrete.

In small structures with lightly loaded columns, there is little reason for using high-strength concrete if the resulting column is understressed or if, in order to fully utilize concrete strength, the column is so small as to present practical construction difficulties. In addition, smaller columns are more likely to require consideration of the effects of slenderness.

In multistoried buildings, high-strength concrete can be used in columns of the lower stories. Where steel percentage is reduced to the minimum in upper stories, concrete strength can be reduced to the minimum used on the project, usually $f_c' = 3$ ksi. To help avoid construction errors, the same strength concrete should be specified for all columns in each story.

In meeting objectives of the project other than strength (see the discussion in Chapter 1), it is sometimes necessary to keep the size of columns to a minimum. For this puporse, use of high-strength concrete is advantageous.

In proportioning columns, engineers should consider the relative cost of carrying the load on concrete versus the cost of carrying the load on reinforcing steel.

Example 9.8   What is the cost of carrying 100 k of load through 1-ft length of column on concrete and on reinforcing steel? Assume

- $f_c' = 3$ and 5 ksi
- $f_y = 60$ ksi
- Eccentricity $e \leqslant e_{min}$
- Average load factor = 1.55
- In-place cost of concrete = \$65/yd^3 for $f_c' = 3$ ksi
  $\qquad\qquad\qquad\qquad\qquad$ = \$70/yd^3 for $f_c' = 5$ ksi
- Cost of reinforcing steel, in place = \$0.30/lb
- Tied column

Solution for Concrete

From Eq. 9.5,

$$\phi P_n = P_u = 1.55P = 0.80 \times 0.70 \times 0.85 \times 3A_{gc}$$

$$P/A_{gc} = 0.921 \text{ for } f_c' = 3 \text{ ksi}$$

Similarly, $P/A_{gc} = 1.535$ for $f_c' = 5$ ksi

Calculate the cost of carrying 100 k on concrete.

For $f_c' = 3$ ksi, cost $= \dfrac{\$65 \times 100}{0.921 \times 144 \times 27} = \$1.82/'$

for $f_c' = 5$ ksi, cost $= \dfrac{\$70 \times 100}{1.535 \times 144 \times 27} = \$1.17/'$

Solution for Reinforcing Steel

Assume that ties add 10% to the volume of steel and laps add 20% to the volume.

$$\phi P_n = P_u = 1.55P = 0.80 \times 0.70 \times 60A_{st}$$

$$P/A_{st} = 21.7 \text{ ksi}$$

Calculate the cost of carrying 100 k on longitudinal reinforcing steel.

$$\text{Cost} = \frac{\$0.30 \times 100 \times 3.4 \times (1 + 0.1 + 0.2)}{21.7} = \$6.11/'$$

Ties

Individual longitudinal bars in columns must be restrained laterally to prevent outward buckling of the bars and a consequent loss of strength. If bars buckle, the face shell of concrete would be spalled. Sizing and spacing of ties is semiempirical based on long practice found to be satisfactory. Requirements in the ACI Code (Code 7.10.5 and 10.14.8) can be summarized as follows.

1. Size of ties must be at least #3 for #10 longitudinal bars and smaller and at least #4 for larger bars and for bundled longitudinal bars. For composite columns, tie bars must have a diameter at least 1/50 times the greatest side dimension of the column but not less than a #3 bar and need not be larger than a #5 bar.

2. Spacing of ties shall not exceed
   - 48 tie bar diameters.
   - 16 longitudinal bar diameters.
   - least dimension of column (one-half the least side dimension of composite columns).

3. Ties must extend full length of the column (see Fig. 9.26).

4. Each longitudinal bar must be enclosed and braced by a tie (see Fig. 9.27). In square or round columns, complete circular ties are also acceptable, if at least four longitudinal bars are spaced uniformly around a circle.

5. Additional ties may be required where longitudinal bars are offset (see the discussion in Chapter 10, Joints).

6. Ends of ties must terminate in a standard hook for anchorage. For #5 tie bars and smaller, a 90° bend plus a 6 tie bar diameter extension at the free end of the bar is sufficient. For seismic design, a 135° bend plus a 10 bar diameter extension is required.

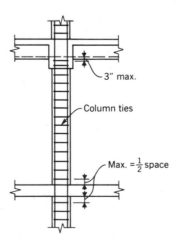

FIGURE 9.26   Tie requirements.

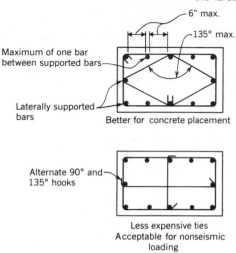

FIGURE 9.27 Tie arrangements.

In addition, it is good practice to arrange ties to give as much clearance as possible in the center of the column to allow easy placement of concrete.*

Spirals

As shown in Fig. 9.4, closely spaced spirals of adequate size result in ductile columns after spalling of the face shell of concrete over the spirals. Because the ACI Code encourages ductile structures, higher axial loads are permitted on spirally reinforced columns (see Eqs. 9.4 and 9.5). Assuming the face shell of a ductile column will spall at loads approaching the ultimate, the amount of spirals required to ensure a ductile column is the amount of spiral steel required to carry the load carried by the face shell of concrete. The axial load carried by the face shell is

$$0.85f'_c(A_g - A_c)$$

where A_g and A_c are the gross area of the column and the area of the column core measured to the outside diameter of the spiral, respectively. Tests show that spiral steel is at least twice as effective as longitudinal steel in supporting axial load, thus axial load carried by spirals is

$$2\rho_s A_c f_y$$

where ρ_s = the ratio of the volume of the spiral reinforcement to the total volume of the column core.

Equating the above two strengths and rounding the constant upward slightly gives the ACI Code equation (Code 10.9.3) for the volume of spirals.

---

*Even the best design will be for naught if the contractor cannot get concrete to the bottom of the column forms.

$$\rho_s = 0.45 \left(\frac{A_g}{A_c} - 1 \right) \frac{f_c'}{f_y} \tag{9.43}$$

Following are other Code requirements for spirals (Code 7.10.4):

1. A spiral must be at least $\frac{3}{8}$ in. in diameter.
2. Clear spacing between spirals shall be at least 1 in. but not more than 3 in. Spacing is usually selected to the nearest $\frac{1}{4}$ in.
3. Spirals are anchored by an extra $1\frac{1}{2}$ turns at each end and spliced by a 48 bar diameter lap, or by welding.
4. Spirals must be spaced evenly and held in position by two to four spacer bars.
5. Spirals must extend from the top of the slab or footing at the bottom of the column to the underside of the lowest horizontal reinforcement in floor members above, or into a column capital to the point where the diameter is twice that of the column.
6. Where beams or brackets do not frame into all sides of a column, ties shall extend above the termination of the spiral to the bottom of the slab or drop panel.

Selection of Tied versus Spiral Reinforcement

From Eqs. 9.4 and 9.5 it can be seen that spirals increase the maximum axial load capacity when eccentricity is less than the minimum.

$$\frac{0.85 \times 0.75}{0.80 \times 0.70} - 1 = 0.138$$

Spirals cost one and one-half to two times as much per pound as do ties. Thus, the use of spirals will be economical only if the cost of spirals is substantially less than 13.8% of the cost of a tied column. In general, this will be true only under the following conditions:

- Columns are large.
- Concrete strength is low.
- Columns are round and not square.
- Eccentricity e is less than e_{min}.
- Longitudinal reinforcement is heavy.
- Use of spirals enables reuse or standardization of formwork.

The effect of some of these parameters is illustrated in Fig. 9.28, which plots the volume of spirals less the volume of ties as a function of column size, column shape, and concrete strength.

Spirals may be used used where necessary to achieve the smallest possible column size. Engineers should also evaluate the difficulty of obtaining a small quantity of spirals of an unusual size.

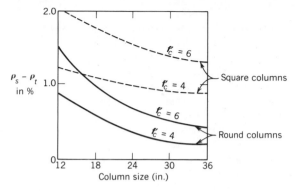

FIGURE 9.28 Volume of spirals less volume of ties, as a function of column size and type and of concrete strength.

Design Procedures

Assuming that concrete outlines have been established and loads have been determined, design of columns can proceed in an orderly manner using the following step-by-step procedure.

Step 1 Decide on the strength of concrete to be used in each story of columns. Sometimes this is an iterative process. Adjustments can be made after Step 3 and the design repeated.

Step 2 Evaluate the effects of slenderness on columns (see Section 9.4).

Step 3 Select longitudinal reinforcement using the procedures in Section 9.3. For the greatest accuracy and economy, design aids in the form of tables, graphs, or computers should be used in the final selection of bars. If the design must be done manually, the simplified procedures presented in Section 9.3 are conservative but sometimes not as economical as those using design aids.

Step 4 Select lateral reinforcement, ties or spirals. If the column resists a significant transverse shear force, its shear strength should be computed and lateral reinforcement selected to function as shear reinforcement, if necessary. Unbraced columns in the lower stories of tall buildings, legs of portal frames and similar members could have shear forces requiring added shear reinforcement.

Step 5 Select splices (see Chapter 5).

Step 6 Select connection details (see Chapter 10).

PROBLEMS

1. What is the maximum factored load permitted on a 12-in.-thick L-shaped column, each leg 2′–6″ long, with $f'_c = 4.5$ ksi, $f_y = 60$ ksi, and 12 #11 longitudinal bars?

2. Using approximate procedures, how much service axial load and moment about each axis alone can be sustained by a 12 × 18-in. tied column with $f'_c = 5.3$ ksi, $f_y = 50$ ksi, and 4 #11 longitudinal bars? Assume that eccentricity is equal to the minimum.

3. What is the maximum service axial load that can be supported by the column in Problem 2 at balanced strain conditions?

4. Select square tied column dimensions and longitudinal reinforcement to support a service axial live load of 380 kips and a dead load of 515 kips, using one of the following sets of parameters. Assume that eccentricity is less than the minimum, and slenderness does not control the design.

Case	f'_c (ksi)	f_y (ksi)	Range of ρ_g (%)
A.	3	60	1 to 2
B.	4	40	2 to 3
C.	4	60	3 to 4
D.	6	60	4 to 6

5. Using approximate procedures, how much moment will each of the columns in Problem 4 resist at minimum eccentricity?

6. What maximum clear story height is permitted without reduction of rectangular column strength for slenderness effects when the thickness of the column is 12, 16, 20, 24, and 36 in. (A) for a braced frame with columns bent in double curvature or (B) for an unbraced frame governed by loads of short duration? Assume that the effective length is equal to the actual length.

7. Select dimensions, longitudinal reinforcement, and concrete strength for a column in a 10-story building. Minimize changes in column dimensions and concrete strength. Use $f_y = 60$ ksi. Factored loads are 110 kips for the roof and 170 kips for each of the 10 floors. Assume that slenderness effects can be neglected and that the factored moment at each floor is 40 ft-kips and at the roof it is 25 ft-kips.

8. Select ties for a 16 × 24-in. column with 10 #10 bars arranged as shown.

9. Select spirals for a 28-in. square column using $f'_c = 4$ ksi, $f_y = 60$ ksi, and 2 in. cover on the spirals.

Selected References

1. ACI Committee 340. *Design Handbook, in Accordance with the Strength Design Method of ACI 318-77*, Volume 2—Columns, Publication SP-17A(78), American Concrete Institute, Detroit, 1978.

2. *CRSI Handbook*, Concrete Reinforcing Steel Institute, Chicago, 1984.

3. A.L. Parme, J.M. Nieves, and A. Gouwens. "Capacity of Rectangular Columns subject to Biaxial Bending," *ACI Journal*, V.63, pp 911–923, September 1966.

4. Boris Bresler. "Design Criteria for Reinforced Columns under Axial Load and Biaxial Bending," *ACI Journal*, V.57, pp 481−490, November 1960.
5. ACI Committee 318. *Commentary on Building Code Requirements for Reinforced Concrete (ACI 318-83)*, American Concrete Institute, Detroit, 1983.
6. *Specification for the Design, Fabrication and Erection of Structural Steel for Buildings*, see also *Commentary* on these Specifications, American Institute of Steel Construction, New York, 1978.
7. James G. MacGregor, John E. Breen, and Edward O. Pfrang. "Design of Slender Concrete Columns," *ACI Journal*, V.67, pp 6−28, January 1970.
8. B.R. Wood, D. Beaulieu, and P.F. Adams. "Columns Design by P-Delta Method," *Proceedings, ASCE*, V.102, No.ST2, pp 411−427 Feburary 1976.
9. B.R. Wood, D. Beaulieu, and P.F. Adams. "Further Aspects of Design by P-Delta Model," *Proceedings, ASCE*, V.102, No.ST3, pp 487−500, March 1976.

CHAPTER 10

JOINTS

10.1 OBJECTIVES

Joints occur in cast-in-place concrete structures for three reasons: (1) members intersect or change direction, for example, a beam-column joint; (2) the structure cannot be cast at one time, requiring construction joints; and (3) movement is expected throughout the life of the structure, requiring expansion joints.

Although joints are a major concern in all structures, they are not usually as time consuming to design in concrete as they are in structural steel or precast concrete structures.

In most cases, concrete outlines and reinforcement of members framing into a joint will have been selected before the joint design is considered. Occasionally, design of members framing into a joint will have to be revised to accommodate design requirements of the joint. Design objectives for the three types of joints are quite different and are stated separately.

For joints formed by intersecting members,

1. Provide for transmission of column loads through floor systems and into foundations.
2. Arrange bars to avoid undue congestion, permit placement of concrete, and allow proper placement of bars to resist tension forces.
3. Anchor flexural reinforcement where beams or slabs end at a joint.
4. Select reinforcement to resist shear forces internal to the joint and to prevent buckling of bars in compression.

For construction joints,

1. Locate joints.
2. Select size of keyways to resist shear, if required.
3. Select size and location of additional shear reinforcement, if required.

For expansion joints,

1. Select spacing and location of joints.
2. Select type of joint and details.

10.2 COLUMN JOINTS

Joints in a column normally occur at three locations: at the foundation, at intersecting levels of floor framing, and at the top level of framing or the roof. Becuase these locations have differing requirements, they will be considered separately.

Foundation-Column Joint

Where a column rests on a concrete foundation, the design bearing strength on concrete ϕP_n must not exceed

$$\phi P_n = \phi 0.85 f_c' A_1 \tag{10.1}$$

where A_1 is the loaded area (Code 10.15).

Note that this is equivalent to the concrete portion of the nominal load strength of a column at zero eccentricity, as given by Eq. 9.3. Capacity reduction factor ϕ is the same as for a tied column, that is, 0.70. This strength computation applies to concrete in the column as well as concrete in the foundation.

From Eqs. 9.4 and 9.5, it is apparent that bearing stress at the end of a column may be higher than that permitted in the column. In a spiral column, the stress increase is $1/0.85 = 1.18$, and in a tied column the stress increase is $1/0.80 = 1.25$. Although the ACI Code is silent as to the distance from the joint interface that this stress increase is permitted, it is reasonable to require that bearing stress be reduced to the stress permitted in the columns within a distance equal to the least dimension of the column.

A condition frequently encountered is a foundation with concrete of moderate strength supporting a column with concrete of higher strength. In such cases, it is convenient to recognize higher bearing strength of foundation concrete when the supporting member of the foundation is larger than the column. When the loaded area is completely surrounded and confined by concrete, the design-bearing strength (given by Eq. 10.1) on the loaded area may be multiplied by $\sqrt{A_2/A_1}$ but not more than 2. A_2 is the maximum area of the portion of the supporting surface that is geometrically similar to and concentric with the loaded area. When the supporting surface is sloped or stepped, A_2 may be taken as the larger area shown in Fig. 10.1.

The above principles can be applied to other situations, such as a steel column bearing on a concrete foundation or supporting a concrete beam.

Force and moment at the base of a column must be transferred to the supporting foundation by bearing on concrete and by reinforcement. Dowels (short reinforcing bars) are usually used as reinforcement because it is difficult to hold long

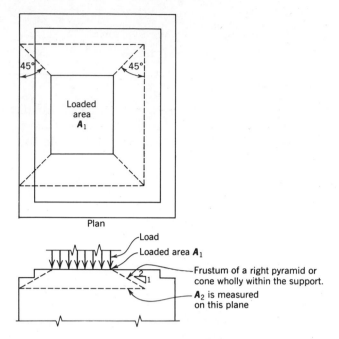

FIGURE 10.1 Application of frustum to find A_2 in stepped or sloped supports. (Courtesy of the American Concrete Institute, from Commentary on Building Code Requirements for Reinforced Concrete.)

vertical bars in exact alignment while casting foundation concrete. Dowels are anchored in the foundation and spliced to column bars just above the foundation where minor adjustment can be made for location and plumbness. Splicing can be done by any method described in Chapter 5. Dowels must carry all compressive force that exceeds concrete-bearing strength of either member and any computed tensile force across the joint interface. Furthermore, the area of reinforcement across the joint shall not be less than 0.005 times the gross area of the column (Code 15.8). Shear due to lateral forces must be transferred to the foundation in accordance with shear-friction (see Chapter 6) or by other means. Shear rarely controls the design of joints in columns, except perhaps in seismic zones.

The diameter of dowels must not exceed the diameter of longitudinal bars by more than 0.15 in. or one bar size. Dowels must not be larger than a #11 bar and must extend into the column and lap with the longitudinal bars a distance not less than the development length of longitudinal bars or the splice length of the dowels, whichever is greater. Dowels must extend into the footing a distance not less than the development length of the dowels. Hooked dowels do not decrease the required compression development length but may be useful if the dowel will be in tension.

Example 10.1   What dowels are required at the foundation for a 20-in. square tied column with 12 #11 longitudinal bars? Column steel f_y = 60 ksi, column concrete f'_c = 5 ksi, and foundation concrete f'_c = 3 ksi. Assume that the foundation is concentric with and much larger than the column and that the column is fully loaded with no moment transferred between foundation and column.

Solution A_g = 20 × 20   = 400in.^{2}

A_{st} = 12 × 1.56 = 18.72in.^{2}

From Eq. 9.5, the column load is

Concrete   0.80 × 0.70[0.85 × 5(400 − 18.72)] =   907 k

Reinforcing bars   0.80 × 0.70[60 × 18.72]   =   629

P_u = 1536 k

Bearing strength of foundation concrete:

0.70(0.85 × 3 × 2) = 3.57 ksi

Bearing strength of column concrete:

0.70(0.85 × 5) = 2.98 ksi

Force transferred by concrete:

2.98 × 400 = 1190 k

Net force to be transferred by dowels   =   346 k

Stress in dowels = 0.8 × 0.7(60 − 0.85 × 5)   = 31.22 ksi

Area of dowels required $= \dfrac{346}{31.22} = 11.08^{□\prime\prime}$

(Minimum dowel area = 400 × 0.005 = 2.00$^{□\prime\prime}$, does not control)

Use 8 #11 dowels, $A_s$prov. = 12.48$^{□\prime\prime}$ (13% over)

or 12   #9 dowels, $A_s$prov. = 12.00$^{□\prime\prime}$ (8% over)

NOTE:

1. The best dowel selection would be 8 #11 bars because there would be less congestion than with 12 bars.
2. In this example, if column concrete f'_c > 6 ksi, the bearing strength of column concrete > 0.70(0.85 × 6) = 3.57 ksi, and the force transferred by bearing on concrete would be controlled by the foundation concrete, rather than the column concrete.
3. If moment is transferred between column and foundation, the bending stress would reduce the force that could be transferred by bearing on concrete and more force would have to be transferred by dowels.

Standard practice is to provide dowels matching the column verticals to avoid lengthy explanations and justifications to owners and contractors. The standard practice also saves time in design. If the area of the dowels is less than that of the column vertical bars through construction error, the reduced area might be justified by the above procedure.

Transmission of Column Load Through Floor

In multistoried buildings, columns commonly have higher strength concrete than intersecting floor systems. Laboratory tests have shown that concrete strength in the column may be 1.4 times that specified for the floor system without significant reduction in column strength. (For example, f'_c = 4.2 ksi in the column and f'_c = 3 ksi in the floor.) When the ratio of strengths of concrete in the column to concrete in the floor system exceeds 1.4, some precautions must be taken to transmit column loads through the floor system.

To transmit column loads, concrete of the strength specified for the column may be placed in the floor at the column location [see Fig. 10.2(a)]. Although simple and inexpensive, this procedure requires inspection to ensure that concrete of the proper strength is used in the proper locations.

Another option is to base the strength of a column through a floor system on the strength of the concrete used in the floor system and provide dowels through the joint extending into the column above and below the joint a distance equal to the compression development length [see Fig. 10.2(b)]. Dowels must be restrained laterally in the same manner as other longitudinal bars. When the joint is confined laterally on all four sides by beams of approximately equal depth or by slabs, strength of the concrete in the joint may be assumed equal to 75% of the column concrete strength plus 35% of the floor concrete strength (Code 10.13). This option has an obvious disadvantage if the column would be congested with added dowels. Bundling the bars may help to alleviate congestion. On the other hand, inspection of dowels is easier than inspection of high-strength concrete.

When deep beams frame into columns on one, two, or three sides, vertical column bars and added dowels within the joint should be restrained using the same ties or spirals as used in the column below (see Fig. 10.3). Where dowels are added, more ties than required in the column below may be necessary.

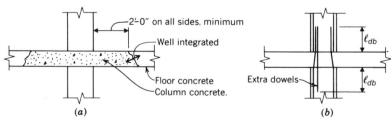

FIGURE 10.2 Transmission of column loads through floor system.

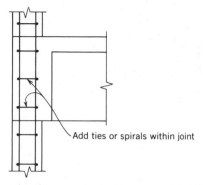

FIGURE 10.3 Lateral ties or spirals within joint.

Bar Arrangement

When columns above and below a joint are the same size and bars are arranged in a rectangular pattern, corner bars and, perhaps, other bars must also be offset if they are lap spliced. The inclined portion of the offset bar must have a slope not greater than 1 in 6 relative to the axis of the column. Offset bars must be bent before placing them in the forms. At offset bends, the axial force in the bar has a horizontal component that is resisted by the column core when acting inward. When acting outward, the horizontal component must be resisted by ties or spirals within 6 in. from the bend point if parts of the floor construction do not provide restraint. Ties or spirals must be designed for 1½ times the horizontal component of the force in the bar. Where a column face is offset 3 in. or more, offset longitudinal bars are not permitted and separate dowels must be provided. Dowels must be lap spliced with the outermost bars and either lap spliced with the innermost bars, or spliced mechanically (Code 7.8) [see Fig. 10.4(a)]. Lap splice lengths are determined in the usual manner, increasing the lap length where spacing between bars is greater than 6 in. The maximum offset of 3 in. has been established by a long tradition of successful usage.

When columns above and below a joint are the same size and bars are arranged in a circular pattern, no offset is necessary if all bars and dowels in the contact splices can be accommodated in one ring with the required clearance between bars. Otherwise, offset bars will still be required. Engineers should examine each joint to determine whether offset bars are needed and how bars should be arranged to assure the necessary clearance between column bars and between column and beam bars.

When bars are offset in the conventional manner, as in Fig. 10.4(a), moment capacity will be reduced if the offset bars are in tension because the effective depth is reduced. In such cases, the member must be designed for the reduced effective depth or mechanical splices should be used without an offset to maintain the effective depth.

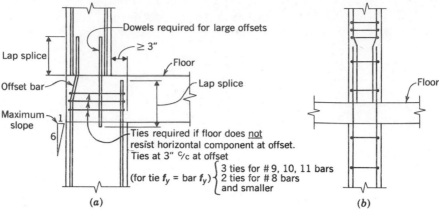

FIGURE 10.4 Offset bars and dowels at column joint.

Another option to provide maximum moment capacity in the column at a joint is to offset-bend the bars in the upper column, as in Fig. 10.4(b). Because this option is rarely used in the United States, when it is selected, engineers should clearly specify the location of bends in vertical bars and the location of ties.*

Vertical bars in columns above a joint are frequently smaller in size or number than in the column below. The area of lower column bars extended above the floor joint as dowels in the upper column must equal the area of bars required in the upper column. Lower column bars not needed in the upper column may be cut off an inch or two below the floor line at the joint, as in Fig. 10.5(a).

When the number or area of bars in the upper column exceeds that in the lower column (e.g., when changing strength of concrete in columns, as at the eighth floor in Fig. 14.3), dowels must be provided at the joint, as in Fig. 10.5(b). In any case, the area of dowels above a joint must equal the area of vertical bars required in the column above.

Reinforcement is frequently congested at beam-column joints. To minimize this congestion, columns should be designed with as few longitudinal bars as possible. This will also minimize the number of ties required. Likewise, beams should be designed with as few negative moment bars as possible, consistent with maximum spacing requirements. Bottom bars in beams pose fewer problems because those not required to extend into the support can be cut off flush with the column or 1 in. inside the column (see Fig. 10.6). Only one quarter of the bottom bars need extend into the support at least 6 in. at continuous supports and one third at end supports. It is good practice to extend all bottom bars into supports unless prevented by bar congestion.

When one or more faces of both the column and the beam are flush, longitudinal bars in both members occupy the same plane because both require the same

---

*Field errors frequently result from unclear instructions. Errors also result when the work desired is contrary to conventional practice, especially if instructions are brief. Special attention should be drawn to unusual or unconventional requirements.

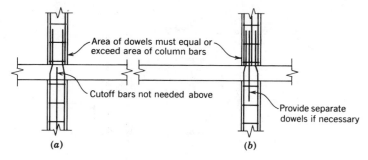

FIGURE 10.5 Column dowels at a joint.

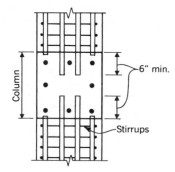

FIGURE 10.6 Plan of beam bottom bars in beam-column joint.

concrete cover over reinforcement. There are several options available to avoid bar interference in addition to the one mentioned above. Column verticals can be offset below the beam instead of just below the floor level, as shown in Fig. 10.7(a). Another option is to move the beam bars in the flush face inside by 1½ or 2 in. to miss the column bars, as shown in Fig. 10.7(b). Beam bars so moved could be offset at the column, or straight bars could be used and moved for their full length, in which case stirrups should be bent narrower to accommodate the bar spacing. This is especially appropriate if the flush face is a spandrel beam with vertical slotted anchors for attaching brick or other facing material. (Concrete cover over bars should be measured at the slotted insert. Thus, the outer face shell of concrete on the beam, equal to the width of the slot, becomes ineffective for structural purposes.) In either case, minimum spacing of the beam bars must be maintained.

The best method for avoiding bar interference is to make beams wider than columns by at least 2 in. on each side so that column bars and beam bars in the corner will pass each other without interference. Figure 10.7 shows beams wider than columns on one side.

In any case, where two bars pass each other, at least ½ in. clearance should be provided. Where clearances are tight, lugs on the bars tend to interlock and prevent one bar from sliding past the other, making erection difficult or even impossible.

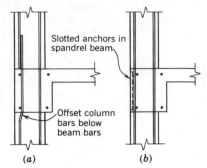

FIGURE 10.7   Bar bending to avoid interference at beam-column joint.

Anchorage of Flexural Bars

At interior supports, anchorage of flexural bars is usually accomplished by extending them into the joint as far as necessary, even into the span on the other side (see Fig. 5.6). At exterior supports, however, flexural bars must be anchored within the joint.

At exterior supports, the strength of flexural tension bars must be developed across the width of the support measured from the inside face of support to a point 2 in. inside the outside face. In most cases, straight bars will not be satisfactory unless the floor framing system extends as a cantilever beyond the outside face of the column [Fig. 10.8(a)]. The column is usually not wide enough to anchor straight bars (see Table A5.1). Hooked bars will normally be required, giving first preference to those with no restrictions (column A of Table A5.9) and then those with 2½ in. side clearance (column B of Table A5.9) and, finally, those with side clearance and also restrained with ties (column C of Table A5.9) [see Fig. 10.8(b)) and 10.8(c)]. If the beam or slab depth is insufficient to permit vertical installation of the hook, it may be rotated to slope as necessary, but the hook will probably then interfere with column vertical bars so that about 3½ in. clearance from the outside face of the column will be required (see Fig. 10.9). Where possible, it is best to place flexural bars with 90° hooks inside the column vertical bars, but the tail end of the hook should not extend

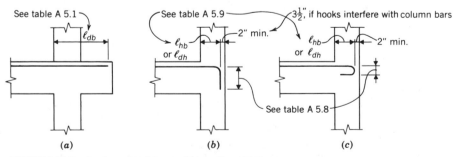

FIGURE 10.8   Anchorage of flexural bars at end columns.

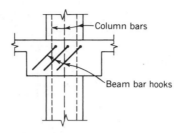

FIGURE 10.9 Installation of long hooks in a shallow beam.

below the bottom of the beam to avoid extending through the construction joint into the column below. If higher strength concrete is placed in the floor system around the column to transmit column loads through the floor system, the higher strength concrete may be used to reduce the hook development length.

If the joint is not large enough to accommodate the required tension bar hooks, there are at least two options open to the engineer.

1. Increase the area of steel provided so that a shorter hook development length is required.

2. Enlarge the beam or column, or both, to reduce the area of steel required (if the beam is made deeper) or permit installation of more bars of smaller size.

Example 10.2 Select top bars at the exterior support for a 24-in.-wide by 20-in.-deep beam resting on a 12-in. square column where ^-A_sreqd. $= 2.51^{□''}$ in the beam. Assume $f'_c = 3$ ksi and $f_y = 60$ ksi.

Solution Two bar selections will be considered.

1. 8 #5 bars with standard 90° hooks.

   $A_s$prov. $= 2.48^{□''}$ (1% under)

2. 4 #8 bars with standard 90° hooks and ties on the tail of the hook.

   $A_s$prov. $= 3.16^{□''}$ (26% over)

Referring to Table A5.9, hooks on #5 bars require a 9.6″ development length with 2½″ side cover, and hooks on #8 bars require 12.3″ development length with both 2½″ side cover and ties enclosing the hooks. With #8 bars, excess steel is provided and the hook development length can be reduced proportionally. Thus, hook length required = 12.3(2.51/3.16) = 9.8″. In both selections, hooks can be anchored in a 12″ column after allowing 2″ clearance at the outside face.

Hooks on both #5 and #8 bars can be extended down vertically after allowing 2″ cover at the top. For #8 bars, the width of a 90° hook is 16″ (from Table A5.8).

Normally, the fewest number of flexural bars would be selected. In this example, #5 bars would be the better selection because use of #8 bars would require ties

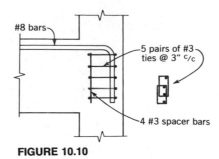

FIGURE 10.10

around the hooks, as shown in Fig. 10.10. Ties should be in pairs to properly restrain the hook extensions.

For thin columns, it may be inconvenient to anchor beam bars across the width of the column. In such cases, it may be advantageous to anchor column bars in the beam. Top beam bars in tension can then be bent down to lap with the column bars. At the top of a column, the longitudinal bars in tension should be bent horizontally. The horizontal bend at the top of vertical bars in a column will probably be lower than the theoretical position because the bar fabricator will cut bars a little short to be sure that they fit inside the concrete outlines. The reduction in effective depth will usually not be critical. The bend in both column and beam bars should have a diameter at least twice as large as the diameter of a standard hook in order to assure that full stress in the bar can be developed around the bend.

Example 10.3 Design the joint between a 24-in.-wide by 20-in.-deep beam and a 24-in.-wide by 8-in.-deep column in Fig. 10.11 to transfer 40′k of moment. The joint is at the top of the column. Assume f'_c = 3 ksi and f_y = 60 ksi for both beam and column. Six #6 longitudinal bars are required in the column.

Solution   Assuming $(\phi f_y j)$ = 4.0 and d = 17.5″,

$$A_s = \frac{40}{4.0 \times 17.5} = 0.57^{\square''}$$

Use 3 #4 top bars in the beam. $A_s$prov. = $0.60^{\square''}$

Referring to Fig. 10.11,

- Extend the 3 inside column bars into the beam a distance equal to the development length in compression.
- Extend the 3 outside column bars to the top of the beam and terminate them in standard hooks with twice the minimum diameter on the bend.
- Bend the 3 beam top bars down and lap with the column bars a length equal to the tension lap length for the #4 top bars. The bend diameter should be twice the minimum diameter.
- Continue the column ties to the top and add an extra set of ties to assure that

a splitting failure will not occur if beam bars are set inside the column bars. Without the ties, a potential crack could follow the bars for the length of the lap with failure occurring along the lap, as shown in Fig. 10.12.

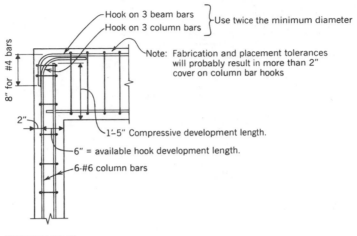

FIGURE 10.11

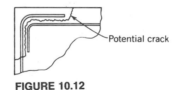

FIGURE 10.12

Another option is to enclose the hooks of the #4 beam bars in ties spaced at 3 times the bar diameter or $1\frac{1}{2}''$ c/c. Six ties would be required to cover the 8" length of the tail on the #4 bar hooks. From Table A5.9, the development length of the #4 hook is 6.1". Because excess steel area is provided, the development length can be reduced to $6.1(0.57/0.60) = 5.8''$. The available anchorage length = 6". For this option, column bars need not be hooked horizontally, and the bend diameter for hooks on the #4 bars can be the minimum diameter, but many closely spaced ties are required.

When bars from one member are bent into another member or anchored by a hook, the hook should be turned in such a way as to resist the diagonal component of the compressive force in the concrete. For example, consider a freestanding column with a cantilever beam near mid-height, as in Fig. 10.13(*a*). If the beam top bars are hooked down, the diagonal compressive forces at the inside radius of the hook will be counteracted by the diagonal component of the compressive forces from beam and column meeting at the joint, as shown in Fig. 10.13(*b*). It is a good practice to make the radius of the bend about twice that in a standard hook. Of course, the bend should be contained completely within the joint. If the hook were turned

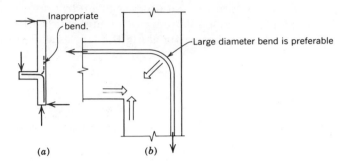

FIGURE 10.13   Direction of bent bars.

up, the path of diagonal compressive forces would be less direct and would need to be resisted by shear strength of concrete in the joint or by additional shear reinforcement.

Concrete Stress in Beam-Column Joints

Triaxial compression increases the compressive strength of concrete in any one direction by confining the concrete. Thus, the presence of compression from floor members perpendicular to the axis of the column does not reduce column capacity; nor is any increase in column capacity recognized as it is generally impractical to provide the triaxial compression and confinement necessary to obtain an increase in column strength.

When large moments are transferred between beam and column, especially moments due to seismic forces, shear stress on the concrete within the joint may be so high as to require reinforcement. ACI-ASCE Committee 352 has prepared recommendations for joint design in which the column width is equal to or greater than the beam width and normal weight concrete is used [1]. When beams are wider than the column, the joint is confined and reinforced by concrete on all sides. Engineers will rarely need to consider shear stress in a joint if design for earthquake does not control.

Offset Connections

Occasionally, for architectural reasons, the edge of floor framing will be at or even behind the inside face of the column, as in Fig. 10.14. Such connections should be designed for moment due to eccentricity of the connection as well as frame moments. Moments parallel to the slab edge will be transferred to the column by torsional shear in the interfacing plane. A conservative design would provide reinforcement to resist all moment and transverse and torsional shear without assigning any value to the concrete. Structural steel shearheads might also be required.

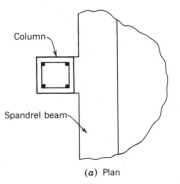
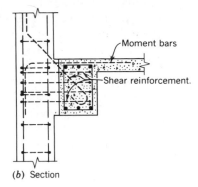

FIGURE 10.14   Offset spandrel beam.

Composite Columns

Structural steel cores of composite columns are spliced in a manner similar to columns in structural steel frames. End-bearing splices may transfer up to 50% of the total compressive force in the steel core, and the remainder must be transferred by splice plates. Ends must be accurately finished at end-bearing splices with positive provision for alignment of one core above the other.

At the foundation, the base plate of the steel core may be as large as the entire composite column, or larger. In this case, the steel base plate may be designed to transfer the total load on the composite column to the foundation using appropriate mechanical connectors to transfer load in the longitudinal bars to the steel base plate. A second option is to design the base plate to transfer the load carried by the structural steel section and concrete to the foundation and provide holes in the base plate to pass the longitudinal bars. Another option is to enlarge the column near its base so that an ample concrete section is available for transfer of the portion of the total load carried by the reinforced concrete section to the footing by compression in the concrete and by reinforcement (Code 7.8.21).

10.3 OTHER MONOLITHIC JOINTS

Joints between intersecting beams and between beams and slabs or joists are generally accomplished without additional design considerations if design of individual members is performed correctly. That is, shear in individual members must consider the conditions of support, bottom bars must be properly anchored by extending into the support 6 in. or more, all bars at the ends of members must be properly anchored to develop the design stress, and shear reinforcement and lateral ties must continue to the face or even into supporting members to the point where they are no longer needed.

When slabs or joists intersect a beam, top bars from the slab will usually cross bars from the beam without interference because only ¾ in. concrete cover is required on slab bars, whereas 2 in. cover (1½ in. required cover plus ½ in. for stirrups) is generally provided for beam bars. When slabs have exterior exposure, 1½ in. concrete cover is required, and slab bars larger than #4 will interfere with intersecting beam bars. Likewise, longitudinal top bars in intesecting beams will interfere with each other as the same cover is required in all beams.

Bottom bars in the supported beam can usually be placed on top of bottom bars in the supporting beam. Where bars in both beams carry either tensile or compressive stresses, it is necessary to select which beam to favor and place bars in the outer layer in one beam and in the inner layer in the other beam, as in Fig. 10.15. Of course, a reduced effective depth must be used in designing the beam with bars in the inner layer.

Knee Joints

Where flexural members are bent so that the tension face is on the concave side, there is a tendency for tension bars to straighten and pull out of the member by spalling the face shell of concrete. Where the bend is a gentle radius, the outward component of tensile force must be resisted by stirrups uniformly spaced in the curved portion. Where the bend is at a point, as in Fig. 10.16, tension bars are usually extended straight past the bend to the compression face and anchored by extending them parallel to the compression face. The distance from the crossing point of the tension bars to the end of each bar should be at least as long as the tension development length of the bars. The minimum beam width must allow for offsettting the bars at the bend.

Compression forces at the bend have an outward component T_r, as shown in Fig. 10.16(b), that tends to split the beam. The component T_r may be resisted by an

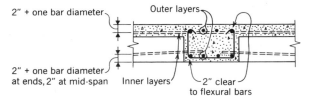

FIGURE 10.15   Bar placement at intersecting beams.

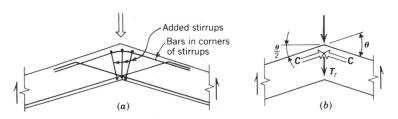

FIGURE 10.16   Reinforcement of bent beam or "knee."

external load in some cases. In a heavily loaded beam with a significant bend, T_r must be resisted by stirrups extending as close to the compression face as possible and wrapped around the tension bars. These stirrups should be in addition to those required for shear. Where the bend is not significant and bending stresses are low, as in lightly loaded slabs, it is common practice to omit the stirrups. Design of stirrups to resist the splitting force T_r may use a capacity reduction factor the same as that used for flexure, $\phi = 0.90$.

To determine when stirrups at the bend are required, consider the splitting force $T_r = 2C \sin \theta/2$, as in Fig. 10.16(b). Assume that the force T_r is resisted by tension in the concrete over a length of the beam equal to its effective depth d and that tensile strength of concrete $= \phi 5\sqrt{f'_c} b_w d$, as permitted for flexure in plain concrete. Equating the tensile strength to the force T_r and substituting $C = T = \rho_w b_w d f_y$ gives

$$\phi 5\sqrt{f'_c} b_w d = 2\rho_w b_w d f_y \sin \frac{\theta}{2}$$

Solving for θ,

$$\text{Maximum } \theta = 2 \text{ arc sin} \frac{2.5\ \phi\sqrt{f'_c}}{\rho_w f_y} \tag{10.2}$$

where $\phi = 0.65$ for plain concrete in tension.

Because of the assumptions used in developing Eq. 10.2, the factor 2.5 is approximate. Large, important members should be designed conservatively by assuming a value of 1.0 instead of 2.5, for the factor in Eq. 10.2. Stirrups should be designed to resist the entire splitting force and be distributed uniformly in the knee joint.

When solid stair slabs with redundant strength, such as that described in Ex. 3.5, are lightly loaded with $\rho \leqslant 0.5\%$, $f'_c = 3$ ksi, and $f_y = 60$ ksi, the maximum bend without reinforcement $\theta = 35°$. For beams, it is conservative to provide reinforcement to resist the splitting force when the bend angle θ is more than 40% of that given by Eq. 10.2. If the bend angle θ is more than 45°, careful detailing of reinforcement is required (see Ref. 2).

Where a flexural member is bent so that the compression face is on the concave side, the flexural tension and compression forces have equal and opposite radial forces acting inward; hence, there is little tendency to split the beam. In large, heavily loaded knee joints, to avoid high concentrations of stress inside the bend, the radius of bend on tension bars should be larger than the minimum permitted for the bar size but not more than the effective depth d. In sharp bends, such as a 90° corner, a fillet in the compression face is desirable.

Example 10.4 A 16-in.-wide by 24-in.-deep beam is bent upward, as in Fig. 10.16, so that the angle between the bottom faces is 145°. The beam is reinforced with 3 #8 tension bars. Select extra stirrups at the bend point. Assume $f'_c = 3$ ksi and $f_y = 60$ ksi.

Solution $C = T = 3 \times 0.79 \times 60 = 142$ k

$$\frac{\theta}{2} = 0.5(180 - 145) = 17.5°$$

$T_r = 2C \sin 17.5° = 2 \times 142 \times 0.30 = 85$ k

For stirrups, $A_s = \dfrac{85}{60} = 1.42^{\square''}$

More simply, $A_s = 3 \times 0.79 \times 2 \sin 17.5° = 1.42^{\square''}$

Use 4 #4 stirrups, $A_s$prov. $= 4 \times 2 \times 0.2 = 1.60^{\square''}$

In computing A_s, a capacity reduction factor is not needed because the compressive force C has already been magnified by $1/\phi$.

10.4 CONSTRUCTION JOINTS

All concrete structures, except perhaps for isolated footings or pads, are cast in sections with construction joints between them. The engineer responsible for design of the structure should specify the location of construction joints, preferably as general criteria to give the contractor as much leeway as possible to manage the work.

Construction joints should be located at points of zero or minimum shear in flexural members and at points where there is a minimum congestion of steel. In addition, it is advisable to have some compressive force across the joint to hold the joint in contact to transfer shear not anticipated by design.

Joints in columns will generally be located at the top of foundations and floor slabs and at the underside of the lowest beam or slab in floor framing systems or at the underside of column capitals in flat slabs.

Joints in beams and slabs are located at the point of zero shear under total load, generally at mid-span. If pattern loading or partial loading causes significant shear under some loading condition at all points in the beam, the joint should be designed specifically for the conditions.

Construction joints should be perpendicular to the compressive force acting across the joint; thus, in most cases, the joint will be perpendicular to the axis of the member. All reinforcing bars intersected by the joint should extend through the joint without interruption. A transverse keyway is commonly built into each joint regardless of the computed value of shear across the joint, except in slabs so thin that keyways are impractical. The key should be about ¼ to ⅓ of the depth of the member. When keys are made of lumber, contractors may put a slight taper in them to permit easy removal after casting concrete. Figure 10.17 illustrates a typical construction joint normally specified in flexural members. Diagonal shear reinforcement across the joint should be designed to resist shear in excess of one half the shear permitted on concrete because the full depth of the section may not be effective in . resisting shear.

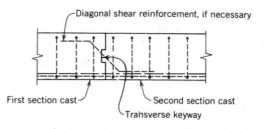

FIGURE 10.17 Construction joint in flexural members.

The location and details of construction joints in thick mat foundations should be specified due to their importance and because location of lines of zero shear are not apparent to construction workers. Multiple shear keys may be used, and diagonal shear reinforcement across the joint will normally be required.

10.5 EXPANSION JOINTS

When changes in the length of a structure occur after it is built, the movement may cause damage to portions of the structure resisting movement or to nonstructural elements. In order to minimize damage, a structure can be separated into two parts by a joint that permits movement in any direction. Such joints are commonly called "expansion joints," although contraction movements are also accommodated. The Portland Cement Association has issued an excellent booklet on "Building Movements and Joints [3]. Expansion joints are expensive and should be used sparingly.*

Significant fluctuations in the length of concrete structures can occur due to changes in temperature. If prestressing is present, the length will shorten due to creep. Volume changes from concrete shrinkage generally need not be considered because embedded reinforcement resists the shrinkage and cracks relieve shrinkage stresses, resulting in little change in overall length. Pedestrian bridges or other connections between individual buildings that can sway independently in the wind or during an earthquake usually require expansion joints separating the bridge from one or both of the buildings.

In computing thermal movement, changes in temperature from the average temperature over a 2- or 3-day period in the warmest season to a similar average in the coldest season should be used. Use of daily temperature extremes would be too conservative because it takes a day or longer to heat or cool a structure to the ambient air temperature. The temperature of roof structures exposed to the sun may rise above the ambient air temperature due to solar heating. Enclosed buildings continually heated in the winter and air-conditioned in the summer will experience little, if any, fluctuations in length from temperature.

---

*Conventional wisdom among experienced engineers is that a lack of expansion joints may cause problems but the presence of expansion joints is sure to cause problems.

Computation of volumetric changes due to prestressing is covered in any standard text on prestressing.

Thermal expansion and contraction of a structure will be resisted by its foundations through columns, walls, and other vertical elements. Although this resistance is frequently disregarded in computing length changes, the engineer should at least recognize that computed free thermal expansion and contraction will probably be more than will be measured in the completed building.

Expansion joints are normally not necessary in enclosed, heated and air-conditioned buildings up to 200 ft long or more. Expansion joints are normally not necessary in open structures exposed to temperature extremes unless the length exceeds about 100 ft. Structures may exceed these lengths without expansion joints if the engineer computes the fluctuations in length and considers their effect on the structure and on nonstructural elements.

For structures longer than 300 to 400 ft, a pour gap 2 to 4 ft wide can be left open across the structure at mid-length. After the structure is enclosed and heated, the pour gap can be concreted, making a monolithic structure for the full length. Before concreting, each portion of the structure can move independently to accommodate thermal and shrinkage movements during construction. The pour gap should be only as wide as necessry to lap-splice reinforcement extending into it [4].

Joints should extend straight across a building. They should be located at mid-length of the building, at its narrowest point if the width is variable. Joints are normally constructed initially 1 in. wide but should be wider if the expected movement would close the joint to less than ½ in. Joints in floors need covers that require careful architectural detailing to present a smooth floor surface. Joint covers are expensive and are seldom as satisfactory as a floor without joints; therefore, joints should be placed in locations as unobrusive as possible, preferably where covers are not needed.

A common method of constructing an expansion joint is to support slabs and beams from one side of the joint on brackets on the other side of the joint as in Fig. 10.18(a). The supported slabs and beams are allowed to move on sliding bearings or elastomeric pads. Design of brackets should give special attention to the horizontal force caused by movement. The force can be computed as the reaction of the supported slab or beam multiplied by the coefficient of friction of the bearing surface. The most critical design condition will usually occur when the expansion joint has opened to maximum width and bearing width is at a minimum.

A second common method is to build the structure on each side of the joint, independent of the other, with only the required clearance between them, as in Fig. 10.18(b). At the joint, double columns and double beams are constructed. Columns on each side should be at the same location, and beams on each side should be the same size to minimize differential deflection of floor members.

A third method used occasionally is similar to the second. Columns on each side of the joint are separated by a few feet. Slabs and beams are cantilevered from the columns to the joint, as in Fig. 10.18(c). If the columns are separated by a full bay dimension, the cantilevers will be one-half bay long and deflection will be excessive unless special precautions are taken. Smooth, sliding dowels are frequently used to prevent differential deflection from disturbing the floor surface.

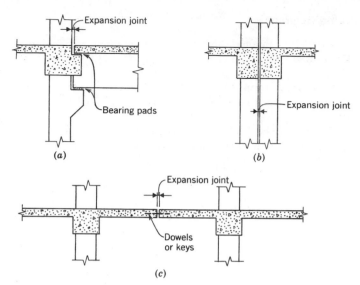

FIGURE 10.18   Expansion joints.

PROBLEMS

1. Select dowels for the foundation of the following columns. Assume that the columns are fully loaded, eccentricity is less than the minimum, no moment is transferred between the column and the foundation, and the foundation is much larger than the column on all sides.

Size (in.)	Col. Reinf.	Col. f'_c (ksi)	f_y (ksi)	Found. f'_c (ksi)
A. 8 × 24	6 #9 tied	3	60	3
B. 8 × 24	6 #8 tied	4	40	3
C. 30 round	27 #11 spiral	5	60	4
D. 36 × 36	30 #14 spiral	7.5	60	4
E. 16 × 20	4 #10 tied	5	60	4

2. Select dowels for a 24-in. square tied column with 10 #11 longitudinal bars bearing concentrically on a 30-in. square pedestal, 3 ft high. Assume f'_c = 4 ksi for both column and pedestal, f_y = 60 ksi, and no moment is transferred between the column and the pedestal.
3. Select dowels for transmitting the column load through a floor system where the column is 20 in. square above and below the joint. The column below is reinforced with 12 #11 bars and the column above is reinforced with 8 #11 bars. Column concrete f'_c = 6 ksi and floor concrete f'_c = 3 ksi. The floor system surrounds the column at a constant depth.
4. Select negative moment bars at the exterior support for a 12-in.-wide by 24-in.-deep beam framing into an 18-in.-wide by 10-in.-deep column when ^-A_s = 1.40$^{□''}$ in the beam. Assume f'_c = 3 ksi and f_y = 60 ksi for both the column and the beam.

Selected References

1. ACI-ASCE Committee 352. "Recommendations for Design of Beam-Column Joints in Monolithic Reinforced Concrete Structures," *ACI Journal*, V.82, pp 266−283, May−June 1982.
2. Erik Skettrup, Jørgen Strabo, Niels Houmark Andersen, and Trols Brøndum-Nielsen. "Concrete Frame Corners," *ACI Journal*, V. 81, No.6, pp 587−593, November−December 1984.
3. *Building Movements and Joints.* Portland Cement Association, Chicago, 1982.
4. "Palatial Hall Design Spans the Centuries," *Engineering News-Record*, pp 32−36, December 6, 1984.

CHAPTER 11

FOUNDATIONS

11.1 OBJECTIVES

The structural frame must carry all horizontal loads, as well as gravity loads, to the ground in a safe and serviceable manner. Because the surface of the ground is usually not suitable for supporting building loads, the structural frame must carry building loads into the ground to a level or bearing stratum that can support the building loads without further assistance from construction. For example, most buildings must be founded below the level of topsoil, below surface disturbances such as plowed land, below organic material subject to decay, below the zone of surface material subject to heaving from frost in northern climates, and below the effects of erosion such as might occur at a seashore. To meet these limitations, most foundations are at least 2 to 3 ft below the surface of the soil. Suitable soil-bearing strata are frequently available only at still greater depths.

The foundation is that part of a structure that receives loads from columns, piers, walls, and other members and transmits them to soil of sufficient strength to support the loads without excessive settlement. The foundation includes the supporting soils. Differential settlement from one part of a foundation to another part must be limited. In addition to gravity loads, the effects of wind, earthquake, loads from other sources, moment, and uplift on the foundation must be considered.

A footing is an individual member of a foundation supporting one (or, sometimes, more than one) column, pier, or wall.

The substructure is that part of a structure below the grade line. It includes the foundation and sometimes a basement and other portions of the structure. The superstructure is that portion of a structure above the grade line.

Foundations may be broadly classified as shallow, or spread, foundations and deep foundations. Spread foundations are those bearing on soils a short distance below grade and below the lowest floor level; they are discussed in Sections 11.3, 11.4, 11.5, and 11.6. Deep foundations employ special construction techniques to carry loads deeper into the soil to reach a bearing stratum strong enough to support the loads; design of deep foundations is covered in Section 11.7.

Objectives in the design of foundations are as follows:

1. Select the type of foundation to be used.

For Shallow or Spread Foundations

2. Select the size, shape, and location of individual footings to satisfy soil-bearing requirements.
3. Select the thickness of footings to provide adequate footing strength.
4. Select the number, size, spacing, and arrangement of reinforcing bars.
5. Design auxiliary features such as dowels, mud slab, top bar support system if bars are heavy and the footing is thick, and concrete casting procedures if individual footings are very large.

For Deep Foundations

6. Select the type and capacity of pile or caisson.
7. Select the number of piles or caissons to support each column, wall, or load.
8. Select reinforcement in individual piles and caissons, if required.
9. Select the shape, size, and reinforcement for pile caps and tie beams, if required.

11.2 LOADING AND PROPORTIONING

Skillful design of foundations blends knowledge of structural engineering and geotechnical engineering because soil and structure interact. Knowledge of this interaction is necessary to determine loads, pressures between soil and structure, and other design criteria. In most cases, a structural engineer will be advised by a geotechnical engineer on allowable soil pressures, load capacities, depth to bearing strata, and other relevant matters. Geotechnical considerations are beyond the scope of this text, and appropriate typical values will be assumed in explaining the design of foundations.

A competent geotechnical engineer should recommend an allowable soil-bearing pressure based on either 1) the load imposed on the footing, or 2) the total load of the column, pier or wall plus the weight of the footing (or net weight of the footing, i.e., the weight of concrete less the weight of displaced soil) and its surcharge (superimposed soil, slab, or other construction directly over the footing). Surcharge loads are generally uniform loads. If consideration of surcharge and weight of the footing is required, the easiest procedure in the design of spread footings is to subtract the unit footing weight and surcharge from the allowable soil-bearing pressure and to proportion the footing for the net remaining bearing capacity.

The building code controlling design of a building will require minimum depths to the bottom of footings below the final outside grade to prevent frost heave, erosion, and the entrance of vermin.

The size and shape of footings are based on unfactored service loads, rather than factored loads, so that soil pressures will be as uniform as possible throughout the foundation. Thus, the required area **A** of a footing is

$$A = \frac{D + L}{p} \quad \text{or} \quad \frac{D + L + W}{1.33p} \quad \text{or} \quad \frac{D + L + E}{1.33p} \tag{11.1}$$

whichever is greater, where **p** is the design soil pressure.

Frequently, the weight of the footing and its surcharge may be disregarded. In other cases, the geotechnical engineer will require that surcharge be considered in determining the size of footings. Surcharge is frequently disregarded in design of basement wall footings where there is very little superimposed load on the footing on the inside of the wall and a high superimposed load from the backfill material on the outside of the wall. Wall footings are usually placed symmetrically under the wall.

If a footing is subject to overturning or uplift, only those live loads that contribute to overturning or uplift should be included. Dead loads (including weight of the footing and surcharge directly over the footing) that resist overturning or uplift must be multiplied by 0.9 to allow for the possibility of actual loads lighter than those assumed in design. A factor of safety against overturning or uplift of at least 1.5 should be maintained. The building code controlling design of the structure should be checked for more restrictive requirements.

In some cases with compressible soils, footings must be proportioned using the dead load and only a portion of the live load, because full live load is rarely applied permanently. A more uniform long-term settlement is achieved by equalizing bearing pressure under permanent loads only. Soil pressure under footings with the largest percentage of live load must still be equal to or less than the allowable pressure when loaded with both live load and dead load. In effect, this will require larger footings for members supporting a small live load. The soil pressure under total dead and live load will be less for members with a small live load than for members with a large live load.

After the footing size and shape have been selected, the remainder of the design is completed using factored loads and bearing pressures.

Eccentrically loaded footings should be avoided. Eccentricity will either cause moment in the member being supported or nonuniform bearing pressure under the footing, or both. If the pressure is nonuniform, the footing may settle differentially across its width and may tilt and damage other parts of the building.

If a footing must be loaded eccentrically or if a bending moment in addition to a vertical load is applied at the top of the footing, the moment may be distributed between moment caused by a nonuniform bearing pressure and moment in the supported member. In cases where a distribution must be made, it is in proportion to the stiffness (unit rotation) of the column or member being supported to the stiffness (unit rotation) of the subgrade soils (see Chapter 13). In practice, most engineers assume that the nonuniform soil pressure resists all moment if the soil is stiff and the column is flexible. Alternatively, the column or member being supported is designed to resist all moment. In some cases it is convenient to conservatively assume that both nonuniform soil pressure and the supported member resist the full moment and thus avoid computations to distribute the moment. Such computations are inexact at best.*

The pressure resulting from vertical load P and moment M must not exceed the allowable bearing pressure p, or $1.33p$ for loading cases that include wind or earthquake. If the eccentricity M/P is within the kern of the footing (middle third for rectangular footings with moment parallel to one side, see Fig. 11.1), the pressure is given by the familiar formula

$$p = \frac{P}{A} \pm \frac{M_x c}{I_x} \pm \frac{M_y c}{I_y} = \frac{P}{A}\left(1 \pm \frac{6e_x}{\ell_x} \pm \frac{6e_y}{\ell_y}\right) \qquad (11.2)$$

where M_x and M_y = moments in the x and y direction, respectively.

e_x and e_y = eccentricity in the x and y direction, respectively.

$$= \frac{M_x}{P} \text{ and } \frac{M_y}{P}, \text{ respectively.}$$

ℓ_x and ℓ_y = lengths of the footing in the x and y direction, respectively.

I and c = section properties from strength of materials for the interface area between bottom of the footing and the soil.

I = moment of inertia.

c = distance from the neutral axis to the extreme fiber.

Weight of the footing and its surcharge can be considered in using Eq. 11.2, but all stabilizing dead loads not contributing to the overturning moment must be reduced by 10%, as noted earlier.

Because the interface between footing and soil can take no tension, Eq. 11.2 is not applicable when the eccentricity places the resultant vertical load outside the

---

*Assessing the stiffness of foundation soils is fraught with many uncertainties and questionable accuracy. In case of doubt, prudent engineers elect to "play it safe."

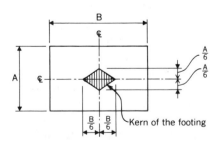

FIGURE 11.1 Footing kern.

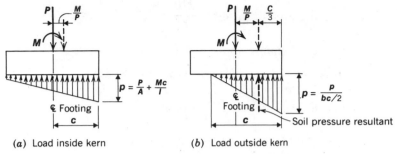

FIGURE 11.2   Soil-bearing pressure under eccentrically loaded footings.

kern. Soil pressure shown in Fig. 11.2(*b*) for rectangular footings when the moment is parallel to one of the sides is given by

$$p = \frac{p}{bc/2} \tag{11.3}$$

where **b** = the footing width, and,

   c = three times the distance from the resultant vertical reaction to the edge of the footing.

When the moment is not parallel to one of the sides, it is necessary to check footing pressures by trial and error.

Footings loaded outside the kern are sensitive to small increases in moment or decreases in stabilizing dead load. The footing should be checked for overturning by assuming that the moment can increase at least 50% while the stabilizing dead load remains constant. In this case, an increase in soil pressure equal to the increase in moment would be permitted. As noted earlier, only 90% of the stabilizing dead load should be considered.

Footings are generally not designed for moment caused by the eccentricity of a footing placed within normal tolerances.

Example 11.1   Determine the size of a footing required to support a service dead load of 120k and a service live load of 90k with an overturning moment due to wind, independent of vertical loads, applied in any direction, of 175′k. (Wind load must be considered to apply in any direction for footings supporting isolated columns. When a column is part of a frame, wind load, in some instances, may be assumed to act only in orthogonal directions.) Allowable soil-bearing pressure is 5 ksf at the bottom of the footing. Assume that the bottom of the footing is 3 ft below grade.

Solution

Case I for Dead Loads Plus Live Loads only.

   P = **D** + **L** = 120 + 90 = 210 k

$$\text{Footing area required} = \frac{210}{5} = 42^{\square\prime}$$

Use 6'-6" square footing, area = $6.5^2 = 42.2^{\square\prime}$

Case II for Dead Load Plus Live Load Plus Wind

$$\text{Eccentricity} = \frac{M}{P} = \frac{175}{210} = 0.83'$$

To keep load inside the kern, $\ell_{min} = 6e$. Use a square footing because wind can blow from any direction.

Try 8'-0" square footing, area = $8^2 = 64^{\square\prime}$
Moment acting in the diagonal direction is most critical.
The diagonal kern distance

$$= \frac{8.0}{6\sqrt{2}} = 0.94' \text{ from the column centerline}$$

$0.94 > 0.83$; therefore use Eq. 11.2.

On the diagonal, $I/c = S = \dfrac{8.0^3}{6\sqrt{2}} = 60.3 \text{ ft}^3$

$$\text{Soil pressure} = \frac{P}{A} + \frac{Mc}{I}$$

$$= \frac{210}{64.0} + \frac{175}{60.3} = 3.28 + 2.90 = 6.2 \text{ ksf} \qquad \underline{\text{OK}}$$

Allowable soil pressure under loading combinations that include wind = $5 \times 1.33 = 6.67$ ksf

Case III for Dead Load Only Plus 150% Wind (Overturning)—Wind Forces Parallel to Sides of Footing

Try 8'-0" square footing and assume that the footing is 3' thick.

$$\text{Footing weight} = 8.0^2 \times 3 \times 0.15 = 28.8 \text{ k}$$

$$\text{Column dead load} = \underline{120}$$

$$\text{Total dead load} = 148.8 \text{ k}$$

$$\text{Eccentricity} = \frac{175 \times 1.5}{148.8 \times 0.9} = 1.96'$$

Because the reaction lies outside the kern, use Eq. 11.3.

$c = 3(4.00 - 1.96) = 6.12'$

$$\text{Soil pressure} = \frac{148.8 \times 0.9}{8.00 \times 6.12/2} = 5.47 \text{ ksf} \qquad \underline{\text{OK}}$$

Permitted soil pressure under conditions in Case III is 150% of the permitted soil-bearing pressure in Case II, or 10.0 ksf.

Case IV for Dead Load Only Plus 150% Wind (Overturning)—Wind Diagonal to Sides of Footing

If wind forces can act in directions diagonal to the footing, the reaction is well beyond the diagonal kern distance of 0.94' for a 8'-0" square footing. Try 9'-0" square footing, 3' deep.

Footing weight = $9^2 \times 3 \times 0.15 =$ 36.4 k

Column dead load   = 120.0

Total load   = 156.4 k

Conservatively, reduce the load by 10%,

Footing load = $156.4 \times 0.9 = 140.8$ k

Because the computations become quite laborious in making an exact analysis, assume that the neutral axis (line of zero soil-bearing pressure) is on the diagonal of the footing and then compute axial load capacity and moment-resisting capacity.

Referring to Fig. 11.3, magnitude of the soil-bearing pressure resultant R is the volume of the pressure pyramid, and its location is at the centroid of the pressure pyramid. Thus,

$$R = \frac{9 \times 9}{2} \times 10 \text{ ksf} \times \frac{1}{3} = 135 \text{ k} < 140.8$$

Location of resultant, $x = \dfrac{9}{\sqrt{2}} \times \dfrac{3}{4} = 4.77'$

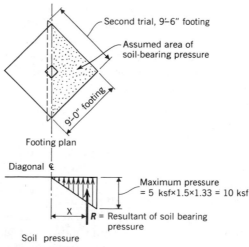

Footing plan

Diagonal ℄

Soil pressure

Second trial, 9'-6" footing

Assumed area of soil-bearing pressure

9'-0" footing

Maximum pressure = 5 ksf×1.5×1.33 = 10 ksf

R = Resultant of soil bearing pressure

FIGURE 11.3

Potential overturning resisting moment

$$= 135 \times 4.77 = 644'\text{k} >> 175 \times 1.5 = 263' \text{ k}$$

Even though the computed resultant of soil-bearing pressure is less than actual dead loads (reduced by 10%), the footing is satisfactory because the potential overturning resisting moment is much greater than the applied moment. To verify this statement, there is not much error introduced by assuming a 9'-6" square footing to determine the magnitude and location of the soil pressure resultant but only a 9'-0" square footing to determine the overturning resisting moment. Thus,

$$\boldsymbol{R} = \frac{(9.5)^2}{2} \times 10 \text{ ksf} \times \frac{1}{3} = 150.4 \text{ k} > 140.8 \text{ k}$$

Location of resultant, $\boldsymbol{x} = \dfrac{9}{\sqrt{2}} - \dfrac{9.5}{\sqrt{2}} \times \dfrac{1}{4} = 4.68'$

Potential overturning resisting moment

$$= 140.8 \times 4.68 = 660'\text{k} >> 175 \times 1.5 = 263' \text{ k}$$

A 9'-0" square footing is conservative. The savings in construction cost realized by a smaller footing would be less than the cost of additional computations to select a smaller footing, if only one footing is needed.

To provide protection against corrosion, the ACI Code requires 3 in. minimum concrete cover on all bars where concrete is cast against and permanently exposed to earth. Where concrete is exposed to earth or weather but not cast against earth, the cover over reinforcement can be reduced to 2 in. for #6 bars and larger and to 1½ in. for #5 bars and smaller (Code 7.7.1). For large footings where muddy or loose soils are expected, a thin slab, 1 to 3 in. thick, is sometimes cast on the ground before placing reinforcement. It is called a "mud" slab because the purpose is to provide a clean surface for workers and prevent muddy soil from contaminating the reinforcement or the bottom layer of concrete. Mud slabs in good condition can be used to reduce the concrete cover required over reinforcement; however, to ensure that concrete can surround the bars, it is good practice to provide at least as much cover between the mud slab and the bars as the cover required where concrete is not exposed to weather or in contact with earth.

11.3 SPREAD FOUNDATIONS

Spread foundations distribute load directly to a shallow soil-bearing stratum. Soil-bearing pressure is normally assumed to be uniform, as in Fig. 11.4(a), although actual pressure may be higher at the edges of a footing where it is bearing on a cohesive soil [Fig. 11.4(b)] or higher near the middle of the footing where it is bearing on a granular soil [Fig. 11.4(c)]. A geotechnical engineer will advise the structural engineer if it is necessary to design footings for nonuniform soil-bearing pressure.

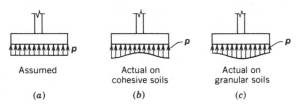

FIGURE 11.4 Soil-bearing pressures.

To simplify construction and to promote repetition of similar footing elements, it is desirable to design the thickness of individual footings in multiples of 4 in. (e.g., 12, 16, 20, 24 in.), with a minimum thickness of 12 in. Likewise, the widths and lengths of footings should normally be in multiples of ½ ft (e.g., 4, 4½, 5, 5½ ft). The widths of wall footings should be varied by increments of 4 in. With steps of this magnitude, footings will vary in capacity by steps of 30% for small footings and 10% or less for large footings.

The thickness of concrete in footings on soil must not be less than 6 in. above reinforcement. Many engineers consider 12 in. as the minimum total thickness for a footing. In some cases, footings thicker than that required for shear or flexure may be necessary to develop column dowels for concrete columns or anchor bolts for steel columns.

The bottoms of footings must always be cast against the soil on which they bear. The edges of footings may also be cast against soil if a neat cut can be made in the soil and if it will remain vertical until concrete is cast. Where soils are loose or ravel with exposure, the edges of footings must be formed. If the soil must resist lateral forces, casting the edge of the footing against soil may be desirable or necessary.

Spread foundations can be grouped as

1. Isolated footings under columns and piers,
2. Strip footings under walls or three or more closely spaced columns on a straight line,
3. Combined footings under two or three columns, piers, or walls.
4. Mat foundations, which are continuous footing slabs under four or more columns, piers, or walls arranged in a grid pattern.

Design of each of these foundations will be considered separately.

Design Procedures

Spread foundations can be designed efficiently by using the following step-by-step procedure.

1. Determine the required area of footing in contact with the ground using service loads and allowable soil-bearing pressure.
2. Select the size, shape, and location of the footing.

3. Convert service loads to factored loads. Divide the total factored loads by the footing area to compute factored soil pressure. Use factored loads and soil pressure in design for shear and for flexural reinforcement.

4. Determine the thickness required for shear, preferably without the use of shear reinforcement.* In most cases, the minimum thickness to resist shear will be sufficient to develop dowels for concrete columns and anchor bolts for steel columns and to provide adequate moment resistance. If it is not, increase the depth as necessary.

5. Select shear reinforcement, if required.

6. Select reinforcement to resist moment, if required.

As with beams and slabs (see section 2.2), where *positive* flexural reinforcement (resisting positive moment) is required by analysis in a foundation member, the ratio $\rho = A_s/bd$ provided shall not be less than that given by $\rho_{min} = 200/f_y$, where f_y is in psi.

Alternatively, the area of flexural reinforcement provided at every section, positive or negative, must be at least one-third greater than that required by analysis (Code 10.5). Although not strictly required by the ACI Code, these provisions are conservatively applied to negative reinforcement also, as it is the primary reinforcement in most footings.

Positive moment is the moment caused by the bending of a member between two supports with flexural tension in the member on the face opposite to the loaded face. In a footing with positive moment, flexural tension is in the top. Negative moment is the moment caused by the bending of a member at a support with flexural tension in the loaded face. In a footing with negative moment, flexural tension is in the bottom.

Code provisions for temperature reinforcement do not apply to foundations because footings are much less subject to large temperature variations and to shrinkage stress from drying. Foundations are buried in the ground and they usually have more massive proportions than slabs and beams. Nevertheless, it is advisable to use somewhat more reinforcement than the minimum when ρ is much less than that required for slabs (see Section 3.2).

For economy, fewer bars of larger sizes should be used. Maximum spacing of bars should be about 18 in., although the ACI Code has no specific limits. Bars must be developed on each side of each critical section for moment by embedment length, a hook (for tension only), or a mechanical device. Straight bars are normally used, but hooked bars can be used if necessary.

Construction workers will usually place heavier and longer bars in the bottom layer. Design should be based on this assumption and the placing sequence specified where it is critical.

For simplicity, bar spacing should be rounded down to the next lower inch. Bars parallel to the edge of a footing should be placed with at least 4 in. cover, as in Fig. 11.5, to allow space for tying intersecting bars.

---

*One mark of maturity in a structural engineer is a reluctance to use finicky details, such as stirrups in footings.

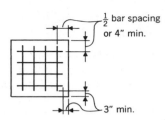

FIGURE 11.5 Plan of reinforcing bars in isolated footings.

The external moment on any section of a footing is determined by passing a vertical plane through the footing and computing the moment of the forces acting on the footing on one side of that vertical plane. The maximum factored moment for an isolated footing must be computed at critical sections located as follows:

1. At the face of the column, pedestal, or wall, for footings supporting a concrete column, pedestal, or wall [see Figs. 11.6(a) and 11.7].

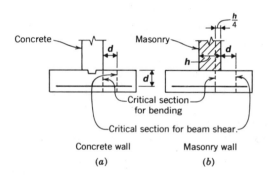

FIGURE 11.6 Location of critical sections for wall footings.

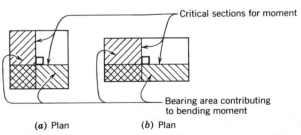

FIGURE 11.7 Critical sections for moment in isolated footings.

2. Halfway between the middle and the edge of the wall or pier, for footings supporting a masonry wall or masonry pier (see Fig. 11.6(b)].

3. Halfway between the face of the column and the edge of the steel base plate, for footings supporting a column with a steel base plate [see Fig. 11.8(c)]. The column could be a precast column as well as a steel column.

For footings supporting a column, pedestal, or wall, location of the critical section for one-way or beam action shear is a plane extending across the entire width of the footing at a distance d from the face of the column, pedestal, or wall, as in Figs. 11.6 and 11.8(b). Shear is computed the same as for one-way slabs or beams (see Chapter 6).

For two-way action or punching shear, the critical section is located around the column, pile, or reaction area so that the critical section perimeter b_o is a minimum but need not approach closer than $d/2$ to the face of the column, pile, or reaction area, as in Fig. 11.8(a). When the column is bearing on a steel base plate, the distance $d/2$ may be measured from the critical section for bending, as in Fig. 11.8(c). Shear is computed in the same manner as for two-way slabs (see Chapter 7).

When moment is transferred between the column and the footing, a portion will be transferred by shear in a manner similar to moment transfer in two-way slabs (see Section 7.3 for a discussion of calculation procedures). The ACI Code does not require calculation of shear in footings caused by moment transfer, but it seems prudent to do so when large moments are transferred. Most practicing engineers disregard the increase in shear due to transfer of small moments but select thicker footings than are required for shear.

Shear force around a column is computed by reducing the total load on the column by the soil reaction under the column within the critical perimeter b_o. Using total load on the footing (without reduction) as the shear force will be conservative, by as little as 3% for thin footings with low soil pressure and by as much as 40% or more for thick footings with high soil pressure. Shear force around a pile can be conservatively taken as the pile reaction.

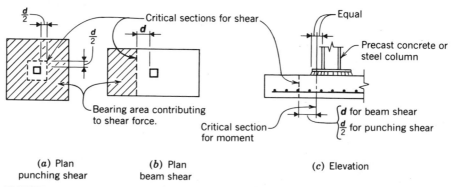

(a) Plan
punching shear

(b) Plan
beam shear

(c) Elevation

FIGURE 11.8   Location of critical sections for shear in footings supporting columns or piers.

The maximum shear stress in concrete due to factored loads and moments is limited to the following:

For beam action, $\qquad\qquad\qquad\qquad\qquad\qquad\qquad 2\phi\sqrt{f'_c}$

For two-way action (punching shear), $\qquad \left(2 + \dfrac{4}{\beta_c}\right)\ \phi\sqrt{f'_c}$

but not greater than $\qquad\qquad\qquad\qquad\qquad\qquad 4\phi\sqrt{f'_c}$

where $\boldsymbol{\beta_c}$ is the ratio of the long side to the short side of concentrated load or reaction area.

Punching shear will control for square footings, and beam shear will control for greatly elongated footings. Either beam shear or punching shear may control in footings intermediate between square and elongated, and both shear conditions must be checked.

Wherever possible, footings should be made thick enough to avoid using shear reinforcement, which may be unreliable and expensive in footings. Footings are important members of a structure, but they are husky and even crude. Delicate measurements, tolerances, reinforcement, and construction procedures are not appropriate for footings.*

Plain Concrete Footings

Due to their massive size, many footings are reinforced with only minimum reinforcement. Frequently, not even minimum reinforcement is required, so footings can be built of plain concrete. Plain concrete has no reinforcement or less than the minimum reinforcement required for reinforced concrete. Plain concrete is designed assuming that there is no contribution to strength from any reinforcement that may be present [1].

The use of plain concrete is limited to the following types of members:

1. Continuously supported by soil or by other structural members capable of providing continuous vertical support.
2. Not used to resist earthquake or blast loading.
3. Not used for pile caps.
4. With concrete strength $\boldsymbol{f'_c} \geq 2.5$ ksi.
5. With a minimum total thickness of at least 8 in. for footings.

Most spread footings will meet these limitations. However, most engineers will not use plain concrete for large or important footings, regardless of the level of stress. For footings with tension in the top surface, engineers will be more conserva-

---

*During the next rain or drizzle, try digging a hole at least 3 ft square in the ground accurately located and sized to the nearest inch and have your measurements independently verified by someone else a few days later. The odds of the second measurement confirming your own are low.

tive in using plain concrete because shrinkage cracks due to drying may initiate flexural cracks. The bottom of the footing in contact with the ground is less likely to dry, shrink, and crack. Where the use of plain concrete footings is permitted and appropriate, reinforcement should be used only where it will result in lower construction costs. In evaluating costs, consider the quantity of backfill above the footing, as well as concrete, steel, and formwork.

Flexural stresses in plain concrete are computed by the usual methods of structural mechanics for homogeneous materials ($f = Mc/I = M/S$), although the actual distribution of stresses in plain concrete is somewhat more complex. The maximum fiber stress in plain, normal weight concrete due to factored loads and moment must not exceed $5\phi\sqrt{f'_c}$ in tension nor $\phi f'_c$ in compression. Because concrete is much weaker in tension than in compression, compression controls only when the member is subject to large axial compression load. The strength reduction factor $\phi = 0.65$ for flexure, compression, shear, and bearing of plain concrete.

Where concrete is cast against soil, as is usually done in footings, the overall thickness h used in design must be taken as 2 in. less than the specified thickness to allow for unevenness of excavation and for some contamination of concrete next to soil. If a mud slab in good condition is used, the full specified thickness of a footing may be considered in design.

Load factors for plain concrete are the same as those for reinforced concrete. Critical sections for moment in plain concrete are located the same as those for reinforced concrete. Critical sections for shear in plain concrete are located similarly to those for reinforced concrete except that the effective thickness h (total thickness less 2 in. as noted above) is used rather than the effective depth d. The maximum shear stress in plain concrete due to factored loads and moments is the same as that for reinforced concrete.

Because the principal tensile stress in a homogeneous material is higher than the average shear stress (see Fig. 11.9), shear stress in plain concrete rectangular members under beam action is computed using

$$v_u = \frac{3V_u}{2bh} \tag{11.4}$$

For two-way action or punching shear, substitute b_o for b in Eq. 11.4. Note the similarity to equations and permissible stresses for reinforced concrete (chapters 6 and 7; Eqs. 6.2 and 7.1).

Shear stress rarely, if ever, controls the design of plain concrete footings of rectangular cross section. The relationship of flexural and shear stresses to permissible stresses can best be visualized by plotting the ratio of computed stress to permit-

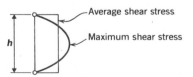

FIGURE 11.9   Shear stress in homogeneous, rectangular member.

ted stress in nondimensional form as a function of the span-to-depth ratio ℓ/h, as in Fig. 11.10.

In a plain concrete footing resisting one-way flexure, the flexural stress f_b is given by

$$f_b = \frac{M_u}{S} = \frac{3w_u\ell^2}{bh^2}$$

where $M_u = w_u\ell^2/2$ and $S = bh^2/6$.

Because the permissible flexural stress $f_{ba} = 5\phi\sqrt{f_c'}$,

$$\frac{f_b}{f_{ba}} = \frac{3w_u\ell^z}{bh^2 5\phi\sqrt{f_c'}}$$

Rearranging, inserting a factor of $(12\ \text{in./ft})^2$ to maintain compatibility of units, and letting $b = 1$ ft (12 in.),

$$\frac{f_b}{f_{ba}} \times \frac{\phi\sqrt{f_c'}}{w_u} = \frac{(\ell/h)^2}{240} \tag{11.5}$$

where $\sqrt{f_c'}$ is in psi

$\quad w_u$ = soil pressure in psf,

$\quad \ell$ and h are in the same units,

$\quad h$ = actual depth less 2 in.

In a plain concrete footing resisting beam shear, the shear stress v_u is given by

$$v_u = \frac{3V_u}{2bh} = \frac{3w_u(\ell - h)}{2bh}$$

where $V_u = w_u(\ell - h)$.

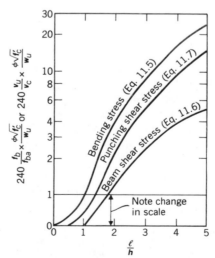

FIGURE 11.10 Stresses in a rectangular plain concrete members.

Because the permissible shear stress $v_c = 2\phi\sqrt{f_c'}$,

$$\frac{v_u}{v_c} = \frac{3w_u(\ell - h)}{2bh2\phi\sqrt{f_c'}}$$

Rearranging, inserting a factor of $(12 \text{ in./ft})^2$ to maintain compatibility of units, and letting $b = 1$ ft,

$$\frac{v_u}{v_c} \times \frac{\phi\sqrt{f_c'}}{w_u} = \frac{(\ell/h - 1)}{192} \qquad (11.6)$$

In a plain concrete footing resisting two-way flexure, the ratio of computed flexural stress to permitted flexural stress is the same as for one-way flexure. Referring to Fig. 11.11, the shear stress v_u is given by

$$v_u = \frac{3V_u}{2b_oh} \leqslant \frac{3}{8}\, w_u\, \frac{(4\ell^2 - h^2)}{h^2} = \frac{3}{2}\, w_u[(\ell/h)^2 - 0.25]$$

where $V_u = w_u(b2\ell - h^2)$
minimum $b_o = 4h$
and $b = 2\ell$ at the most critical condition.

Because allowable shear stress $v_c = 4\phi\sqrt{f_c'}$ when other conditions are most critical,

$$\frac{v_u}{v_c} \leqslant \frac{3w_u[(\ell/h)^2 - 0.25]}{2 \times 4\phi\sqrt{f_c'}}$$

Rearranging, inserting a factor of $(12 \text{ in./ft})^2$ to maintain compatibility of units, and letting $b = 1$ ft,

$$\frac{v_u}{v_c} \times \frac{\phi\sqrt{f_c'}}{w_u} \leqslant \frac{(\ell/h)^2 - 0.25}{384} \qquad (11.7)$$

Equations 11.5, 11.6, and 11.7 are plotted in Fig. 11.10. For rectangular sections, it is obvious that shear stress never controls. Engineers may design plain concrete rectangular members by considering flexural stresses only, if there is no significant tensile stress on the member from shrinkage or external forces, and if no significant bending moment is being transferred between column and footing.

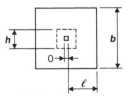

FIGURE 11.11 Critical section in footing with small column.

11.4 WALL FOOTINGS

When loads on a wall do not vary along its length or vary only by small amounts, the wall footing should be designed as a strip of uniform width. Concrete walls can distribute concentrated loads and other nonuniform loads over a considerable distance (see Chapter 12). Masonry walls will also distribute some nonuniform loads along the length of the wall. The required width of the footing is simply the service load per unit of length divided by the soil-bearing pressure, using compatible units. The minimum width of a wall footing should be somewhat wider than the wall itself to allow for inaccuracies in placement. Most engineers will project the footing at least 4 in. beyond each face of the wall.

Footings for cantilevered retaining walls are discussed in Section 12.5.

The depth of a wall footing is almost always selected to resist shear without use of shear reinforcement, using the same procedures as for one-way slabs.

Selection of reinforcement to resist moment can proceed in the same way as for one-way slabs, considering the footing inverted and cantilevered from the wall on each side. If the footing has a short cantilever on each side of the wall, reinforcement is frequently omitted when flexural stress in plain concrete is less than $5\phi\sqrt{f_c'}$.

In footings under masonry walls and other fragile or nonstructural walls, it is common practice to design longitudinal reinforcement so that the footing will carry part of the wall load over hypothetical soft spots in the soil. The amount of wall load to be carried and the span is based on engineering judgment considering the strength and rigidity of the wall and the strength and plasticity of the soil. Usually, the minimum reinforcement provided is that required for shrinkage and temperature reinforcement or 0.0018 times gross concrete area for grade 60 deformed bars and 0.002 times gross concrete area for grade 40 or 50 deformed bars (Code 7.12).

It is good practice to enlarge masonry wall footings at the ends of walls and at corners. The enlargement will tend to tip the corner inward toward the center of the wall, thus minimizing cracking in the wall due to outward movement if a soft spot in the foundation soil is encountered near the corner.

Where wall footings make large steps from a high elevation to a low elevation, the wall footing is frequently omitted at the step and the footing designed for increased load on each side of the gap. The wall, of course, must be designed to span over the gap (see Fig. 11.12).

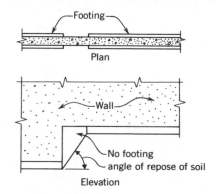

FIGURE 11.12 Stepped wall footing.

Example 11.2 Design a 12-in.-thick footing with $f'_c = 3$ ksi to support a 12-in. concrete wall carrying a dead load of 6 k/′ and a live load of 3.5 k/′. Use a soil pressure of (A) 2 ksf and (B) 3 ksf. In each case, the soil pressure need not include the weight of the footing or surcharge.

Solution A   For soil pressure of 2 ksf,

$$\text{Width of footing} = \frac{(6 + 3.5)}{2} = 4.75'$$

Use width = 5′-0″ with 2′-0″ overhang on each side of the wall.
Factored wall load = $1.4 \times 6 + 1.7 \times 3.5 = 14.35$ k/′

$$\text{Factored soil pressure} = \frac{14.35}{5} = 2.87 \text{ ksf}$$

Check flexure as plain concrete footing:

$$\boldsymbol{M_u} = \frac{\boldsymbol{w\ell^2}}{2} = \frac{2.87 \times 2^2}{2} = 5.74'\text{k/}'$$

Effective footing thickness = $12 - 2 = 10''$

$$\text{Section modulus, } \boldsymbol{S} = \frac{\boldsymbol{bh^2}}{6} = \frac{12 \times 10^2}{6} = 200 \text{ in.}^3/'$$

$$\text{Bending stress} = \frac{\boldsymbol{M_u}}{\boldsymbol{S}} = \frac{5.74 \times 12}{200} = 344 \text{ psi} \qquad \underline{\text{NA}}$$

$$\text{Permissible bending stress} = 5\boldsymbol{\phi}\sqrt{f'_c}$$
$$= 5 \times 0.65\sqrt{3000} = 178 \text{ psi}$$

Therefore, footing must be reinforced or made thicker.

Check shear as a reinforced concrete footing:
The critical section is distance $\boldsymbol{d} = 8.75'' = 0.73'$ from face of the wall.

$$\boldsymbol{V_u} = (2.0 - 0.73)2.87 = 3.65 \text{ k/}'$$

$$\boldsymbol{v_u} = \frac{3.65}{12 \times 8.75} = 35 \text{ psi} \qquad \underline{\text{OK}}$$

$$\text{Permissible shear stress} = 2\boldsymbol{\phi}\sqrt{f'_c} = 2 \times 0.85\sqrt{3000} = 93 \text{ psi}$$

Select transverse reinforcement for footing:

$$\boldsymbol{M_u} = 5.74'\text{k/}', \ \boldsymbol{d} = 8.75''; \text{ assume } (\boldsymbol{\phi f_y j}) = 4.3$$

$$\boldsymbol{A_s} = \frac{5.74}{4.3 \times 8.75} = 0.15^{\square''}/'$$

Minimum reinforcement for flexure is conservatively ⅓ greater than that required by analysis,* or $0.15 \times 1.33 = 0.20^{□''}/'$

Use #4 bars at 12" c/c. $\mathbf{A_s}$prov. $= 0.20^{□''}/'$

Select longitudinal reinforcement:

No temperature or flexural reinforcement is needed.
use 3 #4 bars to hold transverse bars in position.

Solution B   For soil pressure = 3 ksf,

$$\text{Width of footing} = \frac{6 + 3.5}{3} = 3.17 = 3'\text{-}2''$$

Overhang on each side of wall = $1'\text{-}1''$

$$\text{Factored soil pressure} = \frac{14.35}{3.17} = 4.53 \text{ ksf}$$

Check flexure as plain concrete footing:

$$\mathbf{M}_u = \frac{\mathbf{w}\ell^2}{2} = \frac{4.53 \times 1.08^2}{2} = 2.66'\text{k}/'$$

Section modulus $\mathbf{S} = 200$ in.$^3/'$ (from Solution A)

$$\text{Bending stress} = \frac{\mathbf{M}_u}{\mathbf{S}} = \frac{2.66 \times 12}{200} = 160 \text{ psi} \qquad \underline{\text{OK}}$$

Permissible bending stress = 178 psi (from Solution A)

Use plain concrete footing.

In this example, note the following:

1. Higher bearing pressure permits a smaller footing and a shorter cantilever past the face of the wall; therefore, a plain concrete footing is more likely to be permissible.

2. Even though a plain concrete footing is used, dowels must be inserted to match vertical reinforcement in the wall (see Chapter 12).

11.5 ISOLATED SPREAD FOOTINGS

Columns, piers, and similar load-carrying members are usually supported on square, isolated footings, unless a rectangular shape is necessary for a footing to clear underground obstructions. Extremely elongated columns and rectangular piers that occupy a high percentage of the footing area are supported on rectangular footings.

*For this and subsequent examples in Chapter 11, requirements for minimum reinforcement are conservatively applied. The ACI Code does not require such application, and engineers may disregard it.

In such cases, the footing overhang beyond the edge of the column or pier is made equal on all four sides (see Fig. 11.13). Footings of other shapes, such as round, hexagonal, trapezoidal, and so forth, are generally avoided as they are more expensive than rectangular footings to excavate, form, and reinforce. For the remainder of this chapter, reference to columns includes piers, pedestals, and other similar load-carrying members.

Isolated footings are almost always placed concentrically under the column load. Normally, the only exception is where an eccentric footing is required to clear underground obstructions. In such cases, the column must be designed for the eccentric moment—an expensive procedure. The design of isolated footings generally assumes that the footings have no fixity so that no moment is transferred between column and footing due to frame action. This assumption obviously has limits, and experienced engineers will sometimes design isolated footings for applied moment due to frame action, as well as for vertical loads and horizontal forces.

Footings on soils with very high bearing capacity, such as rock, may be so small that the distance $d/2$ to the critical section for shear would fall outside the footing for all but the thinnest footings. In such cases, the footing should be at least deep enough to develop column dowels in compression. Minimum penetration into the bearing stratum may control the footing depth. The footing should then be reinforced with vertical bars and lateral ties as required for a column (see Fig. 11.14).

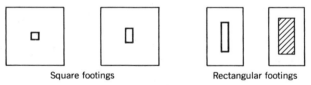

Square footings   Rectangular footings

FIGURE 11.13 Plan of typical isolated footings.

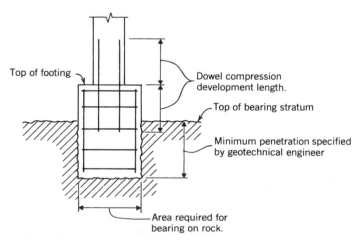

FIGURE 11.14 Footing bearing on rock.

In a two-way footing, moments in each direction must be computed using the full soil pressure, just as in a flat slab where moments in each direction are computed using full load on the slab.

Area of steel, anchorage, and other requirements for flexural reinforcement in footings are computed in the same manner as for other flexural members. For square footings, bars should be distributed uniformly across the entire width of the footings using the same bars and spacing in each direction to minimize field errors. In footings where bending moment is nearly the same in each direction, the effective depth is taken as the average depth for the two directions. Theoretically, steel in one direction will be slightly overstressed, and steel in the other direction will be slightly understressed, but overall strength of the footing will be adequate.

Example 11.3 Design a square footing to support a 16 × 24-in. column carrying a 460 k service dead load and a 370 k service live load. Allowable soil-bearing pressure is 5 ksf. Design footing for two conditions: (A) disregarding the weight of the footing and the soil above it, and (B) including the weight of the footing and the soil above it plus 100 psf live load on the basement floor. Assume that the bottom of the footing is 4'-6" below the basement floor. Use f'_c = 3 ksi and f_y = 60 ksi in the footing.

Solution A

$$\text{Area of footing required} = \frac{460 + 370}{5} = 166 \text{ ft}^2$$

Because the footing must be square, the side dimension = $\sqrt{166}$ = 12.9'

Use 13'-0" square footing, area = 169$^{\square'}$

Factored column load = 460 × 1.4 + 370 × 1.7 = 1270 k

$$\text{Factored soil pressure} = \frac{1270}{169} = 7.53 \text{ ksf}$$

Select depth to meet shear requirements:

An iterative trial and error procedure is usually the simplest. Try a 28" deep footing.

d = 28" − 3" cover − 1" bar size = 24"

The shear perimeter is a rectangle: (16 + 24) × (24 + 24) = 40" × 48"

b_o = 2(40 + 48) = 176"

Subtracting the force of soil pressure inside the critical shear perimeter,

V_u = 1270 − 7.53(3.33 × 4.00) = 1270 − 100 = 1170 k

$$v_u = \frac{1170}{176 \times 24} = 277 \text{ psi} \qquad \qquad \underline{\text{NA}}$$

Allowable shear stress = $\phi 4\sqrt{f'_c}$ = 0.85 × 4$\sqrt{3000}$ = 186 psi

Try a 36″ deep footing

$d = 36 - 3 - 1 = 32''$

The critical shear perimeter is a rectangle 48″ × 56″

$b_o = 2(48 + 56) = 208''$

$V_u = 1270 - 7.53(4.00 \times 4.67) = 1270 - 140 = 1130 \text{ k}$

$$v_u = \frac{1130}{208 \times 32} = 170 \text{ psi} < 186 \text{ psi} \qquad \underline{\text{OK}}$$

Select flexural reinforcement:

Assume average effective depth d in each direction, and use the same bars in each direction. The critical direction for bending is at the wide face of the column. Therefore, the critical section for bending is half the column width, or 8 in., from the centerline of the column and the cantilever span of the footing is $(13'/2) - 8'' = 5'\text{-}10''$

$$M_u = \frac{w_u \ell^2}{2} = \frac{7.53 \times 13 \times 5.83^2}{2} = 1660' \text{k}$$

Assume $(\phi f_y j) = 4.3$ because only minimum reinforcement will be required.

$$A_s = \frac{M_u}{(\phi f_y j)d} = \frac{1660}{4.3 \times 32} = 12.10^{\square''}$$

Check for minimum reinforcement:

$$\rho = \frac{12.10}{13 \times 12 \times 32} = 0.24\% < = 0.33\% = \frac{200}{f_y}$$

Therefore, conservatively increase reinforcement by one-third, or 0.33%, whichever is less. The required reinforcement is then

$$13 \times 12 \times 32 \times 0.0033 = 16.60^{\square''}$$

$$\text{or } 12.10 \times 1.33 = 16.10^{\square''}.$$

Use 13 #10 bars, $A_s$prov. = $16.51^{\square''}$

The bar development length is measured from the critical section for moment to the end of the bar

$$= (5'\text{-}10'') - 3'' = 5'\text{-}7'' = 67''$$

Referring to Table A5.1, the development length of a #10 bar is 56 in. $\underline{\text{OK}}$

Solution B It is conservative to assume that all material between the bottom of the footing and the basement floor slab has the same weight as concrete, except in calculation for overturning or sliding.

Gross soil pressure $= 5.00$ ksf

Surcharge $4.5 \times 0.15 = 0.67$

Live load $= \underline{0.10}$

Net soil pressure $= 4.23$ ksf

Area of footing required $= \dfrac{460 + 370}{4.23} = 196.2^{\square'}$

Footing side dimension $= \sqrt{196.2} = 14.01'$

Use $14'\text{-}0''$ square footing.

Determination of thickness and selection of reinforcement can proceed as in Solution A.

Just as in two-way flat slabs where the reinforcement is heavier in column strips, so too the reinforcement in a rectangular footing supporting square or nearly square columns should be heavier under the columns. Quoting from the ACI Code, Section 15.4.4:

> Reinforcement in the long direction shall be distributed uniformly across entire width of footing. For reinforcement in short direction, a portion of the total reinforcement given by Eq. 11.8 shall be distributed uniformly over a band width (centered on centerline of column) equal to the length of short side of footing. Remainder of reinforcement required in short direction shall be distributed uniformly outside center band width of footing.

$$\frac{\text{Reinforcement in band width}}{\text{Total reinforcement in short direction}} = \frac{2}{(\beta + 1)} \tag{11.8}$$

where β = ratio of long side to short side of footing.

To minimize potential construction errors in spacing bars, it is good practice to space reinforcement in the short direction uniformly. Code requirements will be met if reinforcement is increased by the ratio

$$\frac{2}{1 + 1/\beta} = \frac{2\ell_x}{\ell_x + \ell_y}$$

where ℓ_x = length of the long side of the footing

ℓ_y = length of the short side of the footing.

Application of this requirement is shown in the following example.

Example 11.4 Design a footing to support a 16-in. square column carrying a 260 k dead load and a 190 k live load, using an allowable soil-bearing pressure of 4 ksf, disregarding the weight of the footing and the surcharge. An obstruction to the footing is 3'-6" from the centerline of the column on one side, as shown in Fig. 11.15(a). Use $f'_c = 3$ ksi and $f_y = 60$ ksi in the footing.

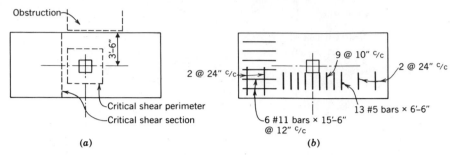

(a) (b)

FIGURE 11.15

Solution

$$\text{Area of footing required} = \frac{260 + 190}{4} = 112.5^{\square\prime}$$

$$\text{Maximum width of footing} = 3.5 \times 2 = 7.0'$$

$$\text{Required length of footing} = \frac{112.5}{7} = 16.07'$$

Use $16'\text{-}0'' \times 7'\text{-}0''$ footing with a bearing area $= 112^{\square\prime}$

$$\text{Factored column load} = 260 \times 1.4 + 190 \times 1.7 = 687 \text{ k}$$

$$\text{Factored soil pressure} = \frac{687}{112} = 6.13 \text{ ksf}$$

Select depth to meet shear requirements:
For Case I, beam shear, use the critical section across full width of the footing. Try 32"-deep footing.

Assume that long bars will be in the bottom of the footing:

$$d = 32 - 3 - \frac{1.4}{2} = 28.3''$$

The critical section for shear is $28.3 + 8 = 36.3'' = 3.03'$ from the centerline of the column.

$$V_u = \left(\frac{16}{2} - 3.03\right) 7 \times 6.13 = 213 \text{ k}$$

$$v_u = \frac{213}{7 \times 12 \times 28.3} = 90 \text{ psi} \qquad \underline{\text{OK}}$$

$$\text{Allowable shear stress} = \phi 2\sqrt{f'_c} = 0.85 \times 2\sqrt{3000} = 93 \text{ psi}.$$

For Case II, slab-punching shear, use the critical shear perimeter around the column.

Average effective depth $d = 32 - 3 - 1.2 = 27.8''$

The shear perimeter is a square $(27.8 + 16) = 43.8''$ on a side.

$b_o = 43.8 \times 4 = 175.2''$

$$V_u = 687 - 6.13 \left(\frac{43.8}{12}\right)^2 = 687 - 82 = 605 \text{ k}$$

$$v_u = \frac{605}{175.2 \times 27.8} = 124 \text{ psi} \qquad\qquad \text{OK}$$

Allowable shear stress = 186 psi (see Ex. 11.3).

Select flexural reinforcement in the long direction of the footing:
The critical section is at the face of the column.

$$M_u = \frac{6.13 \times 7(16/2 - 0.67)^2}{2} = 1150'\text{k}$$

$d = 28.3''$;   assume $(\phi f_y j) = 4.3$,

$$A_s = \frac{1150}{4.3 \times 28.3} = 9.48^{\square''}$$

Check for minimum reinforcement,

$$\rho = \frac{9.48}{7 \times 12 \times 28.3} = 0.004 > 0.0033 \qquad\qquad \text{OK}$$

Use 6 #11 bars, $A_s$prov. = $9.36^{\square''}$ (1% under)   OK

Bars extend $\left(\dfrac{16 \times 12}{2} - 8 - 3\right) = 85''$ beyond the critical section for moment.

From Table A5.1, the development length of #11 bars is 69''.   OK

Select flexural reinforcement in the short direction of the footing:

$$M_u = \frac{6.13 \times 16(7/2 - 0.67)^2}{2} = 394'\text{k}$$

$d = 36 - 3 - 1.4 - 1.0/2 = 31.1''$,   Assume $(\phi f_y j) = 4.3$,

$$A_s = \frac{394}{4.3 \times 31.1} = 2.94^{\square''}$$

$$\rho = \frac{2.94}{16 \times 12 \times 31.1} = 0.0005 << 0.0033 = 200/f_y$$

Therefore, conservatively increase the computed steel area by one-third.

$A_s$reqd. $= 2.94 \times 1.33 = 3.92^{\square\prime\prime}$

Alternatively, for bars uniformly spaced,

$A_s$reqd. $= \dfrac{2.94 \times 2 \times 16}{16 + 7} = 4.09$

Use 13 #5 bars uniformly spaced at 15″ c/c

$A_s$prov. $= 4.03^{\square\prime\prime}$ (1% under) OK

The distance available for developing bars

$= \dfrac{84}{2} - 8 - 3 = 31″$

From Table A5.1, development length of #5 bar = 15″. OK

Check the quantity of bars under the column.

From Eq. 11.8, $A_s$reqd. in the column bandwidth

$= \dfrac{2.94 \times 2}{1 + 16/7} = 1.79^{\square\prime\prime}$

$\left(\dfrac{7}{16}\right)$ 13 = 5.7 bars will be in the column band width

$A_s$prov. $= 5.7 \times 0.31 = 1.77^{\square\prime\prime}$ (1% under) OK

For a more conservative design, combine the requirements of minimum reinforcement and concentration under the column. The short bars would then be distributed as shown in Fig. 11.15(*b*). Alternatively, for uniformly spaced bars,

$A_s$reqd. $= \dfrac{3.92 \times 2}{1 + 1/2.28} = 5.45$ in.^{2}

Use 13 #6 bars uniformly spaced at 15″ c/c $A_s$prov. $= 5.72^{\square\prime\prime}$

In actual practice, design of isolated square spread footings is usually standardized and selected from a handbook rather than designed individually.

11.6 COMBINED SPREAD FOOTINGS

Sometimes a nearby property line or underground obstruction makes it impossible or impractical to center an isolated footing under a column or a wall. Also, sometimes adjacent, isolated footings of normal proportions would overlap or nearly overlap. In such cases, a common solution is to support two or more columns on one combined footing. When three or more columns are closely spaced on a straight line, the footings can be combined in a strip footing. Footings supporting four or more columns not on a straight line are called mat foundations. Common types of combined spread footings are shown in Fig. 11.16.

The soil pressure under combined footings supporting two columns, or three columns not on a straight line, is statically determinate. The soil pressure is not statically determinant under strip footings and mat foundations. Instead, they must be analyzed for actual column loads by methods that take into account the stiffness of the foundation and nonuniform bearing pressure distribution [2].

Where soil-bearing pressure is low and footings are large, combined footings reduce the reinforcement in footings because both negative and positive moment can resist the total static moment. Low-bearing pressure is frequently associated with compressible soils and large foundation settlements. Strip footings and mat foundations tend to reduce differential settlement by their continuity and rigidity. Strip footings can be used in one or both directions to cover a larger area of foundation soils. When the total area of footings exceeds about 50% of the area of the building, it is generally more economical to use a continuous mat foundation covering the entire foundation area or to use a deep foundation. The latter type of foundation is covered in Section 11.7.

As with isolated footings, it is essential that the center of gravity of the bearing area of a combined footing coincide with the center of gravity of loads on columns and walls supported by the footing. Lack of congruity will result in unequal soil-bearing pressures which, in turn, may overload the soil on one side of the footing, cause unequal settlement and, in extreme cases, overturn the footing.

Combined footings are proportioned and located based on total loads. For buildings with low live load, the bearing pressure under partial live load does not control, even though the live-to-dead-load ratio is different from one column to the other. In buildings with heavy live load, particularly those in which full live load could occur on one column while little or no live load is on the other column, it may be necessary to proportion the combined footing for loading cases other than full load. Normally it is not possible to have a wide disparity of live load between adjacent columns due to continuity of floor framing members.

The wide range of possible shapes and arrangements of combined footings

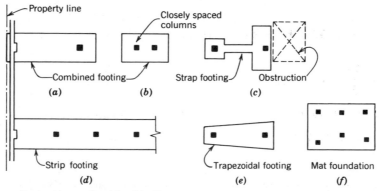

FIGURE 11.16 Combined footings.

allows engineers to exercise their ingenuity in selecting the most appropriate and economical foundation.*

Example 11.5 Select a footing size to support loads on the two columns listed below. Columns are spaced 30 ft c/c, as in Fig. 11.17. Check the bearing pressure under zero live load on one column and 50% live load on the other column. Allowable soil-bearing pressure = 6 ksf.

	Column A (kips)	Column B (kips)
Live load =	200	300
Dead load =	200	100
Total load =	400	400

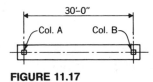

FIGURE 11.17

Solution

$$\text{Area of footing required} = \frac{400 + 400}{6} = 133.3^{\square\prime}$$

Try 4'-0" × 33'-4" footing centered on columns.
Area = 133.3$^{\square\prime}$

By inspection, 50% live load on column A and zero live load on column B would be the most severe condition.

Location of center of gravity measured from column A

$$\frac{100 \times 30}{300 + 100} = 7.5'$$

The kern is $\dfrac{33.33}{6} = 5.55'$ from the centerline of the footing.

$$\text{Eccentricity} = \frac{30}{2} - 7.5 = 7.5'$$

The center of gravity of the loads is outside the kern; therefore, Use Eq. 11.3.

$$c = 3(16.67 - 7.5) = 27.5'$$

*Both "engineer" and "ingenuity" come from the same Latin word meaning talent. Engineers should work hard to justify the lexicographer's etymology.

Maximum soil-bearing pressure

$$= \frac{300 + 100}{4 \times 27.5/2} = 7.3 \text{ ksf} > 6.0 \text{ ksf}$$

The footing is not satisfactory for this loading case. It should be larger, preferably longer rather than wider. The engineer can select a satisfactory size, shape, and location most easily by trial and error.

The thickness of a combined footing is controlled by the most critical shear condition, considering each of the columns supported by the footing individually. Footings of uniform thickness throughout are usually the most economical.

After selecting the size, shape, location, and thickness of a combined footing, design to resist moment can proceed in a manner similar to the design of an inverted, thick slab on two or three supports with known loads at each support. Bottom reinforcement under columns will usually be required as in isolated footings, and top reinforcement will be required in the span connecting supports.

For most concrete members, requirements for bar supports are given by standard specifications and require no design by the structural engineer. In contrast, supports for bars in combined footings and mat foundations require special attention because they are usually heavy (measured in tons rather than pounds) and far above the base (measured in feet rather than inches). For thick footings, engineers should specify how top bar supports must be constructed.

Example 11.6 Design a rectangular combined footing to support two 18-in. square columns 25 ft c/c when the centerline of one column is 1'-2" from a property line beyond which the footing may not extend as shown in Fig. 11.18(a). Column A is integral with a wall perpendicular to the axis of the combined footing. Service loads on the two footings are

	Column A (kips)	Column B (kips)
Live load =	125	250
Dead load =	320	300
Total load =	445	550

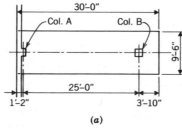

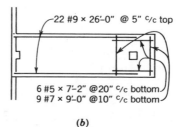

(a)   (b)

FIGURE 11.18

Allowable soil-bearing pressure is 3.5 ksf and need not include the weight of the footing or surcharge. Use f'_c = 3 ksi and f_y = 60 ksi for the footing.

Solution

$$\text{Area of footing required} = \frac{445 + 550}{3.5} = 284^{\square'}$$

Taking moments about column A, the center of gravity of total loads is located $550 \times 25/(445 + 550) = 13.82'$ from column A. Adding distance to the property line, the length of the footing can be $2(13.82 + 1.17) = 30.0'$.
Use $30'\text{-}0'' \times 9'\text{-}6''$ footing; area = $30 \times 9.5 = 285.0^{\square'}$

Factored column loads:

Column A	*Column B*
(k)	(k)
$125 \times 1.7 = 212$	$250 \times 1.7 = 425$
$320 \times 1.4 = \underline{448}$	$300 \times 1.4 = \underline{420}$
$P_u = 660$	$P_u = 845$

$$\text{Factored soil pressure} = \frac{660 + 845}{285} = 5.3 \text{ ksf}$$

Select depth to meet shear requirements:
Because column A is integral with a wall, location of critical shear section is at column B.

For Case I, beam shear, using the critical section across full width of the footing, try $3'\text{-}8''$ deep footing.
For top bars,

$$d = 44 - 1.5 - 1.4/2 = 41.8''$$

The critical section is $(41.8 + 18/2) = 50.8'' = 4.23'$ to the left of column B. Shear on this section is the column load less the load to the right of the section.

$$V_u = 845 - 5.3 \times 9.5(3.83 + 4.23) = 439 \text{ k}$$

$$v_u = \frac{439}{9.5 \times 12 \times 41.8} = 92 \text{ psi} \qquad\qquad \text{OK}$$

Allowable shear stress = 93 psi (see Ex. 11.4).

For Case II, slab-punching shear, use the critical shear perimeter around the column. Use the average effective depth of bottom bars.

$$d = 44 - 3 - 1 = 40''$$

The critical shear perimeter is a square $(40 + 18) = 58''$ on a side.

$$b_o = 58 \times 4 = 232''$$

$$V_u = 845 - 5.3 \left(\frac{58}{12}\right)^2 = 721 \text{ k}$$

$$v_u = \frac{721}{232 \times 40} = 78 \text{ psi} \qquad \qquad \underline{\text{OK}}$$

Allowable shear stress = 186 psi   (see Ex. 11.3).

Select flexural reinforcement in the long direction:

For bottom bars at column B,

$$w_u = 5.3 \times 9.5 = 50.4 \text{ k/}'$$

Clear cantilever span = 3.83 − 1.5/2 = 3.08′

$$^-M_u = \frac{w_u \ell^2}{2} = \frac{50.4(3.08)^2}{2} = 239'\text{k}$$

$$d = 40''; \qquad \text{assume } (\phi f_y j) = 4.3$$

$$A_s = \frac{239}{4.3 \times 40} = 1.39^{\square''}$$

Check for minimum reinforcement,

$$\rho = \frac{1.39}{9.5 \times 12 \times 40} = 0.0003 \ll 0.0033$$

Therefore, $A_s$reqd. = 1.39 × 1.33 = 1.85 in.^{2}

Use 6 #5 bars at approximately 20″ c/c $A_s$prov. = 1.86$^{\square''}$

From Table A5.1,

Development length of #5 bars = 15″. $\underline{\text{OK}}$

For top bars between columns A and B, conservatively compute moments using spans center-to-center of columns,

$$^+M_u \sim \frac{w_u \ell^2}{8} - \frac{^-M_u}{2} = \frac{50.4(25)^2}{8} - \frac{239}{2} = 3820'\text{k}$$

$$d = 41.8''; \qquad \text{assume } (\phi f_y j) = 4.2$$

$$A_s = \frac{3820}{4.2 \times 41.8} = 21.7^{\square''}$$

Check for minimum reinforcement,

$$\rho = \frac{21.7}{114 \times 41.8} = 0.0046 > 0.0033 = \frac{200}{f_y}$$

Use 22 #9 bars at 5″ c/c in the top, $A_s$prov. = 22.0$^{\square''}$

From Table A5.6, the minimum clear span length in which #9 bars can be developed is

$$15.5 \times 1.4 = 21.7' < 23.5' \qquad\qquad \text{OK}$$

To hold top bars in position, use #6 bars at 4'-0" c/c.

Select flexural reinforcement in the short direction:

At column A, assume that the wall distributes all the load transversely. Therefore, no transverse reinforcement is needed at column A.

At column B, the clear cantilever span $= \dfrac{9.5 - 1.5}{2} = 4.0'$

Uniform load $\boldsymbol{w}_u = \dfrac{845}{9.5} = 88.9\text{k}/'$

$$\boldsymbol{M}_u = \frac{\boldsymbol{w}_u \ell^2}{2} = \frac{88.9 \times 4.0^2}{2} = 712'\text{k}$$

$\boldsymbol{d} = 40"$;   assume $(\boldsymbol{\phi f_y j}) = 4.3$

$$\boldsymbol{A}_s = \frac{712}{4.3 \times 40} = 4.14^{\square''}$$

Check for minimum reinforcement,

$$\boldsymbol{\rho} = \frac{4.14}{3.83 \times 2 \times 12 \times 40} = 0.0011 \ll 0.0033$$

$\boldsymbol{A}_s\text{reqd.} = 4.14 \times 1.33 = 5.52^{\square''}$

Use 9 #7 bars at 10" c/c in the bottom, centered under column B.
    $\boldsymbol{A}_s\text{prov.} = 5.40^{\square''}$ (2% under)
From Table A5.1,
Development length of #7 bars = 27"   OK
Reinforcement is shown in Fig. 11.18(b)

Strap Footings

Strap footings under two columns can be used in situations where there would still be a small eccentricity of load under the best placement of an isolated footing under one or both of the columns. Straps are also useful if underground obstructions prevent use of a rectangular combined footing. Strap footings can also be used where the distance between columns creates a large moment if a combined footing is used. For simplicity, the strap (narrow portion of the footing) is usually isolated from foundation soils and used to resist the eccentric moment only. If the strap bears on foundation soils, moments become much higher and the footing becomes a combined footing. To resist moment if the strap bears on the soil, the strap might need to be wider (and thereby resist more load, more moment) or deeper (and increase the depth of the entire footing). Furthermore, proportioning a strap footing may require far more engineering time than potential construction cost savings would warrant.

Example 11.7 Proportion a strap footing and design the strap to support two 16-in. square columns when the footing area is restricted around one of the columns, as shown in Fig. 11.19(*a*). Service loads are 290k on column A and 480k on column B. The average load factor is 1.55 for both columns. Allowable soil-bearing pressure is 6 ksf. Use f'_c = 3 ksi and f_y = 60 ksi for the footing.

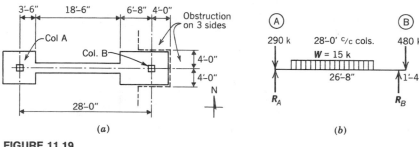

(*a*)

(*b*)

FIGURE 11.19

Solution Assume that positive measures are taken to prevent the strap from bearing on the soil. Its weight must be added to the load on each footing. If the clear distance between footings is 20′ and the strap is 2′ wide by 2′-6″ high, its weight is $2 \times 2.5 \times 20 \times 0.15 = 15$ k.

Referring to Fig. 11.19(*b*), the footing load at column B = R_B

$$R_B = \frac{480 \times 28}{26.67} + 7.5 = 504 + 7.5 = 512 \text{ k}$$

$$\text{Required footing area} = \frac{512}{6} = 85.3^{□′}$$

Use 8′-0″ × 10′-8″ footing, area = $85.3^{□′}$

Footing load at column A = R_A = 290 + 7.5 + (480 − 504) = 274 k

$$\text{Required footing area} = \frac{274}{6} = 45.7^{□′}$$

Use 7′-0″ square footing; area = $49.0^{□′}$

Referring to Fig. 11.19(*b*), the eccentricity of the footing at column B is 1.33′. The strap may be designed for a positive moment (tension in top of the strap) of $480 \times 1.55 \times 1.33 = 992′$k. The maximum moment occurs near column B. Assume a 2′-wide strap so that flexural bars in the strap will not interfere with dowels for the column. Try a 2'-6" deep strap. Effective depth $d = 30 - 2 - 1.4/2 = 27.3"$.

Because the strap is heavily loaded, assume $(\phi f_y j) = 3.65$.

$$A_s = \frac{992}{3.65 \times 27.3} = 9.96^{\square''}$$

$$\rho = \frac{9.96}{24 \times 27.3} = 0.0152 < 0.016 = 0.75\rho_b \qquad \text{OK}$$

Use 8 #10 bars, $A_s$prov. $= 10.16^{\square''}$

Conservatively assume that maximum moment occurs at column B; therefore, bars should extend beyond column B a distance equal to the tension development length of $55.6 \times 1.4 = 78"$ (Table A5.1). Because this is longer than the distance to the obstruction, standard 90° hooks should be used, which require only 28" development length (Table A5.9). Hooks should bend down near the obstruction east of column B. The other end of the 8 #10 bars should extend to column A.

The strap should be designed for a factored shear force of $(504 - 480)1.55 + 15 \times 1.55/2 = 48.8$ k. Conservatively assume that shear is constant for full length of the strap.

Factored shear stress v_u

$$v_u = \frac{48.8}{24 \times 27.3} = 75 \text{ psi} < 93 \text{ psi} \qquad \text{OK}$$

Use minimum shear reinforcement for the full length of the strap.

The square and rectangular footings at columns A and B can be designed as isolated footings. Shear capacity of footings shallower than 2'-6" might be satisfactory, but it would be advisable to use the same depth as the strap to avoid expensive formwork for the strap above the column footings.

Design of the footing at column B could take advantage of the support provided by the strap so that the East-West and North-South reinforcement need only resist the moment of a cantilever span $= (4'-0")-(0'-8") = 3'-4"$

Mat Foundations

When a single footing supports four or more columns arranged in a grid pattern, it is called a mat foundation. As with all other footings, the center of gravity of the footing must coincide with the center of gravity of the loads. Mat foundations usually have a uniform thickness for simplicity of analysis and construction, although sometimes the slab is thickened around the columns to increase shear resistance. The thickening or drop panel could occur either above or below the slab, as shown in Fig. 11.20. In either case, the drop panel should be cast with the slab. If the drop panel is built above the slab, forms must be built and supported in mid-air before and during the casting of concrete, but negative moment reinforcement can

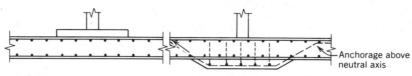

FIGURE 11.20   Drop panels in a mat foundation.

continue straight through below the column. Drop panels built below the slab must be excavated with less efficient machinery than that used for the general excavation, or by hand, at greater expense than the general excavation, and negative moment reinforcement must be specially bent, anchored, and placed.

A mud slab is frequently used under a mat foundation to protect it because its very size requires workers and small equipment to use the soil-bearing surface for foot traffic. Patterns of reinforcement in mats are frequently more conservative than those in flat slabs or flat plates to allow for uncertainties and variables in foundation soil support, settlement, column loading, and bar placement. For example, a grid of both top and bottom bars may be placed throughout the mat. When bars are cut off in the span, more length than that required by analysis should be provided.

A mat foundation is similar to an inverted flat plate with soil pressure as the load on the plate, except that mat loads are much heavier than flat plate loads and the mat slab is much thicker than most flat plates. If a mat foundation is analyzed as a flat plate, computed reactions will rarely coincide with actual column loads. This is because column loads divided by tributary area give bearing pressures that are similar to, but not exactly the same as, the average soil-bearing pressure. Instead, mat foundations must be analyzed for actual column loads by methods that take into account the stiffness of the foundation mat and nonuniform bearing pressure distribution [2].

Because mat footings are large and thick, they may contain more concrete than can be cast at one time, and construction joints become necessary. Due to the size and importance of construction joints in mat footings, the joints should be specifically designed by the structural engineer (see Section 10.4).

Example 11.8   Select the size, shape, and location of a mat foundation to support four columns, located as shown in Fig. 11.21(a), with service loads as follows:

	Columns (kips)				
	A1	*A2*	*B1*	*B2*	*Total*
Dead	250	225	300	350	1125
Live	150	125	175	225	675
Total	400	350	475	575	1800

Allowable soil-bearing pressure is 4 ksf. Weight of the footing and surcharge may be disregarded. Assume $f'_c = 3$ ksi and $f_y = 60$ ksi.

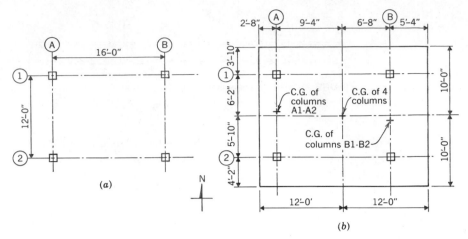

FIGURE 11.21

Solution: The center of gravity (C.G.) of total loads in the East-West direction is

$$\frac{(475 + 575)}{1800} \ 16 = 9'\text{-}4'' \text{ east of line A.}$$

The C.G. of total loads in the North-South direction is

$$\frac{(350 + 575)}{1800} \ 12 = 6'\text{-}2'' \text{ south of line 1.}$$

$$\text{Footing area required} = \frac{1800}{4} = 450^{\square'}$$

Use 20 × 24-ft footing, area = 480$^{\square'}$

The footing is located, as shown in Fig. 11.21(*b*), so that the C.G. of the footing coincides with the C.G. of the loads.

Diagonal columns A1 and B2 with the heaviest loads cause a torsional moment in the footing. On line A, the C.G. of columns A1-A2 is 12 × 350/750 = 5.6′ south of line 1. This causes a torsional moment M_T = 750(6.17 − 5.6) = 425′k.

Alternatively, at line B, the C.G. of columns B1-B2 is 12 × 575/1050 = 6.57′ south of line 1. This causes an equal and opposite torsional moment M_T = (1050 (6.57 − 6.17) = 425′k.

Using Eq. 6.31 and assuming that the minimum footing thickness is 3′, the minimum torsional moment minT_u that need be considered in the design of the footing is

$$\text{Min}T_u = \phi 0.5\sqrt{f'_c}x^2y$$

$$= \frac{0.85 \times 0.5\sqrt{3000} \times 36^2 \times 240}{12,000} = 603'k > 425'k \qquad \underline{\text{OK}}$$

The highest torsional shear stress occurs in the middle of the long sides, that is, the top and bottom surfaces of the footing where flexural reinforcement is normally placed. Thus, even if the torsional moment exceeded the minimum, the footing might still be satisfactory if the flexural reinforcement is increased to provide torsional reinforcement.

A theoretically more accurate shape and location of the footing will result if the East-West centerline of the footing is placed to coincide with the C.G. of columns A1-A2 and with the C.G. of columns B1-B2. Likewise, the North-South centerline of the footing should be placed to coincide with the C.G. of columns A1-B1 and with the C.G. of columns A2-B2. The footing will be skewed with respect to the building coordinates and may be shaped as a parallelogram. These two complications would be confusing to construction workers and may result in field errors. Skewed parallelogram footings should be avoided if rectangular footings placed parallel to building coordinates are satisfactory.

If the percentage of live load on each column might vary significantly from one column on the footing to the next (a rare occurrence in normal building construction), the footing size and shape should be checked for each loading case. Maximum bearing pressure must be kept within the allowable bearing pressure. For each loading case, the highest bearing pressure will occur on the side of the footing nearest the column with the highest percentage of live load.

Design of the footing is completed by converting service loads to factored loads and computing the thickness required to keep shear stresses at each of the four columns within the stress permitted. The critical section perimeter for shear at each column may extend to the edge of the footing rather than enclose the column. In the top of the slab, positive flexural reinforcement should extend from edge to edge for simplicity. Its required area is computed using the full factored average soil-bearing pressure in each direction. In the bottom of the slab, some negative flexural reinforcement should also extend from edge to edge of the slab, sized to meet requirements at lines A and 2. Because the cantilever span at line B is larger than that at line A, additional bars across line B would be necessary. In all cases, clear spans to the face of columns may be used. In computing required positive flexural reinforcement, most practicing engineers would conservatively disregard the benefit of negative moment restraint.

11.7 DEEP FOUNDATIONS

For economic, aesthetic, and other reasons, it is sometimes necessary to build structures on sites that do not have soil conditions suitable for the most economical foundations. That is, soil strata strong enough to support the building do not exist just below the surface or just below basement level. Since prehistoric times, mankind has solved this problem by supporting homes, bridges, and other structures on timber piles (tree trunks turned upside down) driven to "resistance" (to further penetration), that is, deep enough to reach a soil strata of sufficient strength.

Today there are many types of piles and drilled piers available in addition to timber piles. Selection is based primarily on soil conditions (as advised by a geotechnical engineer), magnitude of loads being supported, and economy. Following are the principal differences between types:

1. Material (timber, steel, concrete, or a combination of these materials).
2. Method of installation (piles are usually driven,* whereas drilled piers are excavated and the hole filled with concrete).
3. Size (piles are generally about 12 in. in diameter but can be smaller or larger, whereas drilled pier shafts are usually at least 18 in. in diameter).
4. Cost.

Piles and drilled piers must carry building loads from the top of the member to the bearing strata. The bearing strata might pick up the load in bearing at the point (bottom) of the pile or pier (Fig. 11.22) or by friction along the sides of the member [Fig. 11.22(b) and 11.22(c)]. In addition, soil above the bearing stratum may increase the load on the pile or drilled pier. Soft, compressible upper layers of soil may settle and "hang up" or transfer part of their weight to the piles or piers that do not settle because they are founded on denser materials below. This phenomenon is called negative skin friction and results in a higher load on the pile or pier than that applied at the top [Fig. 11.22(d)].

In certain areas expansive clays may overlay the building site. These clays tend to expand with great force when exposed to moisture. As the clays expand they may heave the pile or pier upward [Fig. 11.22(e)].

For these reasons soil conditions must be considered in designing individual piles and piers. All loads must be carried by the pile or pier to a depth where the loads are transferred to the foundation soils.

There are only a few sizes of each type of pile. Each size is normally assigned one maximum load capacity (ranging from 15 to 400 tons or more). Thus, more than one pile is usually required to support each column. A pile cap similar to a spread footing is constructed to distribute the column load to individual piles.

Drilled piers can be constructed in a wide range of sizes, custom designed to support loads as required. Thus, each column is usually supported by only one pier.

*Certain proprietary piles are excavated with an auger and filled with a cement grout as the auger is withdrawn. They could be considered small drilled piers.

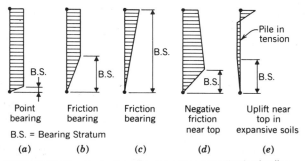

Point bearing   Friction bearing   Friction bearing   Negative friction near top   Uplift near top in expansive soils

B.S. = Bearing Stratum

(a)   (b)   (c)   (d)   (e)

FIGURE 11.22   Typical idealized axial load distribution in piles and drilled piers.

As a check on the overall design, the total capacity of all piles or piers used under a building must be equal to or greater than the total weight of the building. Ten or 20% excess capacity is normal because the capacity of piles or piers at individual columns and load points is usually in excess of the load.

By the nature of their method of installation, piles and drilled piers may have a significant error in their location at ground surface. The specified tolerance on mislocation is normally 3 in. for individual piles or piers and 1 in. for a cluster of piles. Mislocation within tolerance limits is usually not critical for piers but must be considered in design of piles and pile caps.

Soil having a standard penetration value $N \geq 2$ may be considered to provide lateral support and effectively prevent buckling of piles and drilled piers at service loads. Therefore, buckling or length effects on concrete piles and piers need not be considered unless a portion of the member is in weaker soils or is exposed in air or water. In the latter case, corrosion and ice forces may also need to be considered in design.

Dowels should always be used between piers and concrete columns and between pile caps and concrete columns. Dowels may also be needed between concrete piles and pile caps and anchors between steel piles and pile caps. If a column exerts an uplift force on its foundation, the upward force must be transmitted by dowels or other anchors to the pier or piles. These members, in turn, must resist the tensile force and transmit it to a lower stratum where the uplift force can be resisted by weight of the member, skin friction, or other anchorage. Above the point where an uplift force is transmitted to the soil, the force must be resisted by the dead weight of soil. To ensure an adequate factor of safety, only 90% of the dead load should be assumed to resist uplift. The total dead load thus computed should be at least 50% more than the uplift force.

Drilled Piers

Drilled piers (sometimes called caissons) are round because they are excavated by rotating equipment. They may be straight-shafted, as in Fig. 11.23(a), and receive their support either by end bearing or by skin friction along the sides, or both. When conditions warrant, the end of the shaft may be enlarged in a bell, as in Fig. 11.23(b). Cutting tools and design tradition limit the diameter of the bell to three times the shaft diameter. The shaft diameter is usually a minimum of 18 in. or, preferably, 30 in. Engineers should consult contractors on the limitations of their excavating equipment before designing drilled piers. The full height of the bell must be in soil that will stand open during excavation without the use of a casing (a cylindrical steel lining). Construction equipment and procedures will not permit the use of a casing in the bell. Shafts of drilled piers may be excavated without a casing if the hole will stand open. Where necessary, a casing may be used during drilling or afterward to keep the hole open and to prevent the entrance of water. If a casing is used, it may be left in the ground if removal is impractical or if it is needed for reinforcement. In many cases, the casing is removed as the concrete is being cast.

The drilled pier shaft may or may not be reinforced but the bell rarely, if ever, is reinforced, except by extension of the shaft reinforcement.

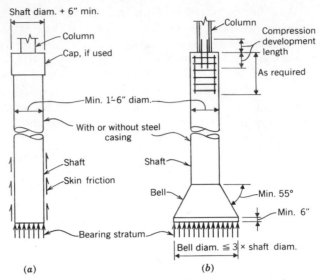

FIGURE 11.23 Drilled pier construction.

It is important to keep water from the excavation and to remove loose material from the bottom of end-bearing piers to achieve suitable bearing conditions. If water cannot be eliminated, special procedures must be used to place concrete under water. A suitable reduction in design strength of concrete should be considered to allow for possible contamination of the concrete.

As with spread footings, drilled piers are proportioned using service loads and allowable soil pressure recommended by the geotechnical engineer. In addition to loads from the structure, design of drilled piers should consider negative skin friction (down drag) and excess weight of the pier foundation over the weight of excavated soil. Negative skin friction may be disregarded if it is less than 10% of the total axial load acting downward on the pier. Excess weight of the pier may be disregarded if it is less than 5% of the total axial load.

Drilled pier shafts penetrating soft, cohesive soils can be protected from additional load due to negative skin friction by placing a casing around the shaft, at least 6 in. larger than the shaft, and filling the annular space with gravel or sand. Pier shafts penetrating expansive soils can be protected from uplift forces in the same manner.

After determining drilled pier outlines, design may proceed by strength design procedures using factored loads. The shafts of many piers are designed as unreinforced or plain concrete, except for dowels near the top. Minimum strength concrete ($f_c' = 3$ ksi) is frequently satisfactory. When drilled piers are numerous, heavy, and long, smaller shafts requiring higher strength concrete or reinforcement may be more economical.

Plain concrete drilled piers may carry an axial factored load ϕP_n of

$$\phi P_n = 0.60 \phi f_c' A_g \qquad (11.9)$$

where $\phi = 0.65$ for plain concrete

and A_g = the effective gross cross-sectional area of pier shaft (see the discussion below on effective shaft size.)

Reinforcement is used in drilled piers if the pier is (1) required to resist lateral loads, (2) in tension, (3) unbraced laterally by firm soils, or (4) required to provide a higher axial load capacity than that provided by concrete alone. Reinforcement need not necessarily extend the full length of the pier.

If reinforcement is required, the shaft should be designed as a reinforced column (Section 9.3). If a continuous, noncorrugated steel casing having a thickness at least 0.0075 times the diameter of the shaft extends the full length of the shaft, it may be included in calculations of area and moment of inertia of the pier shaft section. Welds between casing sections must be full penetration butt welds. Where a drilled pier is unsupported in weak soils, air, or water, the effective column length may be taken as the clear unsupported length plus two shaft diameters.

In calculating the strength of either reinforced or unreinforced pier shafts, the inside diameter of a cased shaft may be used. If there is no permanent casing, a shaft 8 in. smaller in diameter than the nominal shaft diameter must be used in strength computations. The reduction in effective shaft diameter is to allow for contamination and squeezing of the shaft when no casing is present.

Because drilled piers are usually much larger than the columns they support, eccentricity of placement within construction tolerances usually does not control their design. However, the top of highly stressed piers should be checked for combined axial load and bending. The bending moment may be taken as the axial load times the maximum permitted eccentricity in the most critical direction plus other frame moments transmitted to the pier by the column.

Sometimes it is desirable to cast a cap on the pier shaft to facilitate accurate placement of column dowels or anchor bolts. In such cases, the cap should be at least 6 in. larger than the shaft diameter. The depth should be sufficient to develop all reinforcement within the cap.

When two or more drilled piers support one column, a cap is also required—designed in the same manner as a pile cap (see the discussion on pile caps below)—except that pier sizes and capacities are larger than pile sizes and capacities.

When lateral load from the structural frame is transmitted to a drilled pier, the lateral force is resisted by soil pressure along the sides of the shaft. The distribution of soil pressure and the bending moment depends on soil properties and flexural stiffness of the shaft (see Ref. 3 for a description of design procedures).

Example 11.9 An uncased 24-in.-diam. pier is drilled into a bearing stratum that supports 1.2 ksf skin friction. Service load on the pier is 400k, and the average load factor is 1.55. How far must the pier extend into the bearing stratum and for what part of its length must it be reinforced? Assume $f'_c = 3$ ksi for concrete in the pier.

Solution

The required length in the bearing stratum is

$$\frac{400}{1.2 \times 2\pi} = 53'$$

Factored load on the pier $P_u = 400 \times 1.55 = 620$ k. The usable part of the pier is $24 - 8 = 16''$ in diameter. Using Eq. 11.9, the factored axial load capacity of the plain concrete pier

$$\phi P_n = \frac{0.6 \times 0.65 \times 3 \times \pi 16^2}{4} = 235 \text{ k}$$

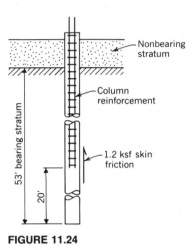

FIGURE 11.24

The pier must be reinforced as a column for all but the lower part. A distance of $53 \times 235/620 = 20'$ may be unreinforced. The pier is shown in Fig. 11.24.

Piles

Piles are made of timber, steel H sections, steel pipe, precast concrete, and thin corrugated steel shells driven with a mandrel and filled with concrete. Some piles are drilled with an auger and filled with concrete grout through the hollow stem of the auger as it is withdrawn. Individual concrete-filled shell piles and augered piles may be designed in the same manner as drilled piers. In addition, the design of tapered piles must consider loading and pile capacity at each section along the length of the pile. The capacity at sections near the tip must not be less than the load on the pile, as shown in Fig. 11.22. Many codes and specifications require that a portion of the shell be disregarded in computing the strength of a pile, to allow for corrosion. The remainder of the shell (some shells may be thinner than the corrosion allowance) may be included in strength calculations, giving due consideration to the direction of corrugations. With horizontal corrugations, the shell will not carry axial load. A geotechnical text should be consulted for other considerations in soil-pile interaction.

After piles are driven to the required resistance (and shell piles are filled with concrete), they are covered with a pile cap on which the column or other member being supported rests. To reduce the size of the cap to a minimum, the spacing of piles is held to a minimum, usually three times the butt (top of the pile) diameter but not less than 2ft.-6 in. c/c. Specified spacing is frequently 3 ft c/c for piles of 30- to 75-ton capacity. Sometimes the geotechnical engineer will require greater spacing. The tips of very long piles may drift laterally during driving and require a wider spacing to avoid interference with each other.

The location of piles is subject to construction tolerances similar to drilled piers, usually 3 in. for individual piles and 1 in. for a cluster. If the first piles driven in a cluster are mislocated, the remaining piles can be relocated before driving to correct for the eccentricity. Thus, the tolerance for the location of the center of gravity of all piles in a cluster can be smaller than for the location of a single pile. The possibility of piles located out of position, within tolerance limits, should be considered in the design of pile caps. The increase in load on individual piles due to misplacement of piles within tolerance limits can be disregarded because the increase in load is less than 7%. For example, consider a two-pile cap with piles a nominal 30 in. c/c, misplaced by 1 in., as in Fig. 11.25. The increase in load on one of the piles is $2 (16/30) - 1 = 0.067$.

Where low capacity (15- to 20-ton) piles support a wall 12 in. thick or more, all pile caps are sometimes omitted. Piles should be staggered on each side of the centerline of the wall, as shown in Fig. 11.26(c), and should be spaced no more than 10 ft c/c to permit balancing the wall between rows of staggered piles. Top and bottom of the wall must be continuously supported laterally. Individual piles must be designed for eccentric load for the case of piles misplaced, within tolerances, 3 in. toward the outside of the wall.

Axial service load capacity of individual piles, when limited by soil-bearing capacity, is given by the controlling building code or by the geotechnical engineer. The weight of the pile cap and surcharge above it and negative skin friction, but not the weight of the piles themselves, are considered load on the piles. Negative skin friction may be disregarded if it is less than 10% of the total axial load acting downward on the pile. The number of piles required to support a column or wall is simply the service load on the column or wall, plus the weight of the pile cap and surcharge, divided by the capacity of an individual pile. At least three piles in a cluster under one column are required for stability of isolated pile caps.

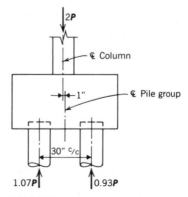

FIGURE 11.25 Two-pile cap.

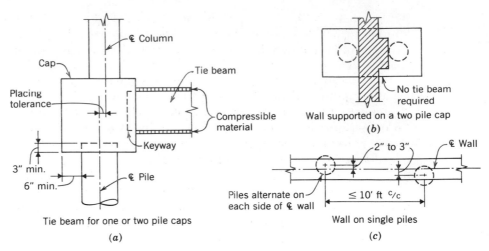

FIGURE 11.26 Eccentric moment on pile cap.

If only one or two piles are required to support the load, tie beams are commonly used to resist the eccentric moment caused by possible misplacement of the pile, if walls or other construction are not available to resist the eccentric moment (see Fig. 11.26). Tie beams in two directions are required for one-pile caps. Tie beams should be designed to carry their own weight plus any other loads on them in addition to resisting moment due to eccentric placement of piles. To keep costs to a minimum and reduce tie beam size as much as possible, it is common practice to avoid placing other loads on tie beams. This is accomplished by placing a soft compressible material above and below the tie beam.

Tie beams are frequently required in design for seismic lateral loads, even for pile caps with three or more piles. For seismic design, longitudinal tie beam reinforcement may not be lap-spliced but must be spliced by welding or by using mechanical splices.

Example 11.10. Design a tie beam between two pile caps of two 40-ton piles each. Assume that the average load factor equals 1.6 and the maximum eccentricity of each pile group due to construction tolerances is 1 in. Use $f'_c = 3$ ksi and $f_y = 60$ ksi in tie beams. Assume that the tie beams carry no other loads. Span of the tie beam is 23 ft.-6 in. clear between pile caps.

Solution

Assume that the pile cap size is 5'-6" × 3'-0" × 2'-6"

The weight of the pile cap = 5.5 × 3.0 × 2.5 × 0.15 = 6 k.

Conservatively, the maximum column capacity

$$= 2 \times 40 \times 2 - 6 = 154 \text{ k}$$

Check shear using loading Case I in Fig. 11.27.

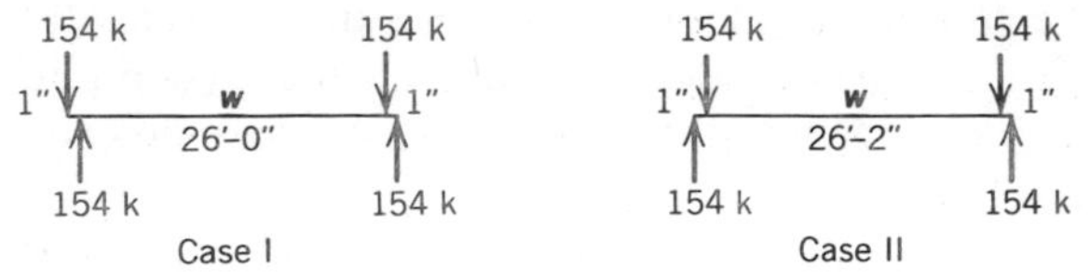

154 k   154 k   154 k   154 k

1" ↓   **w**   ↓ 1"   1" ↓   **w**   ↓ 1"

↑   26'-0"   ↑   ↑   26'-2"   ↑

154 k   154 k   154 k   154 k

Case I   Case II

FIGURE 11.27

Assume a tie beam is $12'' \times 12''$, weight $= 0.15$ k/'

Space between columns $= 23.5 + 2.5 = 26'$

$$\text{Max } \boldsymbol{V}_u = \frac{1.6 \times 154 \times 2 \times 0.083}{26.0} + \frac{1.4 \times 0.15 \times 23.5}{2} = 4.0 \text{ k}$$

Effective depth $\boldsymbol{d} = 12 - 2 - \frac{1}{2} = 9.5''$

$$\boldsymbol{v}_u = \frac{4.0}{12 \times 9.5} = 35 \text{ psi} < 93 \text{ psi} \qquad\qquad \underline{\text{OK}}$$

    Use nominal shear reinforcement #3 at $12''$ c/c closed ties.

    Calculate reinforcement to resist moment use loading Case II in Fig. 11.27.

$$\boldsymbol{M}_u = 1.6 \times 154 \times 0.083 + \frac{1.4 \times 0.15 \times 26.2^2}{8}$$

$$= 20.5 + 18.0 = 38.5' \text{k}$$

$\boldsymbol{d} = 9.5''$; assume $(\boldsymbol{\phi f_y j}) = 4.0$

$$\boldsymbol{A}_s = \frac{38.5}{4.0 \times 9.5} = 1.01^{\square''}$$

Use 4 #7 bars, 2 top + 2 bottom, $\boldsymbol{A}_s$ prov. $= 1.20^{\square''}$

See Fig. 11.28 for arrangement of the tie beam and its reinforcement.

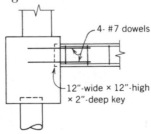

    4- #7 dowels

12"-wide × 12"-high × 2"-deep key

FIGURE 11.28

    Pile caps must cover all piles in a cluster. In width and length, caps should extend at least 6 in. beyond the edge of the piles for 40- to 50-ton piles. The extension should be 9 in. for 100-ton piles and 15 in. for piles up to 200-ton capacity. In practice, the nominal diameter is assumed to be at least 12 in. so that the edge of the cap is at least

1'-3" from the nominal center of the nearest pile after allowing 3 in. for construction tolerance. Larger caps are required for larger piles and piles of greater capacity. Anchorage of flexural reinforcement may also require larger caps, as will be discussed later.

Piles should extend at least 3 in. into the bottom of the pile cap, as in Fig. 11.26(a), and the reinforcing bar mat may rest directly on top of the piles.

Design of pile caps is similar to design of spread footings. Caps are usually designed for the full capacity of each pile rather than the column load divided by the number of piles, to allow for future contingencies. All pile caps with the same number of piles in a cluster should be identical in design, where possible.* For simplicity, computations for moments and shears may be based on the assumption that the reaction from any pile is concentrated at the pile center.

The depth of a pile cap is usually governed by shear. Both beam shear and punching shear must be considered. The ACI Code recognizes that a pile reaction is applied across its full width. The Code provides that the computation of shear on any section through a footing supported on piles of diameter d_p must be in accordance with the following:

1. Entire reaction from any pile whose center is located $d_p/2$ or more outside the section shall be considered as producing shear on that section.
2. Reaction from any pile whose center is located $d_p/2$ or more inside the section shall be considered as producing no shear on that section.
3. For intermediate positions of pile center, the portion of the pile reaction to be considered as producing shear on the section shall be based on straight-line interpolation between full value at $d_p/2$ outside the section and zero value at $d_p/2$ inside the section (Code 15.5.3).

The shear surrounding an individual pile or pair of piles must also be considered in design of a pile cap. However, this consideration will control only for high capacity piles in small, thin pile caps. In any case, it is good practice to be generous with pile cap depth to allow for inaccuracies in construction and to ensure a stiff cap that will distribute load evenly to all piles.

Pile caps frequently have the proportions of deep beams so that compressive struts form in the cap between column and piles. The increase in shear due to deep beam action may be conservatively disregarded, but the anchorage length of flexural reinforcement should be conservatively measured from the inside edge of a pile to the end of the bar (see the discussion on bar anchorage in deep beams in Section 6.5). Ends of bars may be hooked for anchorage or it may be necessary for the edge of the pile cap to extend farther beyond the outside face of the pile than the minimum distance described earlier.

Example 11.11 Design a foundation for a 16-in. square column carrying a 450-k service load, using 50-ton piles 14 in. in diameter spaced 3 ft c/c. Use $f'_c = 3$ ksi and

*Slightly different designs for conditions that appear identical in the field confuse construction workers and lead to errors.

f_y = 60 ksi in the pile cap and assume that the basement floor is 1 ft above the top of the pile cap. Assume construction tolerances of 3 in. on the location of each pile in any direction.

Solution $\dfrac{450}{2 \times 50} = 4.5$ piles

Use 5 piles = 500 k capacity.

Column load	= 450 k

Assume a 3′ thick cap

Cap load = 7 × 7 × 3 × 0.15 = 22

Assume 12″ concrete + 100 psf live load above the cap

Surcharge load = 7 × 7 (0.15 + 0.10) = 12

Total load = 484 k <u>OK</u>

Referring to Fig. 11.29, compute the depth required for shear without shear reinforcement.

Try 3′-0″ deep cap. d = 36 − 3 − 1 = 32″

Critical punching shear perimeter is (16 + 32) = 48″ square

b_o = 4 × 48 = 192″

The theoretical pile center is 2″ outside the critical shear perimeter in each direction. Allowing an additional 3″ for construction tolerances, the pile centers are 5″ beyond the critical shear perimeter in each direction at the corner. The diagonal distance from the corner is $5\sqrt{2}$ = 7.1″, or more than half the pile diameter. Therefore, the full pile reaction should be used in design for shear. Assume that the average load factor = 1.6.

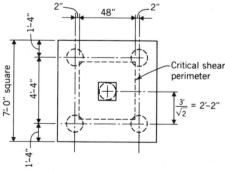

FIGURE 11.29

$$V_u = 100 \times 4 \times 1.6 - 4 \times 4(0.45 + 0.25)$$

$$= 640 - 11 = 629 \text{ k}$$

$$v_u = \frac{629}{192 \times 32} = 102 \text{ psi} < 186 \text{ psi} = 4\phi\sqrt{f_c'} \qquad \underline{\text{OK}}$$

Beam shear does not control because piles are inside the critical section.

Compute the flexural reinforcement required.
Referring to Fig. 11.29 and adding 3″ misplacement of piles to the outside,

$$M_u = 100 \times 2 \times 1.6(2.17 - 0.67 + 0.25) = 560'\text{k}$$

$$d = 32'', \qquad \text{Assume } (\phi f_y j) = 4.3$$

$$A_s = \frac{560}{4.3 \times 32} = 4.07^{\square''}$$

Check for minimum reinforcement:

$$\rho = \frac{4.07}{7 \times 12 \times 32} = 0.0015 < 0.0033 = \frac{200}{f_y}$$

Therefore, increase the reinforcement by one-third.

$$A_s \text{reqd.} = 4.07 \times 1.33 = 5.43^{\square''}$$

Use 9 #7 bars each way, $A_s$prov. $= 9 \times 0.60 = 5.40^{\square''}$
From Table A5.1, for a #7 bar, the tension development length = 27″.
From face of column to end of bar

$$= \frac{7 \times 12}{2} - \frac{16}{2} - 3 = 31'' \qquad \underline{\text{OK}}$$

A more conservative bar selection would be 13 #6 bars each way.

$$A_s \text{prov.} = 13 \times 0.44 = 5.72^{\square''}$$

From Table A5.1, for #6 bars, tension development length = 19.3″, which is slightly less than the distance from the inside face of pile to end of bar = 14 + 6 = 20″.

In actual practice, the design of most pile caps is standardized and selected from a handbook rather than designed individually.

PROBLEMS

1. Design an isolated spread footing to support a 24-in. square column carrying a 640-kip dead load plus 370-kip live load. Soil-bearing pressure = 12 ksf, including weight of the footing and surcharge. Assume that the bottom of the footing is 4 ft below basement floor that has a 100 psf live load on it.

2. Design a footing to support a 16-in. masonry wall with a dead load of 12 klf and a live load of 5 klf. Soil-bearing pressure = 3 ksf, not including the weight of the footing.

3. Design a footing for the wall in Problem 2 using a soil-bearing pressure of 6 ksf.
4. For the same wall and conditions as in Problem 2, design the footing to support 10% of the wall load over an 8-ft-wide soft spot in the foundation soil.
5. Proportion a combined footing to support two columns with service loads as follows (see Fig. 11.30):

	Column A (kips)	Column B (kips)
Dead load	625	500
Live load	415	800
Total load	1040	1300

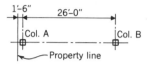

FIGURE 11.30

Soil-bearing pressure is 8 ksf, not including the weight of the footing and surcharge. Assume 80% of the live load can occur on column B simultaneous with 20% live load on column A. No part of the footing may project beyond the property line.
6. Select a satisfactory footing size for Ex. 11.5
7. Select size, and reinforcement if required, for an end-bearing drilled pier to support a column with 1700-kip total service load. The bearing stratum is 55 ft below the surface of the ground and will support a bearing pressure of 20 tons per ft.[2] The soil around the shaft will stand open without casing the pier shaft or the bell. The top 20 ft of soil may exert a negative skin friction of 1000 psf on the pier shaft.
8. Determine the number of 25-ton piles required and the pile cap size and thickness to support an 8 × 8-in. steel column carrying 160-kip dead load and 135-kip live load. The column has a 20-in. square steel base plate.

Selected References

1. ACI Committee 318. *Building Code Requirements for Structural Plain Concrete, (ACI 318-83.1)*. American Concrete Institute, Detroit, Michigan, 1983).
2. ACI Committee 336. "Suggested Design Procedures for Combined Footings and Mats, (ACI336.2R-66)," *ACI Manual of Standard Practice, 1983, Part 4*, American Concrete Institute, Detroit, Michigan, 1983.
3. ACI Committee 336. "Suggested Design and Construction Procedures for Pier Foundations, (ACI336.3R-72)," *ACI Manual of Standard Practice, 1983, Part 4*, American Concrete Institute, Detroit, Michigan, 1983.

CHAPTER 12

WALLS

12.1 OBJECTIVES

The ACI Code defines a wall as a "member, usually vertical, used to enclose or separate spaces." In addition, walls frequently serve other functions, such as retaining earth and liquids, resisting wind pressures, containing bulk materials in storage containers, resisting horizontal forces in the plane of the wall, supporting gravity loads as columns, and acting as beams to carry concentrated loads or uniform loads to points of support. Although some concrete walls serve only as partitions, engineers seeking to reduce construction costs will find additional uses for them.*

Walls are frequently classified by their function. For example,

1. Retaining walls are walls that retain earth or other materials. When they are without lateral support at the top of the wall, they are called cantilever retaining walls.

2. Basement walls are the outside walls of a building, generally below grade, that retain earth outside the building and prevent the entry of water and weather. They are usually supported laterally at both top and bottom of the wall.

3. Wall panels are outside walls that enclose a building above grade, resist wind forces, prevent the entry of water, and provide a thermal barrier, Many wall panels are precast to reduce costs and to facilitate construction of aesthetically pleasing finishes. Cast-in-place wall panels may also serve other functions, such as shear walls.

4. Shear walls are structural elements that resist large horizontal forces in the plane of the wall, such as those generated by earthquake or wind forces, especially in tall buildings. Shear walls may also be used to provide lateral

---

*The more uses, the better. Multiple uses detract little from one another and they make the engineer look good.

support to columns and thus make the construction of columns more economical (see Section 6.8).

5. Bearing walls serve a dual purpose as partition walls and structural support for gravity loads.

6. Transfer walls act as bearing walls and as beams to transfer uniformly distributed or concentrated loads to columns below (see Section 6.5).

7. Fire walls are partitions that divide a large building into sections and deter the spread of fire from one section to another.

The objectives in design of walls are the following:

1. Select wall thickness.

2. Determine the size and spacing of horizontal and vertical reinforcement.

For Walls Serving as Beams

3. Select flexural reinforcement at bottom and top of the wall.

For Cantilevered Retaining Walls,

4. Select the width, thickness, and location of the footing.

5. Determine reinforcement in the footing.

6. Select other details, such as shear keys and expansion joints. Drainage systems must also be considered but are not discussed in this text.

12.2 GENERAL REQUIREMENTS

Walls should be reinforced in two directions with steel of sufficient size, spacing, and strength to control the size and spacing of cracks, as well as to resist moment and shear. The reasons for minimum reinforcement in walls are similar to those for reinforcement in solid slabs (see Section 3.2), but the requirements are different.

The minimum ratio of vertical reinforcement area to gross concrete area must be 0.0012 for deformed bars not larger than #5 and with $f_y \geqslant 60$ ksi and 0.0015 for other deformed bars. The minimum ratio of horizontal reinforcement area to gross concrete area must be 0.0020 for deformed bars not larger than #5 and with $f_y \geqslant 60$ ksi and 0.0025 for other deformed bars. Less vertical reinforcement than horizontal reinforcement is required because the weight of the wall and other gravity loads help prevent horizontal cracking. For walls more than 24 in. thick, it is sufficient to provide reinforcement required in a 24-in.-thick wall [1]. Minimum reinforcement for shear walls is more than that listed here for other walls (see Section 6.8). Shear walls are important structural elements, and reinforcement is essential to ensure ductility.

To facilitate placement of concrete, walls should have a bar curtain (one layer of bars in each direction, tied together) in one face only. Most engineers use only one bar curtain in basement walls and in walls 10 in. thick or less. However, for thicker walls, except basement walls, the ACI Code requires a bar curtain in each face of the wall. One curtain containing between one half and two thirds of the required reinforcement in each direction must be placed not less than 2 in. clear nor more than one third the wall thickness from the exterior face of the wall. The remaining required reinforcement must be placed in a curtain located not less than ¾ in. nor more than one-third the wall thickness from the interior face of the wall (Code 14.3). Basement walls and other walls below grade need have only one layer of steel in each direction, regardless of the thickness of the wall. For massive basement walls (about 2 ft thick or more) two bar curtains may be desirable. Maximum bar spacing requirements apply to each bar curtain separately.

In addition to the above requirements for minimum reinforcement, at least two #5 bars should be placed on all four sides of window and door openings and other large openings. Experience has shown a tendency of concrete around such openings to develop diagonal cracks starting at the corners. Diagonal bars at the corners should not be used as they introduce another layer or two of bars and restrict the placement of concrete. Each end of each #5 bar should extend at least 2 ft beyond the opening to develop the tensile strength of the bar.

Vertical and horizontal bars should be spaced as far apart as possible for economy but not more than three times the wall thickness or 18 in. (Code 14.3.5).

If computations indicate that reinforcement is stressed from loads perpendicular to the wall like a slab (as in a retaining wall), it may be necessary to specify methods for holding the bars in position. Gravity will not hold the bars in the proper location as it does bottom bars in beams and slabs.* For example, vertical #5 bars in a 12-ft-high wall can easily bow 3 or 4 in. toward the center of the wall during concrete placement. Effective depth of reinforcement at the point of maximum moment will be reduced by the amount of the bowing.

Always consider placement of reinforcement in walls for ease of placement of concrete. For example, when heavy bars are required at the top of the wall for flexure, they should be placed one bar above the other, rather than one bar beside the other as in a beam. Horizontal bars in each curtain should be placed in the outside layer unless the design requires the placement of vertical bars in this position (see Fig. 12.1). When vertical bars are in the inside layer, their effective depth is reduced, but control of cracks in the wall is better.

Concrete cover on reinforcement is discussed in Section 1.8 (see Table A1.2).

When walls are waterproofed with an impervious membrane to ensure that no water reaches the wall, concrete cover required for members not exposed to earth or weather may be used even though earth is placed against the membrane.

Exterior basement walls and foundation walls must be at least 7½ in. thick (see Section 12.4) and other walls at least 4 in. thick. Nonbearing walls must be at least 1/30 the least distance between members that provide lateral support, and bearing walls must be at least 1/25 the height or length between supports, whichever is shorter (Codes 14.5.3 and 14.6). Thickness of bearing walls designed as columns can be thinner.

It is difficult to cast and consolidate concrete properly in thin walls. Unless there

---

*Peering into dark spaces between wall forms to see whether rebar curtains are in the correct location is a bit like looking through a sidewalk grating on a sunny day to find a dirty dime at the bottom of the areaway.

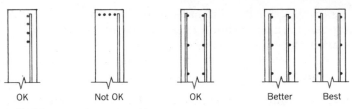

FIGURE 12.1   Placement of bars in walls.

is a special need to reduce the section, it is good practice to make cast-in-place walls at least 8 in. thick for heights of about 8 ft or less and at least 12 in. thick for higher walls.

For simplicity, wall thicknesses should vary by 2-in. increments (8, 10, 12, 14, 16 in., etc.).

Plain Concrete Walls

When loads are light and when location and use of concrete walls minimize the concern for cracking, plain concrete walls can be used. As with footings (see Section 11.3), plain concrete walls have less than the minimum reinforcement required for reinforced concrete or no reinforcement.

Use of plain concrete walls is limited to walls with the following characteristics:

1. Continuously supported by soil, footings, foundation walls, grade beams or other structural member capable of providing continuous vertical support.
2. Not used to resist earthquake or blast loading.
3. With a concrete strength $f'_c \geq 2.5$ ksi.
4. With a thickness of not less than 1/24 the unsupported height or length, whichever is shorter, nor less than $5\frac{1}{2}$ in. for bearing walls. Limits on the thickness of exterior basement walls and foundation walls and of nonbearing walls are the same as those for reinforced concrete.

Even though no other reinforcement is used, two #5 bars should be placed around door and window openings as described above for reinforced concrete walls.

Plain concrete walls must be keyed or doweled to intersecting members where required for lateral stability.

Plain concrete walls without joints will probably have cracks at random spacing and direction. Because reinforcement is not present to control the size of cracks, it is common practice to specify the location and construction of vertical control joints, especially in architecturally exposed walls. Control joints will limit internal stresses due to creep, shrinkage, and temperature effects and minimize unwanted cracks. In areas where the ACI *Building Code Requirements for Structural Plain Concrete* [2] have been adopted by the building code with jurisdiction, control joints in plain concrete walls are required.

Joints will divide the member into flexurally discontinuous elements so that only compressive forces can be transmitted across a control joint. Shear might also be transmitted across a joint by using keyways or sliding dowels. Joints should not be located where bending stress in the plain concrete is required to resist loads or maintain stability of the wall.

In determining the location and spacing of control joints, consider the following:

- Conditions of exposure (indoor vs. outdoor, temperature extremes, humidity of ambient air, and other conditions affecting moisture content and drying of the concrete).
- Shrinkage characteristics of the concrete materials and proportions.
- Curing and other construction procedures.
- Restraint to free movement provided by the foundation, floor framing, and other elements of construction.
- Stresses due to loads on the wall.
- Location of openings, intersecting walls, corners, pilasters, and other features of the building.
- Construction techniques that might affect cracking in the wall.
- Acceptability of some random cracks.

Horizontal joints are generally not required except in multistoried walls. Vertical control joints will normally be spaced not over 20 ft c/c.

Plain concrete walls subject to widely fluctuating temperatures and humidity where random cracks would be unacceptable should have joints at 6 to 10 ft c/c.

If the wall has any reinforcement, it must end at least 3 in. from the joint. The purpose of the joint is to allow cracks to form there rather than elsewhere, whereas bars crossing the joint tend to limit its size and make cracks between joints more likely.

12.3 WALL BEAMS AND WALL COLUMNS

Walls are very efficient as beams due to their great depth and are also usually efficient as columns because much of the load can be supported on concrete rather than steel. If walls are required for partitions or to contain earth pressure, clever engineers will also use them to resist flexure and axial loads, whenever possible.

Walls Acting as Beams

Basement walls are frequently used to distribute load from exterior columns to a strip footing under the wall (see Fig. 12.2). Walls used for this purpose are called grade beams, where the top of the wall (i.e., the beam) is approximately at grade. The cost of footings is reduced, and the cost of walls is increased only slightly, if at all. In most instances, the wall must be designed by deep beam theory (see Section 6.5) due to the proportions of the wall. As with other footings, center of gravity of the footing must coincide with center of gravity of loads. Thus, the width of the footing might vary from one bay to the next, in proportion to the column loads.

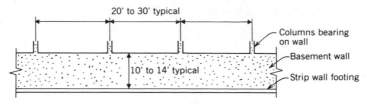

FIGURE 12.2   Elevation of typical basement wall.

Sometimes, a wall can serve as the strap in a strap footing. For example, the function of the strap in the footing shown in Fig. 11.19(*a*) could be served by a wall extending from column A to column B.

Example 12.1   A row of 10 × 20-in. columns bear on a 12-in.-thick basement wall 12 ft high, as shown in Fig. 12.3. The columns are spaced 30 ft c/c and carry service loads as follows:

Dead load = 290 k
Live load = 95
Total load = 385 k

Other uniform service loads bearing on the basement wall at the first floor line are as follows:

Dead load = 2.70 k/′
Live load = 0.95
Total load = 3.65 k/′

Allowable soil-bearing pressure = 6 ksf, not including weight of the footing. Assume f'_c = 3 ksi and f_y = 60 ksi. Select (A) footing size, and (B) flexural reinforcement for the wall as a grade beam.

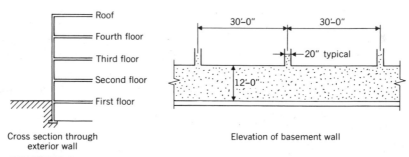

Cross section through
exterior wall

Elevation of basement wall

FIGURE 12.3

Solution A

Distribute the column load uniformly along the footing under the wall:

$$\text{Column load} = \frac{385}{30} \qquad = 12.83 \text{ k/}′$$

Other uniform load $= 3.65$

Basement wall weight $= 12 \times 0.15 = \underline{1.80}$

Total wall load on footing $= 18.3\text{k}/'$

Required footing width $= \dfrac{18.3}{6.0} = 3.05'$

Use 3'-0" wide footing (1.5% under)

(See Section 11.4 for selection of footing reinforcement.)

If column loads were not distributed along the wall, a wall footing 1'-8" wide would still be required and an 8' square footing under each column would be required ($\sqrt{385/6}$). This would be a more expensive solution but might be justified if utilities or other obstructions interfere with a continuous wall footing.

Solution B

Assume effective depth of the wall, $d = 12 - 1 = 11' = 132''$

Span-to-depth ratio $\dfrac{\ell_n}{d} = \dfrac{30 - 1.67}{11} = 2.58$

Because $\mathscr{L}_n/d < 5$, use deep beam design procedures. In this case, the beam is inverted so that reaction points are on the top and the uniform loading is on the bottom. This meets the requirement that loads be on the "top" or opposite face of the beam from the reaction point to justify use of deep beam theory. Locations of positive moment reinforcement and negative moment reinforcement will also be reversed from the normal beam; that is, positive moment reinforcement will be in top of the wall and negative moment reinforcement will be in bottom of the wall.

Column dead load $= 290 \times 1.4 = 406\text{ k}$

Column live load $= 95 \times 1.7 = \underline{162}$

Column total load $= P_u = 568\text{ k}$

Location of the critical section for shear $= 0.15\ell_n$ from face of column, or

$0.15(30 - 1.67) + \dfrac{1.67}{2} = 5.08'$ from the column centerline,

or $15 - 5.08 = 9.92'$ from the centerline of the span.

$V_u = \dfrac{568 \times 9.92}{30} = 188\text{ k}$

$v_u = \dfrac{188}{12 \times 132} = 119\text{ psi}$

From Fig. A6.3a, allowable $v_c \cong 240$ psi $>> 119$ psi. Therefore, no shear reinforcement is required. Use horizontal and vertical reinforcement as required for walls.

For vertical reinforcement,

$$A_s = 12 \times 12 \times 0.0012 = 0.17^{\square''}/'$$

Use #4 bars at 13" c/c, $A_s$prov. $= 0.2 \times \dfrac{12}{13} = 0.18^{\square''}/'$

For horizontal reinforcement,

$$A_s = 12 \times 12 \times 0.002 = 0.29^{\square''}/'$$

Use #5 bars at 13" c/c, $A_s$prov. $= 0.31 \times \dfrac{12}{13} = 0.29^{\square''}/'$

Select the positive moment flexural reinforcement in top of wall:

Conservatively using moment coefficients for end spans (see Chapter 13) to determine positive moment in the grade beam,

$$w_u = 568/30 = 18.9 \text{ k}/'$$

Maximum moment $^+M_u \leqslant \dfrac{w_u \ell_n^2}{14} = \dfrac{18.9(28.3)^2}{14} = 1080'\text{k}$

Because $d/\ell_n = 11/28.3 = 0.39 < 2/5$ (see Section 6.5), nonlinear distribution of strain need not be taken into account, and flexural bars can be located near top and bottom of the wall.

Assume $(\phi f_y j) = 4.3$, and $d = 132''$,

$$A_s = \dfrac{1080}{4.3 \times 132} = 1.90^{\square''}$$

Check minimum reinforcement:

$$\rho = \dfrac{1.90}{132 \times 12} = 0.0012 < 0.0033$$

Therefore, increase both ^+A_s and ^-A_s by one-third.

$$^+A_s\text{reqd.} = 1.90 \times 1.33 = 2.53^{\square''}$$

Use 2 #10 horizontal bars continuous, spaced 13" c/c vertically at top of the wall. $A_s$prov. $= 2.54^{\square''}$. These bars replace 2 #5 wall bars. Splice the #10 bars at the columns. Referring to Table A5.6, the minimum clear span to develop #10 "bottom" bars (positive moment reinforcement) is

$$20 \times 1.4 = 28' < 28.3' \hspace{4cm} \underline{\text{OK}}$$

Select negative moment flexural reinforcement in bottom of the wall:

Conservatively using moment coefficients for end spans to determine negative moment in the grade beam as was done for positive moment,

Maximum $^-M_u \leqslant \dfrac{w_u \ell_n^2}{10} = \dfrac{18.9(28.3)^2}{10} = 1520'\text{k}$

Assume $(\phi f_y j) = 4.3$ and $d = 132''$,

$$A_s = \frac{1520}{4.3 \times 132} = 2.68^{\square''}$$

Use one-third more reinforcement because $^+\rho < 0.0033$.

Therefore, $A_s$reqd. $= 2.68 \times 1.33 = 3.57^{\square''}$

Use 3 #10 horizontal bars continuous spaced 13" c/c vertically at bottom of the wall. $A_s$prov. $= 3.81^{\square''}$. These bars will replace 3 #5 wall bars. Splice the #10 bars at mid-span between columns. Referring to Table A5.6, the minimum span to develop a #10 "top" bar (negative moment reinforcement) is $20' < 28.33'$.   OK

Alternatively, bottom bars could be placed in the footing instead of bottom of the wall. The bars would be easier to place in the footing than in the wall, but extra computations would be required to determine requirements to assure composite beam action, such as roughening the footing–wall-joint interface and providing dowels.

Note that using moments in the end span is conservative when applied to interior spans because moment in interior spans should be somewhat smaller. It is good practice to use the same reinforcement for full length of the wall to avoid field errors.

Although not strictly required, it is good practice to splice the #10 top and bottom bars with a full tension splice to avoid field errors, to assure anchorage of the positive moment bars at the support (see discussion on anchorage of bottom tie bars in deep beams in Section 6.5), and to reinforce for unexpected patterns of moment due to variations in support from the foundation soil.

Walls Supporting Axial Load

Where walls support significant axial load in addition to their own weight, they may be designed as columns (see Chapter 9), subject to all the limitations and requirements for columns, such as minimum area of longitudinal (vertical) reinforcement, size and spacing of lateral reinforcement, distribution of reinforcement, and slenderness. The outer limits of the effective cross section of a column built monolithically with a wall or pier is 1½ in. beyond the outside of the spiral or tie reinforcement (Code 10.8.2). In any case, the horizontal length of wall considered effective in supporting a concentrated load should not exceed the center-to-center distance between loads nor the width of bearing plus four times the wall thickness. The latter requirement might be waived if justified by a detailed analysis (Code 14.2.4).

Regardless of the method of design, vertical reinforcement in walls need not be enclosed by lateral ties or spirals if the area of vertical reinforcement is not more than 0.01 times the gross concrete area or where vertical reinforcement is not required as compression reinforcement.

Walls may be designed to support axial loads by an empirical method if the resultant of all factored loads is located within the middle third of the overall thickness of the wall (Code 14.5). Use of this method is more efficient (in design time) and more economical (in construction cost), than the design of walls as columns.

The design axial load strength ϕP_{nw} of a wall by the empirical method may be computed as follows:

$$\phi P_{nw} = 0.55\phi f'_c A_g \left[1 - \left(\frac{k\ell_c}{32h} \right)^2 \right] \tag{12.1}$$

where $\phi = 0.70$
ℓ_c = vertical distance between lateral supports
h = overall thickness of the wall
A_g = gross area of the wall
and k = effective length factor = 1.0 for most practical cases, where top and bottom of the wall is braced against lateral translation.

For walls not braced against lateral translation but restrained against rotation at top and bottom (an unusual case of a freestanding wall), $k = 2.0$. For walls restrained against lateral translation and restrained against rotation at top or bottom, or both, $k = 0.8$. Large foundations or heavy floor systems at least as stiff as the wall (where stiffness = EI/ℓ) would provide sufficient restraint against rotation to justify $k = 0.8$.

Where columns, beams, or other concentrated loads bear on a wall, the bearing strength is the same as that given in Section 10.2 (see Eq. 10.1). If all portions of a column are not directly over the supporting wall, the wall should include a pilaster at least as large as the column.

Example 12.2 Design the wall in Example 12.1 to support the column loads. Assume that each column has 8 #9 vertical bars and applies minimum moment to the wall and bears on the wall, as shown in Fig. 12.4.

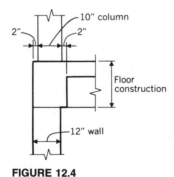

FIGURE 12.4

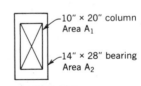

FIGURE 12.5

Solution Referring to Fig. 12.5, the concentric bearing area is $14''/10'' = 1.4$ larger than the column. Increase in bearing stress = $\sqrt{A_2/A_1} = 1.4$ also.

Factored axial load (from Ex. 12.1) = 568 k

Concrete bearing strength =

$0.7 \times 0.85 \times 3 \times 10 \times 20 \times 1.4 \qquad = \underline{500}$ (Eq. 10.1)

Load to be carried by dowels $\qquad = \quad 68$ k

Using Eq. 9.5,

$$\text{Required dowel area } A_s = \frac{68}{0.80 \times 0.7 \times 60} = 2.02^{\square\prime\prime}$$

Use 4 #7 bars $A_s$prov. $= 4 \times 0.60 = 2.40^{\square\prime\prime}$

It would be good practice to use heavier dowels, such as 4, 6 or 8 #9 bars to allow for dowel misplacement, for presence of more moment than computed, and for structural tradition. Owners and construction workers may expect to see dowels matching the column vertical bars and question the engineer if dowels are fewer or smaller.*

Referring to Fig. 12.6, the length of wall effective in supporting the column load = column width plus 4 wall thicknesses = 20 + 4 × 12 = 68″ = 5′-8″

Within this length of wall, the axial loads are

Factored column load $\qquad\qquad\qquad\qquad = 568$ k

Factored uniform load = 2.70 × 1.4 = 3.78 k/′

$\qquad\qquad\qquad + 0.95 \times 1.7 = \underline{1.62}$

$\qquad\qquad\qquad\qquad 5.40 \times 5.67 = \underline{\quad 31 \quad}$

Total factored load $\phi P_{nw} \qquad\qquad\qquad = 599$ k

*Whenever someone raises such a question, the engineer spends a half hour or more in answering it and still risks leaving the questioner less sure of the engineer's competence.

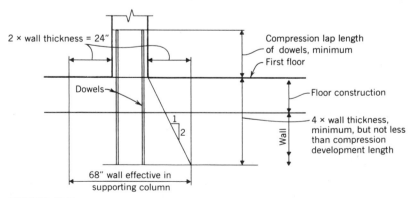

FIGURE 12.6

Check eccentricity of load to determine whether the empirical method may be used. Using Eq. 9.6, the minimum column eccentricity,

$$e_{min} = 0.25\left(10 - 2.5 - \frac{10}{2}\right) = 0.62''$$

Conservatively assume that the footing imparts a similar eccentricity so that at mid-height of the wall, $e = 0.62''$.

Next, assume moment at mid-height of the wall due to soil pressure, $M_u = 8.26'k/'$ (this value will be developed in Ex. 12.3), and that total moment for the length of wall effective in supporting the column $= 8.26'k/' \times 5.67' = 46.8'k$. The eccentricity due to soil pressure

$$e = \frac{46.8 \times 12}{599} = 0.94''$$

Conservatively assuming that both column eccentricity and soil pressure eccentricity are in the same direction, the total eccentricity $= 0.62 + 0.94 = 1.56''$. The resultant of all factored loads is within the middle third of the wall, which extends $2''$ ($12''/6$) from the center of the wall; therefore, the empirical method may be used.

Using Eq. 12.1,

$$\text{Allowable } \phi P_{nw} = 0.55 \times 0.7 \times 3(12 \times 68)\left[1 - \left(\frac{12}{32 \times 1}\right)^2\right]$$
$$= 810 \text{ k} > 600 \text{ k} \qquad \underline{\text{OK}}$$

Dowels at top of the wall need not extend full height of the wall, because the wall itself—reinforced with minimum wall reinforcement—will support the column load. Dowels should extend a minimum of 4 times the wall thickness below the first floor to allow the column load to spread to the width of the wall assumed in design. In practice, most engineers would extend dowels full height of the wall to ensure their proper placement and to avoid lengthy explanations to workers who are accustomed to full-height bars.

12.4 BASEMENT RETAINING WALLS

When a floor is constructed below grade, the outside walls must retain the earth outside the building and resist the earth pressure against the building. For convenience, all such walls are called basement retaining walls. They can be used to support axial loads and transmit loads in flexure while they resist lateral earth pressure.

Basement retaining walls are supported laterally at the foundation and at the first floor (i.e., at bottom and top of the wall) and span between foundation and first floor as a solid slab. Long walls span as a one-way slab. In short walls (length less than two times the height) and at corners of long walls, the wall spans by two-way action.

Intersecting perpendicular walls, heavy pilasters, and other stiff elements may also create two-way bending in a wall. Unless the bending moment is very large, as in a wall several stories below grade where earth pressure is large, it is common practice to design all basement walls as one-way slabs without taking advantage of the reduction in moment for two-way bending. Minimum wall reinforcement will frequently provide adequate reinforcement to resist earth pressures. As with solid slabs, shear rarely, if ever, controls design of basement walls.

Because basement walls are designed to be supported laterally at the top, it is essential that the first floor slab be in place before backfill is placed against the wall. Alternatively, the wall can be braced temporarily before backfilling until after the first floor slab or other support has been constructed.

To ensure that building spaces below grade remain dry, outside walls must be waterproof. Generally, this requires application of a membrane to the outside face of the wall or taking other precautions. Hydraulic and sanitary structures (water tanks and the like) are frequently designed to be watertight without using a membrane. Discussion of measures to waterproof walls is beyond the scope of this text.

Buildings are frequently faced with masonry above grade. Where the grade line is below the top of the basement wall, it may be notched or recessed at the grade line to receive masonry. Because masonry is usually a nominal 4 in. thick, the recess is made 4½ in. wide to allow for construction tolerances and finger space for masons laying bricks. If a 12-in. wall is used, the remaining thickness at the recess is 7½ in. (see Fig. 12.7). Design of the wall for lateral loads from earth pressure must consider the maximum moment occurring at the reduced section, as well as the maximum moment occurring at the full thickness.

Earth pressures and other design conditions are determined by soil mechanics and should be recommended by a competent geotechnical engineer. Following is a brief summary of typical design conditions most commonly encountered in basement retaining walls.

The pressure exerted by earth and other retained material is assumed to be hydrostatic, that is, pressure is equal to the height below the surface multiplied by an equivalent fluid unit weight. The equivalent hydrostatic pressure is generally as-

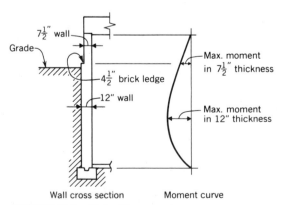

FIGURE 12.7   Wall with recess for masonry.

sumed to be between 30 and 62.5 pcf, the latter being the actual weight of water. The lower value is assigned to dry granular soils, and the higher value to wet soils approaching the properties of a liquid. Certain saturated silts and clays can exert a hydrostatic pressure as high as their unit weight—far in excess of 62.5 pcf. An alternative to constructing the wall to resist higher soil pressure is to remove the soil, replace it with granular material, and provide drainage. If the soil slopes up away from the wall, if there is vehicular traffic on grade near the wall, or if foundations or other heavy loads are on the soil near the wall, additional pressures will be exerted on the wall. Such loads, on or near grade, are called surcharge. For vehicular traffic, lateral pressures from surcharge generally range from 25 to 150 psf. Surcharge pressures are considered constant from bottom of the wall to ground surface or level of application of surcharge pressure, (see Fig. 12.8).

Unless the wall is designed for the full fluid pressure of saturated soil, it is essential that the soil behind the wall be drained to prevent accumulation of water. Drainage can be accomplished by selecting proper backfill material and installing perforated drain pipes at bottom of the wall, sloped to sewers or lower-lying areas. Failure to provide adequate drainage could lead to substantially higher soil pressure against the wall and, ultimately, to failure of the wall.

Sometimes buildings on a sloping site will have earth pressure against basement walls on one side of the building but no pressure against walls in the same story on the other side of the building, as in Fig. 12.9. In such cases, the reaction at top of the wall on the one side must be considered a permanent lateral load on the frame and the frame designed to resist the load in the same manner as wind loads, earthquake loads, and other lateral forces. Alternatively, the reaction at the top of the wall can be resisted by diaphragm action in the first floor, which transmits the load to shear walls parallel to the reaction.

The horizontal reaction at bottom of a basement wall will be resisted by the friction of the footing on foundation soils. If the basement slab is cast tight against the wall, the slab may also resist the horizontal reaction by friction against the soil or by bearing against another wall with an opposing horizontal reaction. (See the discussion on sliding resistance in Section 12.5, Cantilever Retaining Walls.) Resistance to sliding on soils is more than adequate for basement walls unless they support very light gravity loads and resist very large lateral pressure.

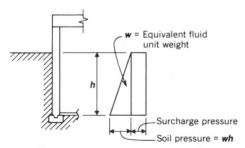

FIGURE 12.8 Lateral pressure on retaining wall.

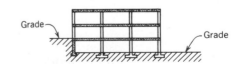

FIGURE 12.9 Cross section of building on sloping site.

Example 12.3   (A) Design the basement wall in Ex. 12.1 for soil pressure the full height of the wall and an assumed hydrostatic pressure of 45 pcf with no surcharge. (B) Design the footing and the wall, assuming the footing may not extend beyond the property line, which is on the outside face of the wall.

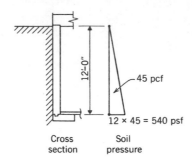

Cross
section

Soil
pressure

FIGURE 12.10

Solution A

Referring to Fig. 12.10, the maximum active soil pressure against the wall at the bottom

$$= 12 \times 45 = 540 \text{ psf}$$

Total pressure against wall, $H = \dfrac{12 \times 0.54}{2} = 3.24 \text{k}/'$

For earth pressure, load factor = 1.7

$$W_u = 1.7H = 1.7 \times 3.24 = 5.51 \text{k}/'$$

Neglecting the restraint of the floor system at the top and the footing at the bottom, it is conservative to assume that the wall is a simple span. For a load increasing uniformly from zero at one end, the maximum moment near mid-span = $0.1283W\ell$. For simplicity, take

$$M_u \simeq \dfrac{W_u \ell_n}{8} \text{ where } \ell_n = \text{clear height of wall} = 12'$$

$$M_u = \dfrac{5.51 \times 12}{8} = 8.26'\text{k}/'$$

Allowing ¾" cover and assuming ¾" bar size,

$$d = 12 - 0.75 - 0.75/2 = 10.8". \qquad \text{Assume } (\phi f_y j) = 4.3$$

$$A_s = \dfrac{8.26}{4.3 \times 10.8} = 0.18^{\text{□}''}/'$$

Use minimum reinforcement #4 at 13" c/c $A_s$prov. = $0.18^{\text{□}''}/'$

Although this selection meets Code requirements, to be conservative and allow some tolerance in the location of vertical bars, assume $d = 10.8 - 1 = 9.8"$.

$$A_s = \frac{8.26}{4.3 \times 9.8} = 0.20^{\square''}/'$$

Use #4 bars at 12" c/c, $A_s$prov. $= 0.20^{\square''}/'$

Solution B

Footing Design

For simplicity, conservatively assume that the average load factor = 1.55. If there is a large quantity of wall with this condition, then it might be economically feasible to compute the actual load factor. The savings in reinforcement could be as much as 3% to 5%.

Referring to Fig. 12.11, the critical section for moment is at the inside face of the wall.

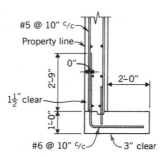

#5 @ 10" c/c
Property line
0"
2'-9"
2'-0"
1½" clear
1'-0"
#6 @ 10" c/c
3" clear

FIGURE 12.11

$$M_u = \frac{w_u \ell^2}{2} = \frac{6 \times 1.55 \times 2^2}{2} = 18.6' k/'$$

Assume a 12"-thick footing, 3" cover, and a ¾" bar size.

$d = 12 - 3 - 0.75/2 = 8.6"$.   Assume $(\phi f_y j) = 4.3$

$$A_s = \frac{18.6}{4.3 \times 8.6} = 0.50^{\square''}/'$$

Use #6 @ 10" c/c, $A_s$prov. $= 0.44 \times \dfrac{12}{10} = 0.53^{\square''}/'$

From Table A5.1, the basic development length of #6 bar = 19.3". Because 100% of the bars are spliced at one point in the wall and

$A_s$prov./$A_s$reqd. < 2, a class C splice is required.

Splice length $= 1.7\ell_d = 1.7 \times 19.3 = 33" = 2'\text{-}9"$

The effective depth of #6 bars in the wall is slightly greater than that in the footing.

Wall Design

Backfill may not always be present against the basement wall and, if present, it may not exert as much pressure as assumed in design. Therefore, moment in the wall due to eccentricity of the footing should not be reduced by moment in the wall caused by soil pressure against it. Likewise, it is good practice to design the wall for moment caused by soil pressure without reducing it for moment caused by eccentricity of the footing. Thus, reinforcement in both faces of the wall is required.

Referring to the moment curve in Fig. 12.12, at 2'-9" above the footing, moment is reduced to

$$\frac{12 - 2.75}{12} \times 18.6 = 14.3'\text{k}/'$$

Effective depth $d = 12 - 1.5 - 0.75/2 = 10.1''$

$$A_s = \frac{14.3}{4.3 \times 10.1} = 0.33^{\square''}/'$$

Select bar spacing to match the spacing of footing dowel bars.

Use #5 @ 10" c/c, $A_s$prov. = 0.37 in.^{2}/'

The #5 bars should extend full height of the wall.

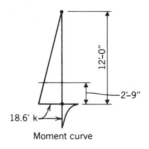

12'-0"

2'-9"

18.6' k

Moment curve

FIGURE 12.12

12.5 CANTILEVER RETAINING WALLS

Retaining walls cannot always be supported laterally at the top, especially where they are isolated from other structures or structural elements. In such situations, the weight of the material being retained can be used to provide the resistance to overturning the wall.* The type of retaining wall to be used in a particular case depends on a number of factors in addition to the cost of construction. The most common types of cantilevered or free standing retaining walls used today are illustrated in Fig. 12.13.

*A form of engineering judo.

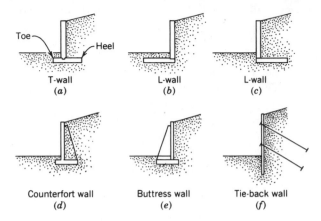

FIGURE 12.13 Types of retaining walls. (a) T-wall; (b, c) L-wall; (d) counterfort wall; (e) buttress wall; (f) tie-back wall.

The inverted T-wall [Fig. 12.13(a)] is by far the most common type of retaining wall for wall heights of about 20 ft or less. Its simplicity recommends it for a wide variety of uses. The edge of the footing on the low side of the wall is called the "toe." The edge of the footing on the high side of the wall is called the "heel."

An L-wall [Fig. 12.13(b) and 12.13(c)] can be used where property lines or other obstruction limits the use of a footing beyond one side of the wall. Its disadvantage is that high soil-bearing pressure is generated on the toe of the footing. For the wall in Fig. 12.13(b), only the weight of the wall and the footing plus any friction of retained material against the wall is available to resist overturning. This disadvantage can be overcome by the use of a wider footing. The wall in Fig. 12.13(c) requires careful detailing to assure adequate shear strength between wall and footing.

A counterfort wall [Fig. 12.13(d)] is similar to an inverted T-wall but uses counterforts (short walls perpendicular to the retaining wall) at regular intervals. The purpose of counterforts is to support the retaining wall. Counterforts are used when the height of a T-wall is more than about 20 ft, to reduce the size and deflection of the wall.

Buttress walls [Fig. 12.13(e)] are similar in design and purpose to counterforted walls but have the buttresses on the low side of the wall instead of counterforts on the high side of the wall. Buttresses are in compression and are somewhat more efficient than counterforts, but they interfere with free use of the space adjacent to the wall.

Tie-back walls [Fig. 12.13(f)] are used for deep excavations with large lateral loads. As the excavation progresses, the wall is supported by tie-backs drilled and anchored into the soil beyond the failure plane of the soil. Ties are made of high-strength steel and prestressed to prevent movement of the wall due to elongation of the ties. After prestressing, ties are grouted for protection against corrosion. When carefully constructed, tie-back walls support the retained soil without significant lateral or downward movement. This is an important consideration where structures are supported on the soil adjacent to the wall on the high side. In addition, tie-back walls leave the low side of the wall unobstructed—an important consideration during construction.

The lateral soil pressure on cantilever retaining walls is the same as on basement retaining walls and is subject to the same considerations. A geotechnical engineer should be consulted on walls higher than about 10 or 15 ft, as they are heavy and expensive. The safety of the wall depends on geotechnical considerations, and the consequences of failure could be serious.

Waterproofing of cantilever retaining walls is usually omitted where they are exposed to the weather. However, drainage is usually assured by installing weep holes through the wall near the bottom at about 10-ft intervals to drain the soil behind the wall. Weep holes are formed by 3- or 4-in. pipes through the wall.

Design of Cantilever Retaining Walls

Because the design of cantilever retaining walls is thoroughly covered in other publications, such as the *CRSI Handbook* [3], it is only reviewed and summarized here. Design will proceed in an orderly fashion if the steps below are followed.

1. Determine the width of the footing and its location with respect to the wall. Considering all gravity and lateral loads, including loads superimposed on top of the wall, the nominal factor of safety against overturning should be at least 2 when the point of rotation is taken as the lower corner of the toe of the footing. The factor of safety is easy to compute in this manner but is unconservatively high because the center of rotation is actually inside the footing near the toe. As the lateral load increases under constant gravity load, approaching overturning, the resultant vertical reaction on the soil approaches the toe. The closer the resultant is to the toe, the higher the soil-bearing pressure (see Ex. 11.1 and preceding discussion). When the bearing pressure reaches the bearing capacity of the soil, failure in the soil occurs and the retaining wall overturns.

   The retaining wall is normally located in the middle third of the width of its footing, but resistance to overturning is relatively insensitive to the exact location.

   Bearing pressure under most retaining wall footings is rather low but should be checked if the allowable soil-bearing pressure is also low. Furthermore, unequal pressure under toe and heel of the footing will lead to rotation of the footing and of the wall. The rotation may not be noticeable in isolated retaining walls. Engineers should be alert to situations where rotation will be objectionable, such as where the retaining wall abuts another wall with vertical lines against which rotation can be compared. Walls can be built with a backward slope equal to the expected movement. However, it is very difficult—almost impossible—to compute rotation of the footing and deflection of the wall precisely.

2. Check the possibility of the footing sliding on the soil. Geotechnical engineers will recommend a coefficient of friction, generally about 0.4 to 0.6. A factor of safety against sliding of 1.5 should be maintained. Passive lateral soil pressure against the toe of the footing can be considered and could be significant if the footing is deeply buried. If resistance to sliding is inadequate, it can be increased by using a key in the footing, which increases the resistance due to passive soil pressure, or by increasing the width of the footing, which increases the weight of soil on the footing and therefore the frictional resistance.

3. Design the wall for flexure. Increase lateral soil pressure by a load factor of 1.7 for use in design of the wall. Be generous on the thickness of the wall to avoid delicate proportions that would be sensitive to inaccurate placement of reinforcement. On high walls, the thickness may be tapered to save concrete, but thickness at top of the wall should be at least 12 in. to permit easy placement and consolidation of concrete. Because formwork for tapered walls costs more than that for walls of uniform thickness, tapered walls should only be used where repetition and reuse of form work will reduce its additional cost to less than the cost of concrete saved.

Placement of dowels in the footing is critical. Misplacement of dowels toward the toe of the footing will reduce the effective depth of reinforcement at the point of highest moment in the wall.

Vertical flexural bars in walls can be cut off near mid-height or at several heights to save weight of steel. Considering the added computational effort, detailing effort, and the confusion and risk of misplacement of steel in the field, short bars should be used only where a large quantity of wall will justify the effort.

Design of reinforcement should take into account two-way slab action in the vicinity of counterforts, buttresses, and other intersecting walls. As in other walls, bars should be placed in the exposed face of retaining walls to control crack widths. In addition, control joints in walls may be constructed to relieve stresses from shrinkage, temperature, and other internal forces that tend to cause cracking.

Shear in the concrete wall rarely controls the design. However, sliding of the wall across the footing could be critical. Top of the footing could be roughened and the joint designed by shear friction concepts (see Section 6.6), diagonal shear bars could be used, or a key provided. Many engineers provide a key in all footing-wall joints.

Considerations in the location and spacing of control joints are similar to those for plain concrete walls, discussed in Section 12.2. Due to the presence of reinforcement, control joints can be spaced farther apart than joints for plain concrete walls—generally 20 to 30 ft c/c.

4. Design the toe of the footing using service load soil-bearing pressure and a load factor of 1.7. Thickness of the footing is usually made the same as the wall, as a minimum.

5. Design the heel of the footing using the weight of soil and surcharge directly above the footing and a load factor of 1.7. Moment in the heel of the footing need not exceed maximum moment in the wall. Moments should be taken about the tensile reinforcement in the wall rather than the face of the wall.

Figure 12.14 shows typical patterns of reinforcement in a cantilever retaining wall. In practice, dimensions and reinforcement of cantilever retaining walls less than 20 ft high are usually taken from tables.

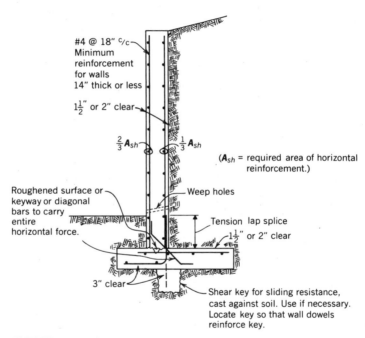

FIGURE 12.14 Cantilever retaining wall (T-wall).

PROBLEMS

1. Select minimum reinforcement for a 8-in. wall.
2. Design a 12-in.-thick grade beam wall, 10 ft high, to distribute column loads to a strip footing under the wall, as in Fig. 12.2. Columns are 14 × 20 in., the long side parallel to the wall, with the outside face 5 in. from the outside face of the wall. Columns are spaced 35 ft c/c. Each column supports an ultimate load of 825 kips. Assume that 8 #11 vertical bars are

used in each column. Assume $f_y = 60$ ksi and $f'_c = 3$ ksi in the wall and $f'_c = 4$ ksi in the columns. Design wall reinforcement for flexure and to support column loads.

3. Select reinforcement for a 12-in.-thick basement wall of 16 ft clear height with a 4½-in. brick ledge at grade, as in Fig. 12.7. Grade is 11 ft above bottom of the wall. Assume hydrostatic soil pressure of 50 pcf plus a lateral soil pressure of 100 psf due to surcharge. Use $f'_c = 3$ ksi and $f_y = 60$ ksi.

Selected References

1. Paul F. Rice. "Structural Design of Concrete Sanitary Structures," *Concrete International Design and Construction*, V.6, No.10, pp 14–16, October, 1984.
2. ACI 318.1-83. *Building Code Requirements for Structural Plain Concrete*. American Concrete Institute, Detroit, Michigan, 1983.
3. *CRSI Handbook*. Concrete Reinforcing Steel Institute, Chicago, Illinois, 1984.

CHAPTER 13

LOAD AND MOMENT DISTRIBUTION

13.1 OBJECTIVES

The first step in the design of a structure after selecting its configuration is to compute the magnitude and location of loads. Loads are frequently, even usually, not uniform nor distributed uniformly, but it is convenient to treat them as such for ease of computation. Meticulous consideration of the exact size and location of many concentrated loads leads to burdensome calculations with little or no increase in accuracy. This chapter discusses when and how to consider nonuniform loads as uniform loads.

The second step in the design of a structure is to compute the maximum moment, shear, torsion, and axial load at each critical section. Normally, only moment and shear are computed for flexural members because small axial load does not affect the design. Likewise, only axial load and moment are normally computed for columns (and frequently only axial load) because small moment and shear do not affect the design. (See Chapter 9 for conditions in which moment, as well as axial load, should be computed for a column. Chapter 6 gives guidelines for conditions in which both shear and axial load should be considered at a section.) This chapter discusses how to calculate moments at the ends of members. From this, shear, axial forces, and moments within the span are determined by statics.

The objectives of the work described in this chapter are to determine the moment, shear, and axial force at each critical section in each member of the structure. These values should be computed by a method that uses the least effort consistent with required accuracy.

13.2 LOAD DISTRIBUTION

Live load on floors includes people, furniture, other movable objects, and movable equipment. Live load on roofs includes rainwater, ice, snow, and casual foot traffic. Dead load includes the weight of concrete, walls, partitions, other building construction, and permanent equipment. People moving in unison, vibrating equipment, and falling objects cause more severe load effects on the structure than the static weight of the people, equipment, or objects. This increase in load is called impact and is normally included in required live loads. For certain equipment and operating conditions, an allowance for impact should be made and treated as live load. The allowance is generally stated as a percentage of the static weight.

Live loads specified in the building code controlling design of the structure are generally of uniform magnitude and are applied uniformly over the full length of the span of a member. (See Chapter 6 for exceptions in the case of shear.) In determining the magnitude of live loads, codes have implicitly recognized that most live loads are actually nonuniform and are applied in concentrated areas (e.g., under the legs of furniture, bookcases, and equipment supports). In most cases, engineers need not be concerned with concentrations of code-specified live load. Exceptions would include vehicular wheel loads on very thin slabs and equipment supports where weight of the equipment divided by its floor area exceeds the specified live load. Live load is not necessarily distributed on all adjacent spans simultaneously to obtain the most severe conditions (see Section 13.3). Live load magnitude and distribution should be computed in the case of warehouse type of loading using the actual unit weight and volume of stored material. For example, it is common practice to compute the live load for floors supporting library shelving or filing cabinets when the spacing and height of the bookshelves or filing cabinets are known. Such live loads are rather heavy and frequently exceed 100 psf.

Most building codes allow a reduction of live load on members carrying a large floor area. Experienced engineers take advantage of this reduction on columns, but many do not on beams, unless it is necessary to investigate the design of a member already constructed.

In northern climates, roofs should be designed for the weight of the maximum snowdrift where drifts may accumulate. Likewise, roofs that are level or nearly so should be designed for the weight of ponded water in areas where deflection of the roof prevents rainwater from flowing directly into drains.

Dead load due to self-weight of the structure is almost always computed as a uniformly distributed load along the span of each member. To justify this assumption, holes or openings through floor and roof slabs are usually conservatively disregarded unless they occupy more than about 10% to 15% of the floor or roof area. The weight of haunches, drop panels, and the like can be conservatively distributed along the length of a member. The weight of joists or ribbed slabs are computed as a uniform load over the floor area. Allowance for distribution ribs and double joists is made by an appropriate additional uniform load.

Normal weight concrete usually weighs about 145 pcf. Because steel weighs considerably more than concrete, the addition of steel reinforcement increases the weight of reinforced over plain concrete. The exact weight of reinforced concrete can be computed by

$$W_{rc} = W_c + \rho(W_s - W_c)$$

where W_{rc}, W_s, and W_c = unit weights of reinforced concrete, steel, and plain concrete, respectively,

and ρ = ratio of steel volume to total volume of concrete.

If weight of steel and plain concrete is taken as 490 and 145 pcf, respectively, and $\rho = 1.5\%$, the weight of reinforced concrete is 150 pcf. This is the value most engineers use for the weight of concrete in computing dead loads. A similar upward adjustment should be made to the weight of plain concrete using lightweight aggregates. Although more heavily reinforced members will weigh more, they are usually a small part of the total concrete volume. Furthermore, holes through floors and the roof reduce the dead load of concrete and tend to offset any unconservative excess weight due to heavy reinforcement.

Openings in walls and partitions for doors, windows, and so forth, are usually conservatively disregarded in computing the dead load unless the area of openings exceeds about 25% to 35% of the total area of the wall or partition.

Exterior walls and walls around stairs, elevator, and duct shafts and other permanent features of the building are treated as line loads. Other interior partitions, even though constructed by permanent methods, are frequently treated as uniformly distributed loads over the entire area of the floor.* Owners may remodel the building and move partitions without consulting an engineer. In some cases, the location of partitions may be unknown when the structure is designed. Furthermore, moments, shears, and axial loads due to uniform loads are easier to compute than those due to irregularly spaced concentrated loads. In establishing a uniform dead load allowance for partitions, engineers should consider the height of partitions, their unit weight, and probable minimum spacing. An allowance of 20 to 50 psf is common.

Allowance should also be made for the dead load of floor finishes, roofing material, ceiling construction, and mechanical work hung from the floor or roof above. When the dimensions but not the weight of a construction material is known, the weight can be easily estimated by using the unit weight from Table A13.1.

The weight of fresh concrete, formwork, and shoring for floor in multistoried buildings, plus construction workers and equipment, is supported by shoring on the floor below. Such construction loads must not exceed the superimposed service dead load plus service live load for which the floor supporting the loads was designed. Because construction loads are frequently higher than superimposed design loads, shores are placed below the support floor to distribute the loads to successively lower levels until no floor carries more than its design load, as in Fig. 13.1. The strength of concrete when construction loads are applied, as well as placement of shores and reshores, should be considered. Most structures are designed assuming that construction loads do not control; construction procedures that overstress the structure are then prohibited. In special circumstances, construction loads may control design of the structure (Code 6.2.2). Recent analytical studies indicate that loads on shores and lower floors are as high as twice the weight of the top story of freshly placed concrete [1]. However, current practice is to assume that loads are distributed, as shown in Fig 13.1.

*This assumption is especially appropriate for solid one- or two-way slabs. It is virtually impossible for a slab strip immediately under a wall to fail without receiving the support of a considerable width of slab nearby.

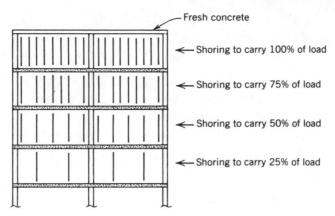

FIGURE 13.1 Example of minimum shoring and reshoring to distribute construction loads to four lower floors.

Wind loads, by their nature, are uniformly distributed but are usually not uniform in magnitude. They will vary with the height above grade, from center to edges of the building, from windward to leeward side to sides parallel to the wind, and depend on roughness and shape of the building skin. For simplicity, most structures of small to moderate size are designed for uniform wind loads applied to the windward side of the building. Engineers should recognize that negative loads, or suction, exists on the leeward side and sometimes elsewhere and that higher-than-average loads may be exerted on individual members, especially if they are small and lightweight.

Other lateral loads include earth pressure and hydrostatic pressure against basement walls and earthquake or seismic forces due to a rapid ground movement. Earthquakes can also have a vertical component that alternately increases and decreases the axial load in columns. A full consideration of earthquake forces requires a study of the structure's dynamic properties.

Hydrostatic pressure may also uplift the foundations of a structure. If the pressure is high enough and unrelieved by drains, the building will float out of the ground.

Temperature changes, concrete shrinkage, and prestress shortening can induce axial forces, moments, and shears into the structure if it is restrained by the ground or other construction. Because thermal and shrinkage stresses tend to be reduced by creep in concrete, they are usually disregarded unless the structure is over about 250 ft in length, is subject to severe temperature variation, or has unusual conditions of restraint or unusually severe limitations on serviceability.

If a portion of the structure is prestressed, the prestressed members will shorten. When the shortening is resisted by nonprestressed members, moments, shears, and axial forces in the nonprestressed members will result.

Loads for each member should be computed independently of other members. For example, loads on beams should be computed by tributary areas without reference to end reactions of slabs that the beams support. If desired or necessary, the slab reactions can later be used to check the beam loading. Similarly, loads on columns should be computed by tributary areas without reference to end reactions of beams or slabs that the columns support. Independent computations are less likely to lead to accumulative error, are usually simpler and more accurate, and provide a means for checking the computations easily by using end reactions of supported members.

When all loads are essentially uniformly distributed, the load on a beam or column can be computed most easily by considering only those loads within the tributary area. The tributary area is the area surrounding the member, extending halfway to the next similar member, as shown in Fig. 13.2. Where a beam spans parallel to a one-way slab, it is customary to include some slab load on the beam to allow for two-way slab action in the vicinity of the beam.

When concentrated loads are equal or nearly equal in magnitude and are spaced uniformly along the length of a beam or slab with loads coinciding with each support*, the beam or slab may be conservatively designed in flexure for a uniform load equal to the concentrated load, divided by the spacing. Using this assumption, the computed maximum shear and moment will always be equal to or greater than the actual shear and moment, as shown in Fig. 13.3(*a*). If stirrups are required,

*These conditions are usually met in buildings that are well designed for maximum economy and efficiency.

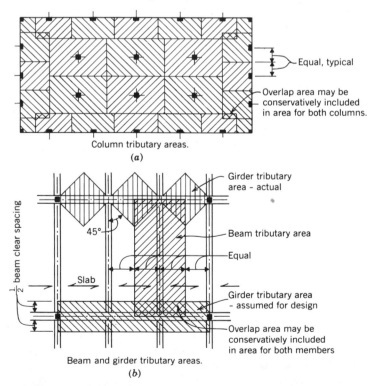

Column tributary areas.

(*a*)

Beam and girder tributary areas.

(*b*)

FIGURE 13.2   Tributary areas.

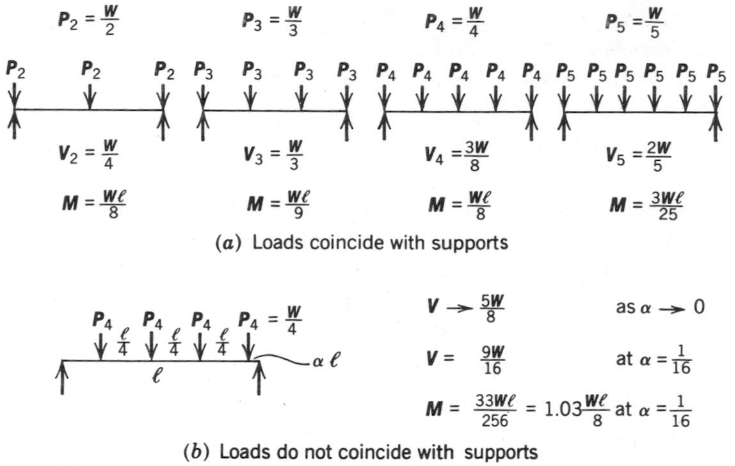

$$P_2 = \frac{W}{2} \qquad P_3 = \frac{W}{3} \qquad P_4 = \frac{W}{4} \qquad P_5 = \frac{W}{5}$$

$P_2 \quad P_2 \quad P_2 \; P_3 \quad P_3 \quad P_3 \quad P_3 \; P_4 \; P_4 \; P_4 \; P_4 \; P_4 \; P_5 \; P_5 \; P_5 \; P_5 \; P_5 \; P_5$

$$V_2 = \frac{W}{4} \qquad V_3 = \frac{W}{3} \qquad V_4 = \frac{3W}{8} \qquad V_5 = \frac{2W}{5}$$

$$M = \frac{W\ell}{8} \qquad M = \frac{W\ell}{9} \qquad M = \frac{W\ell}{8} \qquad M = \frac{3W\ell}{25}$$

(*a*) Loads coincide with supports

$P_4 \quad P_4 \quad P_4 \quad P_4 = \frac{W}{4}$

$\frac{\ell}{4} \; \frac{\ell}{4} \; \frac{\ell}{4}$ $\alpha \ell$

ℓ

$$V \to \frac{5W}{8} \qquad \text{as } \alpha \to 0$$

$$V = \frac{9W}{16} \qquad \text{at } \alpha = \frac{1}{16}$$

$$M = \frac{33W\ell}{256} = 1.03\frac{W\ell}{8} \text{ at } \alpha = \frac{1}{16}$$

(*b*) Loads do not coincide with supports

FIGURE 13.3   Shear and moment caused by concentrated loads with uniform spacing.

beams should be designed for the actual concentrated load spacing because the shear within the span differs from, and is sometimes higher than, the shear caused by an assumed uniform load. When spacing of the concentrated loads is one quarter of the span or less, the beam or slab may be designed for a uniform load without regard to whether or not loads coincide with supports. As shown in Fig. 13.3(*b*), under the most severe conditions, moment will never be unconservative by more than 3%. Although actual shear could be somewhat higher than assumed shear, the area of high shear occurs for only a short distance near the end of the member where both negative moment and deep beam action serve to increase shear capacity of the member. In any case, in design for shear, uniformly spaced concentrated loads of equal magnitude can be treated as a uniform load where the spacing is less than twice the member depth.

13.3   FRAME ANALYSIS FOR GRAVITY LOADS

Joints in structural steel frames and timber frames are usually too flexible to develop full continuity between members unless special precautions are taken. Thus, floor and roof members in steel and timber frames are designed as simple spans under gravity loads, and detailed procedures for determining moments in frames are not required.

In contrast, reinforced concrete frames, by their very nature, normally have rigid joints. Individual members rarely carry a load in simple bending. Instead, both gravity and lateral loads cause moments in a member at the ends, as well as within the span. Moment at the end of a member is transferred through the joint into other members framing into the joint.

In the following discussion, it is assumed that the engineer has had training in analysis of indeterminant structures.

Regardless of the method of frame analysis, it is essential that an engineer be able to visualize the shape of the structure under load. If an engineer's mental picture of the deflected shape is not clear, a freehand sketch, such as those shown in Fig. 13.4, should be prepared and tested for consistency before any calculations are made. The direction of curvature will indicate the direction of the moment and prevent gross errors in which concrete members are reinforced in the wrong face to resist flexural tension. A sketch should indicate the relative magnitude of moments in a frame and assist in the proper location of reinforcement.

Most methods of elastic frame analysis for gravity loads assume that there is no lateral sway of the frame due to asymmetrical loads or frames. In most cases, this is a reasonable assumption and introduces little error. When a frame or its gravity loads are grossly asymmetrical, the frame analysis should take into account the amount of lateral sway and the increase in moments that it causes. (Chapter 9 discusses methods for computing the increase in moment in columns due to P-Δ effects.)

Most methods of frame analysis (and all of those discussed in this text) also assume that the frame is constructed of material that is homogeneous, isotropic, and monolithic.* Reinforced concrete is not homogeneous because it is composed of both steel and concrete, and even concrete alone is not homogeneous. Neither is concrete isotropic because it frequently cracks from flexural, shear, temperature, or shrinkage stresses. Most reinforced concrete frames can be considered monolithic only

*This is a convenient assumption for maintaining the financial solvency of structural design firms, because more accurate assumptions would lead to unreasonably long, laborious computations. Longer computations would not necessarily lead to safer structures.

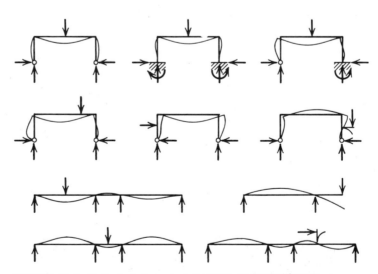

FIGURE 13.4   Deflected shape and reactions on loaded frames.

under temporary lateral loads applied after the completion of the frame. Due to limitations of the construction process, most frames receive gravity loads before completion. Nevertheless, long experience indicates that concrete frames can be safely designed by assuming that flexural stiffness *EI* is proportional to the product of the modulus of elasticity of concrete E_c and the moment of inertia of the gross plain concrete section *I*. (Appropriate exceptions are discussed later in this chapter.)

Another nearly universal assumption in frame analysis is that joints are rigid. That is, the angular relationship between members framing into a joint remains constant throughout the loading history of the frame. As noted in Chapter 10, Joints, this assumption is not necessarily accurate, especially if tension bars are not adequately anchored. The practical effect of lack of joint rigidity is generally to increase mid-span positive moments in beams due to gravity loads and decrease the corresponding negative moments at the ends. For this reason, it is good practice to be conservative with mid-span positive moment in flexural members, especially in end spans where cracking in lightly loaded columns (or torsional cracking in spandrel beams) may reduce their stiffness and, hence, the amount of moment at the discontinuous end.

Any reasonable set of assumptions may be adopted for computing relative flexural and torsional stiffnesses of columns, walls, floors, and roof systems. Assumptions must be consistent throughout the analysis. The effect of haunches and other changes in a cross section should be considered in determining moments.

Rather than try to estimate the exact conditions, it is sometimes advantageous to calculate moments for the extreme conditions and then use the maximum computed moments. For example, moments for a beam framing into a lightly loaded column at the exterior support could be computed, first assuming that the stiffness of the column is equivalent to the uncracked stiffness. This assumption will give the maximum negative moment at the exterior end of the beam. Next, moments could be computed assuming that the stiffness of the column is equal to its cracked section stiffness or assuming that the column has no stiffness. This assumption will result in maximum positive moment at mid-span and maximum negative moment at the first interior support.

Frame analysis should take into account loading patterns (arrangement of live load on some spans but not on other spans) that result in the highest possible moment at each critical section. The loading pattern resulting in maximum positive moment is usually different from the loading pattern resulting in maximum negative moment. This subject is discussed in more detail below, under Moment Distribution. When service live load is less than 3/4 of service dead load, and the nature of the live load is such that uniform live load on all spans is likely to occur, many engineers consider only a loading pattern with full live load on all spans for maximum positive and negative moments [2]. The resulting moments are unconservative for the most severe loading pattern by only a few percent.

The ACI Code requires that gravity service loads be factored before making a frame analysis (Code 8.9).

The common methods of frame analysis are listed below in order of increasing accuracy and engineering time required for their application.

1. Moment Coefficients   This method, based on coefficients specified in the ACI Code, is by far the fastest and simplest method. It is useful for checking moments computed by other methods, for preliminary design, and for final design of members where the potential savings in construction cost does not justify the use of more time-consuming methods. To save engineering time spent in design, moment coefficients should be used wherever possible, including conditions outside ACI Code limitations, when based on the engineer's experienced judgment.

2. Moment Distribution   This manual method can be used over a wide range of loading conditions and frame configurations. It is more accurate than the moment coefficient method for conditions outside the limits of applicability for moment coefficients. It is useful for checking moments generated by a computer, for preliminary design, and for final design, especially where a computer is not available or is more costly.

3. Computer Analysis   Frame moments calculated by computer are especially appropriate for large or tall frames and ones with a complicated configuration. If many loading patterns or complicated patterns must be investigated, a computer frame analysis is very useful. Use of a computer is, of course, subject to its availability. In evaluating whether to use a computer, the cost of an engineer's time in preparing input and output should be considered, as well as the cost of machine time.

4. Two-way Slab Analysis   Chapter 4 discusses the special case of distribution of moments in two-way slab systems.

After end moments in members have been determined by moment distribution or computer analysis, shear, axial forces, and moments within the span are determined by statics.

As with other aspects of structural design, engineers should select a method of frame analysis that will minimize engineering time without sacrificing accuracy or overall economy. Symmetrical structures afford the opportunity for reducing engineering computations. Furthermore, it is rarely necessary to analyze an entire structure consisting of equal spans, equal loads, and the same member sizes. It is usually sufficient to analyze only the end span and the first interior span.

Moment Coefficients

Beams and one-way slabs may be designed for the approximate moments and shears in Table A13.2 (Code 8.3) under the following conditions:

1. There are two or more spans.
2. Spans are approximately equal, with the larger of two adjacent spans not greater than the shorter by more than 20%.
3. Loads are uniformly distributed.

4. Unit live load does not exceed three times unit dead load.

5. Members are prismatic, that is, have the same cross section for the full length of the member.

In using moment coefficients from Table A13.2, ℓ_n is the clear span for positive moment or shear and the average of adjacent clear spans for negative moment.

The use of moment coefficients for preliminary design is very convenient because they do not require a knowledge of member sizes. When moments are calculated by other methods, it is good practice to always compute the equivalent coefficient as a check on the calculations, even though the structure does not fall within the above limitations. By doing so, the engineer will develop an intuitive feeling for the proper magnitude of the coefficient for a broad range of typical conditions.

For an interior span, use of the coefficients results in total design moment 23% greater than static moment based on clear spans.

$$\frac{w\ell_n^2}{16} + \frac{1}{2}\left[\frac{w\ell_n^2}{11} + \frac{w\ell_n^2}{11}\right] = 1.23 \times \frac{w\ell_n^2}{8}$$

For an end span, use of the coefficients results in total design moment about 17% greater than static moment based on clear spans.

$$\frac{w\ell_n^2}{11} + \frac{1}{2} \times \frac{w\ell_n^2}{9} = 1.17 \times \frac{w\ell_n^2}{8}$$

The excess of design moment over static moment makes allowance for pattern loading, for variations in span length, for variations in member sizes, and for somewhat higher static moments based on spans center-to-center of supports.

Moment Distribution

In 1932, Professor Hardy Cross proposed a simple method of distributing moments based on slope-deflection or moment-area concepts. The method quickly became known as the Moment Distribution Method and was used almost exclusively by structural design offices until the advent of computers. Today it is still widely used, especially when a computer is not available or is too expensive for the problem at hand.

Moment distribution is an iterative procedure that consists of three basic steps.

1. Assume all joints in the frame are fixed against rotation, and compute fixed end moments (FEM) at both ends of each member in the frame due to the loading under consideration.

2. One at a time, release each joint to rotate and distribute unbalanced moments to each member framing into the joint in proportion to its stiffness EI/ℓ for prismatic members. When the other end of a prismatic member is pinned (free to rotate), the stiffness may be taken as $0.75EI/\ell$. For nonprismatic

members (e.g., beams with haunches), the sitffness may be slightly different from **El**/ℓ or 0.75**El**/ℓ.*

3. Carryover a portion of the moments distributed in step 2 to the other end of each member. For prismatic members, the carryover factor is 0.50. For nonprismatic members, the carryover factor can be taken from Ref. 3.

Step 2 completes one cycle of moment distribution and steps 3 and then 2 are repeated for additional cycles. In most cases, one or two cycles gives sufficient accuracy as the procedure converges very quickly.

In the analysis of frames and continuous members to determine moments, span length should be taken as the distance center-to-center of supports. For beams cast integrally with supports, moments at faces of supports may be used for design.

As a simplification, solid and ribbed slabs built integrally with supports, with clear spans not more than 10 ft, may be analyzed as continuous slabs on knife edge supports with spans equal to the clear spans of the slab and width of beams otherwise neglected.

When members are not built integrally with their supports (e.g., slabs on masonry walls), the span length should be taken as the clear span plus depth of the member but need not exceed the distance center-to-center of supports.

Moments at the face of support can be computed by considering the beam in Fig. 13.5. The total static design moment in the beam is less than the static moment computed using centerline spans. The amount of reduction is equal to the area of the shear diagram between centerline and face of support, or approximately **V**a/2, where **V** is the shear at the support centerline, and **a** is the width of support. The moment curve is shifted slightly so that negative moment is reduced less than the

*Most texts on indeterminate analysis contain a derivation of flexural stiffness.

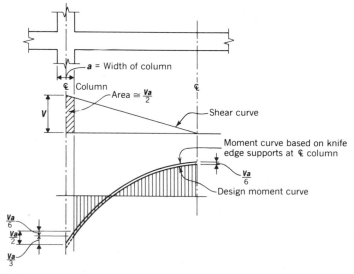

FIGURE 13.5   Reduction in moment at face of support.

total reduction in static moment. Somewhat arbitrarily, it is assumed that negative moment can be reduced by **Va**/3.

It is common practice to conservatively disregard the reduction in positive moment as it is a small amount and the total reduction in moment is actually somewhat less than **Va**/2. Likewise, the reduction in column moments from the neutral axis of the beam to top or bottom face of the beam is conservatively disregarded. The shear in columns is usually much less than that in beams, and the shape of the moment curve in the vicinity of supports is much flatter; hence, the reduction in moment in columns would be much smaller than in beams and is not worth the computational effort.

The above procedure is appropriate for supports with a width less than about 15% of the span length. For wider supports, a more detailed analysis should be performed.

When analyzing a structural frame or continuous members, factored live load, in combination with factored dead load, should be arranged to give the maximum moments at critical sections. Normally, only the three loading conditions illustrated in Fig. 13.6 need be considered.

- Loading Case I will give maximum moment in the beam at column A, maximum positive moment in spans A-B and C-D, and minimum positive moments in spans B-C and D-E. Case I also gives maximum moment in the exterior column A.

- Either Loading Case I or II will give maximum moment in interior columns, depending upon loads, spans, and stiffnesses.

- Loading Case II will give maximum positive moments in spans B-C and D-E and minimum positive moments in spans A-B and C-D.

- Loading Case III will give maximum negative moment at line B.

The minimum negative moment is usually disregarded as it is rarely critical unless there is a stress reversal that would require reinforcement in the bottom of a member at the support for positive moment. This condition sometimes occurs in very short spans adjacent to long spans. The condition of minimum positive moment at mid-span frequently controls the length of top bars for negative moment and, hence, must be computed.

When analyzing frame moments due to gravity loading, it is sufficiently accurate to consider the live load applied only to the floor or roof under consideration, and

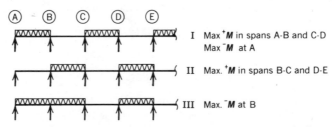

FIGURE 13.6   Arrangement of live load for maximum moment.

far ends of columns built integrally with the structure may be considered fixed (Code 8.9.1). Columns should be designed to resist the axial forces from factored loads on all floors or roof and the maximum moment from factored loads on a single adjacent span of the floor or roof under consideration. Loading condition giving the maximum ratio of moment to axial load should also be considered whether eccentric loading is due to unbalanced floor or roof loads or to other causes (Code 8.8). Figure 13.7 illustrates the type of live load pattern required for maximum moment in interior columns.

In many structures, moments in columns due to gravity loads control the design only for exterior columns with low axial load such as those supporting only the roof or the roof plus one or two floors.

Moment distribution will proceed in an orderly manner if these steps are followed:

1. Compute the relative stiffness at each end of each member in the frame or continuous construction. If concrete strength and type is the same or nearly the same in all members, the relative stiffness is given by I/ℓ (or $0.75I/\ell$ for members pin-connected at the far end). At each joint, compute the ratio of stiffness of each member to the total of stiffnesses of all members framing into the joint.

2. Select the carryover factors for each end of each member.

3. Compute fixed end momens (FEM) at each end of each member. For moment distribution, the sign convention is to assign positive values to moments tending to rotate the joint clockwise and negative values to moments tending to rotate the joint counterclockwise. For members pinned at one end, the FEM is calculated at the fixed end only. Formulas for FEMs are available from several sources. For most practicing engineers, the source most radily available is the AISC *Manual of Steel Construction* [4].

4. Set up a standard form for a systematic recording of the computational data, and record the FEMs.

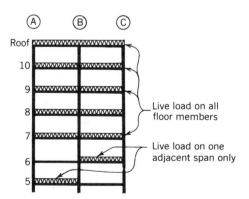

FIGURE 13.7 Distribution of live load for maximum moment in Column B in the fifth story.

5. At each joint, compute the unbalanced moment and distribute it to all members at the joint in proportion to their stiffnesses. The distributed moments should carry a sign opposite to the unbalanced moment so that the algebraic sum of all moments on a joint should equal zero after completion of this step. This step is equivalent to "unlocking" a joint and allowing it to rotate under action of the FEMs. The joint will rotate in the direction of the unbalanced moment until it is balanced by the sum of equal but opposite moments in each member framing into the joint.

6. Multiply the moments distributed in step 5 by the carryover factors and carry over the reduced moments to the far end of each member (except those with pin ends). If the far end is fixed, it absorbs all FEMs and moments carried over to the fixed end. Moments at fixed ends are not carried over or transferred from a fixed end to the other end of the member.

7. Repeat step 5. If greater accuracy is desired, repeat step 6 and then step 5 as often as necessary to achieve the desired accuracy.

8. Add the moments listed for each member at each joint.

9. Check the moments at each joint for numerical error to assure that the sum of moments equals zero.

10. In flexural members, compute the moment at face of the support for loading cases giving maximum negative moments. Compute shear and moments within the span at critical sections. Check the computations by verifying that the absolute sum of positive mid-span moment plus the average of the two negative end moments equals or exceeds total static moment. (For the unusual case of one or both end moments being positive, draw the moment diagram to verify that design moments equal or exceed static moment.)

11. Check moments in flexural members by computing the coefficients $C = w\ell^2/M$ and comparing it to the coefficients given in Table A13.2 or comparing it to the engineer's own experience.

The principles of moment distribution have not changed since the procedure was first proposed. In a brief monograph, the Portland Cement Association explains the application of moment distribution in more detail than is given here [5]. The monograph includes a procedure for determining, in one set of calculations, maximum positive and negative moments from pattern loading.

Example 13.1 Given the second-floor plan and framing plan of a four-story building, shown in Fig. 13.8, calculate:

A. Maximum positive moment in joist J1.

B. Maximum positive moment in joist J2.

C. Maximum negative moment between joists J1 and J2.

D. Maximum moment in North-South direction in column A3.

E. Maximum moments at critical sections in beams B1, B2, and B3.

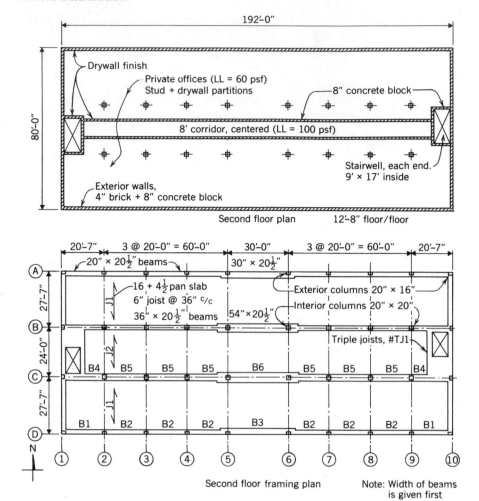

FIGURE 13.8 Floor plans for Example 13.1

Use both moment distribution and moment coefficients, where appropriate. Assume that the same concrete is used in both floor framing and in columns.

Solution A   Compute loading on joists.

Live load: 60 psf × 1.7   = 102 psf

Dead load:

    Joists (see Tables A3.1 and A3.2)   (97 + 6) psf × 1.4 = 144 psf

    Partitions (assumed)   40 × 1.4   = 56

    Ceiling and floor finishes   10 × 1.4   = <u>14</u>

       Total w_u = 316 psf

        or $w_u = 0.316 \times 3 = 0.95 \text{k/}'$ per joist.

Compute concentrated loads on joist J2 from the weight of corridor walls plus the excess of corridor live load over office live load.

The height of the wall is $(12'-8'') - (1'-8'') = 11'$

From Table A13.1, the wall weight is 50 psf for concrete block plus 8 psf for drywall.

$$P_u = 11 \times 58 \times 1.4 + (100 - 60)\frac{8}{2} \times 1.7 = 0.9 + 0.27 = 1.17 k/'$$

or $P_u = 1.17 \times 3 = 3.5$ k per joist for dead + live load

$\qquad = 0.90 \times 3 = 2.7$ k per joist for dead load only

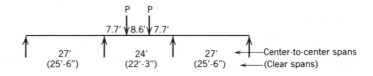

The moment pattern will vary for each individual joist in a bay depending on its proximity to a column. A joist framing directly into a column will be restrained by the column just as a beam would be restrained. For other joists, restraint is provided by the spandrel beam in torsion. The least restraint is provided to the joists farthest from a column. To simplify the calculations, conservatively disregard the stiffness of columns because most joists do not frame into a column and assume that the discontinuous end of joists J1 are pin-connected for Solutions A, B, and C. Joist loading for solution D will provide the critical negative moments at the discontinuous end for joists framing into a column. Note that the assumption of a pin end at the discontinuous end of joist J1 is conservative for negative moment between J1 and J2 and for positive moment in J1 but is unconservative for positive moment in J2.

Because joists have the same concrete outlines in all spans, assume that the moment of inertia of the joist $I = 100$. Conservatively assume that the clear span of joists is measured from face of columns rather than face of beams because beams are the same depth as the joists. Compute stiffnesses and distribution factors:

For J1, $\dfrac{3}{4} \times \dfrac{I}{\ell} = \dfrac{75}{27} = 2.78 \sim 0.40$

For J2, $\qquad \dfrac{I}{\ell} = \dfrac{100}{24} = \underline{4.17} \sim \underline{0.60}$

$\qquad\qquad$ Total $= 6.94 \sim 1.00$

Because members are prismatic, use carryover factor $= 0.50$.

For maximum positive moment, consider live load on joists J1 only. Because partitions are movable and their location is unknown, also consider partition load on joists J1 only.

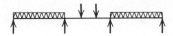

For J1 with one end fixed,

$$FEM_u = \frac{w_u\ell^2}{8} = \frac{0.95(27)^2}{8} = 86.6'k$$

For J2 with two ends fixed and $w_u = (144 + 14)3 = 0.48k/'$ and $P_u = 2.7$ k per joist.

$$FEM_u = \frac{w_u\ell^2}{12} = \frac{0.48(24)^2}{12} = 23.0'k$$

$$\frac{P_uab*}{\ell} = \frac{2.7 \times 7.7 \times 16.3}{24} = \underline{14.1}$$

$$FEM_u = 37.1'k$$

The moment distribution tabulation is shown in Fig. 13.9.

In this case, the additional cycle of moment distribution reduced the moment by about 3%, a trivial amount in most cases. In some cases, the additional cycle will increase moments over the previous total.

*For two symmetrically placed, equal concentrated loads,

$$FEM = \frac{Pab^2}{\ell^2} + \frac{Pa^2b}{\ell^2} = \frac{Pab}{\ell}$$

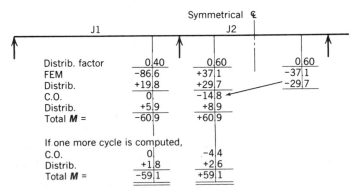

	J1		J2		
Distrib. factor	0.40		0.60		0.60
FEM	−86.6		+37.1		−37.1
Distrib.	+19.8		+29.7		−29.7
C.O.	0		−14.8		
Distrib.	+5.9		+8.9		
Total **M** =	−60.9		+60.9		
If one more cycle is computed,					
C.O.	0		−4.4		
Distrib.	+1.8		+2.6		
Total **M** =	−59.1		+59.1		

FIGURE 13.9 Moment distribution for maximum ^+M in joist J1.

For J1, compute maximum ^+M_u by first computing the reaction at the left (free) end. Thus,

$$R_L = \frac{0.95 \times 27}{2} - \frac{60.9}{27} = 10.6 \text{ k}$$

$$^+M_u = \frac{R_L^2}{2w_u} = \frac{10.6^2}{2 \times 0.95} = 58.8'k$$

Check the coefficient using $\ell_n = 25.4'$, $\qquad 58.8'k = \frac{w_u \ell_n^2}{10.4}$

The continuous joists J1-J2-J1 do not meet the limitations for use of moment coefficients because there are concentrated loads on joists J2. For the purposes of illustration only, moments will be computed using coefficients here and in solutions B, C, and D.

Using moment coefficients from Table A13.2, assume that the discontinuous end is unrestrained.

$$^+M_u = \frac{w_u \ell_n^2}{11} = \frac{0.95(25.4)^2}{11} = 55.7'k$$

Solution B For maximum positive moment in joist J2, consider live load and partition load on joist J2 only.

For J1, dead load $w_u = (144 + 14)3 = 0.48/'$

with one end fixed, $FEM_u = \dfrac{w_u \ell^2}{8} = \dfrac{0.48 \times (27)^2}{8} = 43.7'k$

For J2, $w_u = 0.95k/'$, $\qquad P_u = 3.5 \text{ k}$

With two ends fixed, $FEM_u = \dfrac{w_u \ell^2}{12} + \dfrac{Pab}{\ell}$

$$FEM_u = \frac{0.95 \times (24)^2}{12} + \frac{3.5 \times 7.7 \times 16.3}{24} = 45.6 + 18.3 = 63.9'k$$

The moment distribution tabulation is shown in Fig. 13.10.

Maximum $^+M_u = \dfrac{w_u \ell^2}{8} + P_u a - (^-M_u)$

$$= \frac{0.95 \times (24)^2}{8} + 3.5 \times 7.7 - 54.2$$

$$= 68.4 + 27.0 - 54.2 = 41.2'k$$

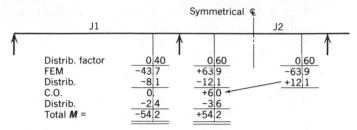

FIGURE 13.10　Moment distribution for maximum ^+M in joist J2.

Check the coefficient using $\ell_n = 22.3$ and by converting concentrated corridor wall loads to equivalent uniform loads. Thus, the partition load

$$w_u \le \frac{1.17 \times 2}{24 \times 0.5} = 195 \text{ psf}$$

For J2, $w_u = (0.316 + 0.195)3 = 1.53$ k/′

$$^+M_u = 41.2 = \frac{w_u \ell_n^2}{18.5}$$

Using moment coefficients from Table A13.2,

$$^+M_u = \frac{w_u \ell_n^2}{16} = \frac{1.53(22.3)^2}{16} = 47.7'\text{k}$$

Solution C　For maximum negative moment between J1 and J2, consider live load and partition load on two adjacent spans but not on the third span.

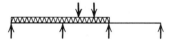

For J1 left,　$FEM_u = 86.6'$k (as in solution A)

For J1 right, $FEM_u = 43.7'$k (as in solution B)

For J2, FEM_u　　= 63.9′k (as in solution B)

The moment distribution tabulation is shown in Fig. 13.11.

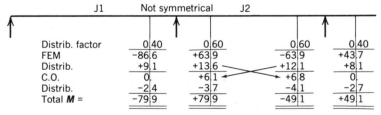

FIGURE 13.11　Moment distribution for maximum ^-M between joists J1 and J2.

In J1, conservatively disregard the increase in shear at the interior support due to the negative moment.

$$V_u = \frac{0.95 \times 27}{2} = 12.8 \text{ k}, \quad \text{and } a = 1.67',$$

$$\frac{V_u a}{3} = \frac{12.8 \times 1.67}{3} = 7.1'\text{k}$$

Design $^-M_u = 79.9 - 7.1 = 72.8'\text{k}$

Check the coefficient using $w_u = \dfrac{1.53 + 0.95}{2} = 1.24 \text{k}/'$

and $\ell_n = \dfrac{25.4 + 22.3}{2} = 23.8'$

$$72.8'\text{k} = \frac{w_u \ell_n^2}{9.8}$$

Using moment coefficients from Table A13.2,

$$^-M_u = \frac{w_u \ell_n^2}{10} = \frac{1.24(23.8)^2}{10} = 70.4'\text{k}$$

Solution D   Compute moment in column A3 by assuming that one-half the joists in the panel are rigidly connected to the column. Referring to Table A3.1, the moment of inertia of the gross concrete section of one joist is 9238 in.[4] Because there are 3-1/3 joists (10 ft/3 ft) in one-half the panel, joists $I = 9238 \times 3.33 = 30,800$ in.[4] for the half panel. Compute stiffnesses and distribution factors at lines A and D.

Stiffness of joists

$$= \frac{I}{\ell} = \frac{30,800}{27 \times 12} = \quad 95.1 \text{ in.}^3 \sim 0.52$$

Stiffness of column above

$$= \frac{I}{\ell} = \frac{20(16)^3}{12 \times 152} = \quad 44.9 \qquad \sim 0.24$$

Stiffness of column below = $\underline{\quad 44.9 \quad} \qquad \underline{\sim 0.24}$

Total stifness at joint   = 184.9 ~ 1.00

Recompute the ratio of stiffnesses at column lines B and C, assuming that joist J1 is fixed at both ends. Actual joist I may be used instead of an assumed value of 100, as in solution A.

For J1, $\dfrac{I}{\ell} = \dfrac{9238}{27} = 342 \sim 0.47$

$$\text{For J2,} \frac{I}{\ell} = \frac{9238}{24} = \underline{385} \sim \underline{0.53}$$

$$\text{Total} = 727 \sim 1.00$$

Live load on exterior spans only (joists J1) will result in maximum moment in exterior columns. In addition to moments due to frame action, add the moment due to eccentric application of the load of the spandrel beams on the exterior columns. Eccentricity equals the distance from centerline of beam to centerline of column.

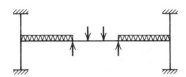

For J1, Eccentric moment equals the beam reaction multiplied by eccentricity. Beam load = 6.2k/' (from solution E)

Beam reaction $P_u = 6.2 \times 20 = 124$ k

Assuming that joists in only half the bay resist the eccentric moment (consistent with the assumption above on stiffnesses) and because eccentricity = $(20'' - 16'')/2 = 2'' = 0.17'$,

$$\text{Eccentric moment per joist} = \frac{(124 \times 0.17)3}{10} = 6.2'\text{k}$$

$$\text{FEM}_u = \frac{w_u \ell_n^2}{12} \qquad = \frac{0.95(27)^2}{12} = \underline{57.7}$$

$$\text{Total moment} = M_u = 63.9'\text{k}$$

For J2, $FEM_u = 37.1'$k (as in solution A)

The moment distribution tabulation is shown in Fig. 13.12.
The moment in column A3 = $16.5 \times 10/3 = 55.0'$k in the North-South direction.

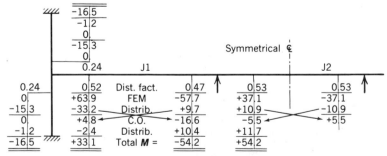

FIGURE 13.12 Moment distribution for moment in Column A3 in North-South direction.

The moment at discontinuous end of joists (lines A and D), after subtracting the external moment due to beam eccentricity, is $(33.1 - 6.2) = 26.9'$k per joist.
As in solution C, $V_u = 12.8$ k, and $a = 1.33'$ at the exterior column.

$$\frac{V_u a}{3} = \frac{12.8 \times 1.33}{3} = 5.7'\text{k}$$

Design $^-M = 26.9 - 5.7 = 21.2'$k

Check the coefficient using $w_u = 0.95$k/$'$ and $\ell_n = 25.4'$, $21.2 = \dfrac{w_u \ell_n^2}{28.9}$

Using moment coefficients from Table A13.2,

$$^-M = \frac{w_u \ell_n^2}{24} = \frac{0.95 \times 25.4^2}{24} = 25.5'\text{k}$$

Note that column moments could have been calculated approximately using the first cycle of moment distribution at the exterior column-joist joint only. Thus, column moment $= 0.24 \times 63.9 \times 10/3 = 51.1'$k or about 6% less than the moment given by two cycles of moment distribution. If the calculated column eccentricity is much less than the minimum eccentricity (as is frequently the case), the unconservative result of the approximate calculation is of little concern.

For simplicity, most engineers in practice would either disregard the moment caused by eccentricity of the spandrel beam framing into the exterior column or would distribute the eccentric moment to the column above, the column below, and to the joists in proportion to their stiffnesses, after the distribution of frame moments. The procedure illustrated in this example is more accurate but is tedious and has little effect on final member design unless the beam is offset from the column centerline far more than is the case in this example.

Solution E   Compute loading on the beams:

Live load: 60 psf $\times \dfrac{27}{2} \times 1.7$ $= 1.38$k/$'$

Joist, finishes, and partitions, 153 psf $\times \dfrac{27}{2} \times 1.4 = 2.89$

Beam load,* $(256 - 97)$ psf $\times 1.67 \times 1.4$ $= 0.37$

Wall, 94 psf $(12.67 - 1.71) \times 1.4$ $= 1.44$

Brick facing on beam, 40 psf $\times 1.67 \times 1.4$ $= \underline{0.09}$

Total $w_u = 6.2$k/$'$

---

*The additional weight of the beam is 150 pcf multiplied by the thickness of 1.71 ft (= 256 psf) less the weight of joists (97 psf). The net unit weight is then multiplied by the beam width.

Because live load is only 22% of the total load (1.38/6.2), pattern loading may be disregarded and one moment distribution may be performed to determine moment at all of the critical sections. Compute distribution factors:

At line 1,

Beam $\dfrac{I}{\ell} = \dfrac{20(20.5)^3}{12 \times 19.75 \times 12} = \;\; 60.6 \sim 0.30$

Column $\dfrac{I}{\ell} = \dfrac{16(20)^3}{12 \times 12.67 \times 12} = \;\; 70.2 \sim 0.35$

$$\underline{70.2 \sim 0.35}$$

$$\text{Total} = 201.0 \sim 1.00$$

At lines 2, 3, and 4,

Beam $\dfrac{I}{\ell} = \dfrac{20(20.5)^3}{12 \times 20 \times 12} \;\;\; = \;\; 59.8 \sim 0.23$

$$59.8 \sim 0.23$$

Column $\dfrac{I}{\ell} = \dfrac{20(20.5)^3}{12 \times 20 \times 12} \;\;\; = \;\; 70.2 \sim 0.27$

$$\underline{70.2 \sim 0.27}$$

$$\text{Total} = 260.0 \sim 1.00$$

At line 5, disregarding the haunch in beam B2,

Left beam $\dfrac{I}{\ell}$ $\qquad\qquad\;\; = \;\; 59.8 \sim 0.23$

Right beam $\dfrac{I}{\ell} = \dfrac{30(20.5)^3}{12 \times 30 \times 12} \;\; = \;\; 59.8 \sim 0.23$

Column $\dfrac{I}{\ell}$ $\qquad\qquad\;\; = \;\; 70.2 \sim 0.27$

$$= \;\; \underline{70.2 \sim 0.27}$$

$$\text{Total} = 260.0 \sim 1.00$$

For B1 and B2, $\textbf{FEM}_u = \dfrac{\textbf{w}_u \ell^2}{12} = \dfrac{6.2(20)^2}{12} = 206.7'\text{k}$

For B3, $\textbf{FEM}_u = \dfrac{6.2(30)^2}{12} = 465.0'\text{k}$

The moment distribution tabulation is shown in Fig. 13.13.

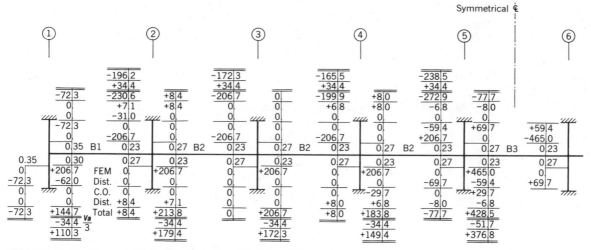

FIGURE 13.13 Moment distribution for beams B1, B2 and B3.

In most instances, in computing moments at the face of supports it is sufficiently accurate to compute shear as the beam load multiplied by one-half the span and disregard the increase or decrease in shear due to unequal end moments.

$$\text{For B1 and B2, } V_u = \frac{6.2 \times 20}{2} = 62.0 \text{ k}, \qquad a = 1.67'$$

$$\frac{V_u a}{3} = \frac{62.0 \times 1.67}{3} = 34.4'\text{k}$$

$$\text{For B3, } V_u = \frac{6.2 \times 30}{2} = 93.0 \text{ k}, \qquad a = 1.67'$$

$$\frac{V_u a}{3} = \frac{93.0 \times 1.67}{3} = 51.7'\text{k}$$

In a design office, the computations could be simplified without significant loss of accuracy by omitting one beam B2 and performing the moment distribution with only two bays of beam B2.

When end moments are nearly equal, positive moments near mid-span can be computed approximately with a very small error by subtracting the average value of negative moments from the simple span positive moment. Thus,

$$^+M_u = \frac{w_u \ell^2}{8} - \frac{^-M_L + {}^-M_R}{2}$$

For beam B1,

$$^+M_u = \frac{6.2(20)^2}{8} - \frac{144.7 + 230.6}{2} = 310.0 - 187.6 = 122.4'\text{k}$$

For beam B2, using the smallest combination of end moments,

$$^+M_u = \frac{6.2(20)^2}{8} - \frac{206.7 + 199.9}{2} = 310.0 - 203.3 = 106.7'\text{k}$$

For beam B3,

$$^+M_u = \frac{6.2(30)^2}{8} - 428.5 = 697.5 - 428.5 \qquad = 269.0'\text{k}$$

Beam B3 does not meet the limitations for use of moment coefficients because it is more than 20% longer than the adjacent span. For purposes of illustration only, moments for all spans will be computed by moment coefficients.

For 20' span, $\ell_n = 18.33'$, $w_u\ell_n^2 = 6.2(18.33)^2 = 2084$

For 30' span, $\ell_n = 28.33'$, $w_u\ell_n^2 = 6.2(28.33)^2 = 4977$

At lines 5 and 6, average $\ell_n \quad = 23.33'$,

$$w_u\ell_n^2 = 6.2(23.33)^2 = 3376$$

Beam B1, at line 1, $^-M_u = \dfrac{w_u\ell_n^2}{16} = \dfrac{2084}{16} = 130.2'\text{k}$

at mid-span, $^+M_u = \dfrac{w_u\ell_n^2}{14} = \dfrac{2084}{14} = 148.9$

at line 2, $^-M_u = \dfrac{w_u\ell_n^2}{10} = \dfrac{2084}{10} = 208.4$

Beam B2, at lines 2, 3, and 4,

$$^-M_u = \frac{w_u\ell_n^2}{11} = \frac{2084}{11} = 189.5$$

at mid-span, $^+M_u = \dfrac{w_u\ell_n^2}{16} = \dfrac{2084}{16} = 130.2$

Beam B3, at lines 5 and 6,

$$^-M_u = \frac{w_u\ell_n^2}{11} = \frac{3376}{11} = 306.9$$

at mid-span, $^+M_u = \dfrac{w_u\ell_n^2}{16} = \dfrac{4977}{16} = 311.1$

Moments at critical sections in beams B1, B2, and B3 calculated by moment distribution and by coefficients are summarized in Fig. 13.14. The equivalent coefficient for moments calculated by moment distribution are also shown. Note that the only moment calculated by coefficients that is not conservative is the end negative moment in beam B3.

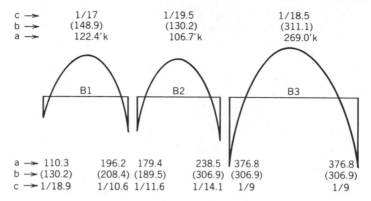

c → 1/17   1/19.5   1/18.5
b → (148.9)   (130.2)   (311.1)
a → 122.4′k   106.7′k   269.0′k

B1   B2   B3

a →→ 110.3   196.2 179.4   238.5 ⎸376.8   376.8⎸
b →→(130.2) (208.4)(189.5) (306.9)(306.9) (306.9)
c →→1/18.9   1/10.6 1/11.6   1/14.1 1/9   1/9

FIGURE 13.14 Summary of moments for Example 13.1. *(a)* Moments at face of support and at mid-span, computed by moment distribution ('k); *(b)* moments at face of support and at mid-span computed by moment coefficients ('k); *(c)* equivalent moment of coefficient = $w_u \ell_n^2$ divided by moment on line *a*.

Even though the beams in bays 2-3, 3-4, and 4-5 have slightly different moments, it is common practice to use the maximum moment found in any of the three bays and specify the same reinforcement for all three beams.*

Computer Analysis

With the widespread availability of powerful computers, there is a temptation to analyze all structures by computer whether appropriate or not. The following brief discussion will help the engineer to decide when to use a computer for frame analysis.

Contrary to common initial perceptions, computer analysis is no more accurate than manual analysis because accuracy is determined by physical conditions and not by the computation method. This point cannot be overemphasized because many engineers rely on pocket calculators with accuracy to eight decimal places or computers with even greater accuracy. Overstressing accuracy in computation leads to inaccurate assessment of structural behavior, fussy details, higher design costs, and higher construction costs. For whatever level of accuracy is necessary, an iterative manual method such as moment distribution can be performed to the desired accuracy. Accuracy within 5% is adequate for most structures (see the discussion in Chapter 1) and is achieved with moment coefficients or one or two cycles of moment distribution.

The advantages of using a computer are threefold.

1. Less intellectual energy is used. Some engineers may not find this to be an advantage. Those that do should be cautioned not to allow their energy level

---

*Identical reinforcement reduces confusion in the field during construction and helps assure the engineer that the proper amount of bars are being placed as the design requires.

to fall so low that errors are made in input to the computer or in using the output.

2. There is less chance for error because manual work is reduced. Errors can still occur in preparing input and using output. Limitations in the hardware and software are potential sources for further error in using a computer.

3. Design costs might be lower (but not necessarily). Design costs should be evaluated for each individual case, just as the cost of one design method versus another is evaluated.

In evaluating whether to use a computer for analysis of moments, consider the following:

1. Availability of hardware and its accessibility at the time needed.

2. Cost of using the hardware.

3. Speed with which results become available from the computer.

4. Time required for preparation of input data and entering the data.

5. Time required for analyzing the output.

6. Documentation of the program. Engineers should be wary of undocumented programs.*

7. Method of analysis used in the program. Is it sufficient for the problem on which it will be used but not overly sophisticated? Programs that are too sophisticated may be more expensive to use. They may also be more complicated, leading to input errors or errors in evaluating output.

8. Consistency of assumptions used in the program compared to the real structure.

9. Availability of sample problems using the program and hardware. Sample problems should cover the range and scope of the problem to be solved by computer and should have been prepared by the engineer or by someone responsible to the engineer.

If a computer is used for frame analysis, the software should be examined for potential difficulties not normally occurring with manual procedures (see the following examples).

Analysis Based on Gross Moment of Inertia I_g   Columns with a high percentage of reinforcement will have a much higher effective moment of inertia than I_g, and columns with light loads and little steel could have a much lower moment of inertia than I_g. Unless tempered with engineering judgment, the use of a single value for I_g in all situations could lead to a dangerous underestimation of moments as, for example, in the positive moment region of end span beams framing into columns near the top of a building. Similar errors can involve beams. For example, analysis based on uncracked torsional stiffness of spandrel beams can lead to an overestima-

*When a building has structural problems, an engineer is far more likely to be held responsible by an unhappy client than is a computer program, which the client can neither see nor understand.

tion of negative moment at the exterior support for slabs supported on the spandrel beam and an underestimation of positive moment in these slabs.

Analysis Based on Moment of Inertia Other than I_g Although theoretically more accurate, using a variable moment of inertia involves so many considerations that academic and research engineers have not attempted it. Practicing engineers should be wary of any program with this feature.

Assumptions Reflecting Well-known Structural Behavior but Not Used in Manual Analysis Because of the Labor Required Such "refinements" are attractive and give an impression of more nearly predicting the true behavior of a structure. However, unless all factors are considered, refinements may lead to false or even unsafe conclusions. For example, a computer program might calculate moments caused by elastic shortening of columns with different load intensities. The actual structure is built over a period of several months, and upper stories will not be affected by deformations that take place before they are constructed.

Analysis Based on the Effects of Differential Settlement and Volume Changes In most cases, moments, shears, and axial loads caused by these effects are relieved by creep. It is inappropriate to analyze structures routinely for the effects of differential settlement and volume changes. Where these may be significant in design, they should be analyzed by an engineer, and a computer should only be used on a case-by-case basis.

Selection of Member Sizes, Bar Combinations, and Other Details Such selections may be impractical or at variance with the engineer's usual practice. It is possible for computers to make a better selection of bars and member sizes than an engineer can, but this feature of a computer program should be carefully checked before use. In any case, engineers should retain control and not use computer output without checking and comparing to past experience.

If used properly, computers have the potential for processing large amounts of data swiftly and accurately, especially if the input data can also be generated automatically.

13.4 FRAME ANALYSIS FOR LATERAL LOADS

In a simple portal frame, as in Fig. 13.15(*a*), a lateral force applied at the upper joint is resisted equally by the two legs if they are the same size. If one leg is stiffer than the other, the stiffer leg resists a higher proportion of the lateral force because it will not deflect as much under a given force. If several portal frames are pinned together, as in Fig. 13.15(*b*), they share equally in resisting a lateral force if the frames are of equal size. These observations lead to an approximate but simple method (called the Portal Method) for estimating the moments in a frame due to lateral loads.

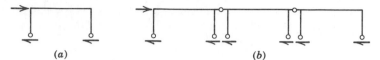

FIGURE 13.15 Portal frames resisting lateral forces.

Portal Method

Use of the Portal Method is based on these assumptions:

1. Loads are applied at joints only. Wind and other loads applied to the building at intermediate places are transferred to the joints by walls or structural elements that are not a part of the frame.

2. The total horizontal shear in the columns in a given story is equal to the total lateral force above that story. The Portal Method is not applicable to frames where shear walls, vertical trusses, or diagonal bracing resist a significant portion of the lateral force.

3. The shear resisted by each exterior column is one-half that resisted by each interior column. This assumption in a multibay frame is equivalent to a series of independent frames pinned together, as in Fig. 13.15(b). Two beams framing into an interior column tend to make it twice as stiff as an exterior column with only one beam framing into it. Also, exterior columns tend to be somewhat smaller than interior columns.

4. A point of inflection is located halfway between joints in both beams and columns. In a multistoried frame with nearly equal bay sizes, story heights, beam sizes, and column sizes, this assumption is reasonably accurate. When columns rest on foundations that provide little resistance to rotation (such as small footings on compressible soils), the point of inflection will move closer to the foundation. In using the Portal Method, the exact location is a matter of engineering judgment. The inflection point could be located from near mid-height to near the bottom of the columns. It is commonly assumed to be 1/4 to 1/3 the column height above the foundation. Because the exact location is unknown, conservative values should be used.

    If the foundation is very stiff (such as pile or mat foundations), it may not permit as much rotation of the column as permitted by beams at the next higher floor line. Under such conditions, the inflection point could be higher than mid-height of the column. An engineer might conservatively assume the inflection point is as much as 2/3 the column height from the bottom for computing column moments at the base but only at mid-height for computing column moments at the top and moments in the beam at the top of the portal frame.

A frame analysis is usually performed using unfactored lateral loads because the load factor to be applied depends on the combination of other loads. After completion of the analysis, moments are multiplied by the proper load factor.

Using these assumptions, all shears and moments are easily computed by statics. Actual computations are illustrated in the following example.

Example 13.2   Given the building frame with a basement in Example 13.1, floor-to-floor heights all 12'-8", a 1'-8" parapet, and 30 psf wind pressure, calculate shear and moments in first- and second-story columns and second-floor joists on line 3.

Solution   A diagrammatic elevation of the building frame is shown in Fig. 13.16(a). Shears and moments are computed as follows:

Wind forces at each joint are based on the tributary area.

Wind force at roof = (6.33 + 1.67)20 ft × 30 psf = 4.8 k

Wind force at floor = 12.67 × 20 × 30 psf   = 7.6 k

Shear in second story = 7.6 × 2 + 4.8   = 20.0 k

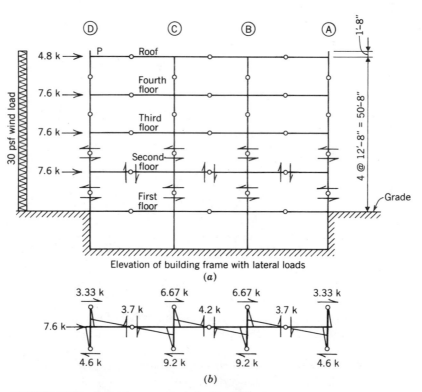

FIGURE 13.16   Portal frame analysis for Ex. 13.2.

Shear in second-story columns,

$$V = \frac{20.0}{6} = 3.33 \text{ k (exterior columns)}$$

$$V = \frac{20.0}{3} = 6.67 \text{ k (interior columns)}$$

Shear in first story $V = 7.6 \times 3 + 4.8$ $= 27.6$ k

Shear in first-story columns $V = \dfrac{27.6}{6}$ $= 4.6$ k (exterior columns)

$$V = \frac{27.6}{3} = 9.2 \text{ k (interior columns)}$$

Moment in second-story columns,

$$M = \frac{3.33 \times 12.67}{2} = 21.1'\text{k (exterior columns)}$$

$$M = \frac{6.67 \times 12.67}{2} = 42.2'\text{k (interior columns)}$$

Moment in first-story columns,

$$M = \frac{4.6 \times 12.67}{2} = 29.1'\text{k (exterior columns)}$$

$$M = \frac{9.2 \times 12.67}{2} = 58.3'\text{k (interior columns)}$$

Moment in joists in bays A-B and C-D,
$$M = 21.1 + 29.1 = 50.2'\text{k per bay}$$

Moment in joists in bay B-C,
$$M = (42.2 + 58.3) - 50.2 = 50.3'\text{k per bay}$$

Shear in joists in bays A-B and C-D, $V = \dfrac{50.2}{27 \times 0.5} = 3.7$ k per bay

Shear in joists in bay B-C, $V = \dfrac{50.3}{24 \times 0.5} = 4.2$ k per bay

Shears and moments due to wind, as calculated above, must be multiplied by the proper load factor when combing with shears and moments due to gravity loads and other forces.

A diagram of the shears and moment pattern is shown in Fig. 13.16(b).

Axial forces in columns can be computed in the same manner. For example, a tensile axial force due to wind exists in column D equal to the sum of the vertical shears in the joists in all floors and the roof in bay C-D.

Computer Analysis

When building frames are tall or are braced laterally in part by shear walls, vertical trusses, or diagonal bracing, analysis by the Portal Method is not appropriate. Other manual methods of analysis are available but they are tedious and laborious. In using these methods, engineers must be very careful to avoid manual errors. Most manual methods make simplifying assumptions that are not valid for tall buildings. For example, changes in the lengths of members are usually ignored, whereas length changes may be significant in tall buildings.

Engineers designing tall buildings today almost always analyze the frame for moments due to lateral loads using a computer. In selecting computer hardware and software, consider the same matters listed above for analysis of moments due to gravity loads.

13.5 REDISTRIBUTION OF MOMENTS

Tests indicate that reinforced concrete has some ductility to transfer moment from one section to another. When tensile reinforcement reaches yield at one section, it will continue to yield while the section rotates, if the concrete does not reach its ultimate strain. Both the steel stress and the moment capacity remain constant, at the same time that the steel yields. A continuous member will carry a load larger than that causing one section to yield. The load increases until all sections reach yield or until the concrete at one or more sections reaches its ultimate strain. At ultimate load level, the moment carried by each section is different from elastic moments. The change in moments is called a "redistribution of moments."

As an approximation, the ACI Code considers that the ductility of a member (its ability to redistribute moments) is a function of the ratio of actual reinforcement used to the balanced reinforcement, or $(\rho - \rho')/\rho_b$ for doubly reinforced members. For singly reinforced members, $\rho' = 0$ and the ratio becomes ρ/ρ_b, where ρ_b is the balanced reinforcement ratio given by Eq. 2.13.

Negative moments calculated by elastic theory at supports of continuous flexural members for any assumed loading arrangement may each be increased or decreased by not more than the ratio **R** (Code 8.4).

$$\boldsymbol{R} = 0.2 \left(1 - \frac{\rho - \rho'}{\rho_b} \right) \tag{13.1}$$

The maximum net reinforcement ratio $(\rho - \rho')$ can be computed easily if the desired moment redistribution ratio is known by solving Eq. 13.1 for $\rho - \rho'$. Thus,

$$\rho - \rho' = (1 - 5\boldsymbol{R})\rho_b \tag{13.2}$$

Equations 13.1 and 13.2 are shown graphically in Fig. A13.1.

Where approximate values for moments are used (e.g., moment coefficients from Table A13.2), redistribution of moments is not permitted because approximate values implicitly assume some redistribution under extreme conditions.

Modified negative moments must be used for calculating moments at sections within the spans. The ACI Code conservatively requires that redistribution of negative moments shall be made only when the section at which moment is reduced is so designed that ρ for singly reinforced members or $\rho - \rho'$ for doubly reinforced members is not greater than $0.50\rho_b$ (Code 8.4).

When negative moment is increased, positive moment will be reduced, and at the positive moment section it is required that $(\rho - \rho')/\rho_b \leq 0.50$. Conversely, when negative moment is decreased, positive moment will be increased, and at the negative moment section it is required that $(\rho - \rho'/\rho_b) \leq 0.50$.

Moments are most easily redistributed in solid slabs in both negative and positive moment regions and in the positive moment region of T-beams because low reinforcement ratios are normally used. In the negative moment region of T-beams, over supports, the reinforcement ratio will frequently be close to $0.75\rho_b$. If necessary, in this case, adequate ductility can be assured by extending bottom bars over the support as compressive reinforcement, taking care to provide a compression lap between the bottom bars from the two adjacent spans. Of course, adequate stirrups acting as lateral ties must enclose the compression bars.

When moments in continuous members are analyzed for several loading conditions, the total moment for which the members must be designed may be reduced by moment redistribution. Moment redistribution may also be used to reduce the maximum amount of reinforcement required in two or more beams that are nearly the same. These two uses are illustrated in Example 13.3.

In actual practice, most engineers will not calculate redistribution of moments unless the construction cost saving is large enough to justify the additional design time or unless it is necessary to investigate the design of a beam constructed accidentally with less than the amount of reinforcement required to resist elastic moments or with misplaced reinforcement.

Negative moments can always be increased or decreased at least 10% if the net reinforcement ratio $(\rho - \rho') \leq 0.5\rho_b$ at all sections in the member.

Example 13.3   Redistribute the moments calculated in Ex. 13.1 to reduce the total design moments (A) in beams B2 and (B) in joists J1 and J2.

Solution A   Moments at centerline and face of support are taken from Fig. 13.13. If all beams B2 are designed for $^-M = 172.3'$k at face of support at each end (^-M at line 3), then $^-M = 179.4$k, at line 2 must be reduced. The amount of reduction is

$$1 - \frac{172.3}{179.4} = 0.04$$

Negative moment regions of beam B2 must then be designed such that $(\rho - \rho')/\rho_b \leq 0.50$. If this condition is not met, ^-M should not be reduced.

Design $^-M = 238.5'$k in beam B2 at line 5 is also higher than $172.3'$k. At line 5, negative reinforcement for beam B3 will reinforce the adjacent beam B2 for the higher moment and no redistribution of moments is necessary.

If both ends of beam B2 are designed for $^-M = 172.3'$k ($^-M = 206.7'$k at the centerline of support), the required design ^+M near mid-span is

$$^+M = \frac{6.2(20)^2}{8} - 206.7 = 103.3'\text{k}$$

The design ^+M can then be reduced to 103.3'k from the 106.7' calculated in Ex. 13.1 for a 3% redistribution of ^+M. Positive moment regions of beam B2 must then be designed such that $(\rho - \rho')/\rho_b \leq 0.50$.

If moments had been calculated in Ex. 13.1 for the most severe pattern of live load distribution, the maximum ^-M and ^+M would have been slightly larger. Design for moments based on live load on all spans can be justified by redistribution of moments up to 10% if all critical sections of the beam are designed such that $(\rho - \rho')/\rho_b \leq 0.50$.*

Solution B   Calculated elastic design moments in joists for each loading condition are shown in Fig. 13.17. Moments not obtained from Ex. 13.1 have been calculated and supplied here. To reduce total design moments, the elastic positive moment in joist J1 for loading Case I, the elastic negative moment at line C for loading Case III, and the elastic positive moment in joist J2 for loading Case II should be reduced.

*Unless an engineer has a lot of spare time, moment redistribution will be used only for investigating completed structures.

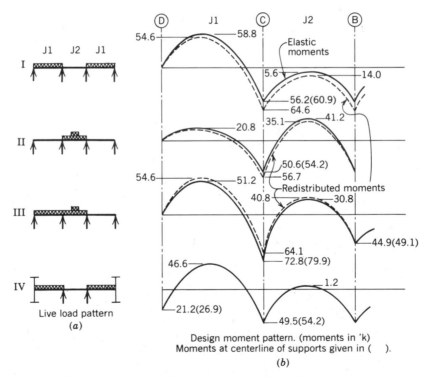

Design moment pattern. (moments in 'k)
Moments at centerline of supports given in (   ).
(b)

FIGURE 13.17   Elastic design moments for Example 13.3.

Loading Case III   Assume $(\phi f_y j) = 4.0$, $d = 19.2''$, $b = 6''$

At the support (C),

$$^-A_s = \frac{72.8}{4 \times 19.2} = 0.95 \text{ in.}^2, \qquad \rho = 0.82\%$$

Assume $f'_c = 3$ ksi and $f_y = 60$ ksi, $0.75\rho_b = 1.6\%$ from Table 2.1, and $\rho_b = 1.6/0.75 = 2.13\%$

$$\rho/\rho_b = \frac{0.86}{2.13} = 0.4 \text{ and the redistribution ratio}$$

$$R = 0.20(1 - 0.4) = 0.12.$$

The maximum moment on line C may be reduced to

$$^-M = 72.8(1 - 0.12) = 64.1'\text{k}$$

From Ex. 13.1, $Va/3 = 7.1'\text{k}$, and the redistributed moment at centerline of support (C) will be

$$64.1 + 7.1 = 71.2 \text{ k}$$

Joist J1 reaction $R_L = 0.95 \times \dfrac{27}{2} - \dfrac{71.2}{27} = 10.2 \text{ k}$

Positive moment near mid-span of joist J1:

$$\frac{R_L^2}{2w} = \frac{10.2^2}{2 \times 0.95} = 54.6'\text{k}$$

By inspection, $\rho_b < 0.50$ at mid-span and ductility is assured.
Approximate positive moment in joist J2:

$$\frac{0.95 \times 24^2}{8} + 3.5 \times 7.7 - \frac{(64.1 + 44.9)}{2} = 40.8'\text{k}$$

Loading Case I   Reduce ^+M_u from $58.8'\text{k}$ to $54.6'\text{k}$ or about 7% ($^+M_u = 54.6'\text{k}$ for loading Case III, after redistribution). The redistribution will increase ^-M_u at line C by about $(58.8 - 54.6)2 = 8.4'\text{k}$ and the mid-span moment in joist J2 by the same amount. Thus, minimum moment (negative) near mid-span of joist J2, $M_u = 5.6 + 8.4 = 14.0'\text{k}$.
Design negative moment at line C

$$^-M_u = 56.2 + 8.4 = 64.6'\text{k}$$

The redistributed negative moment at line C for this case is less than the redistributed moment for Case III.

Loading Case II   Reduce positive moment in joist J2 by increasing negative moment on lines B and C. Because the maximum redistribution of negative moment is 12%

(from Solution B), $^-M = 50.6 \times 1.12 = 56.7'$k, which is still less than the negative moment calculated for loading Case III.

At mid-span of joist J2, the redistributed moment,

$$^+M = 41.2 - (56.7 - 50.6) = 35.1'\text{k}$$

For all loading cases, the maximum moments for which joists J1 and J2 must be designed are summarized in Fig. 13.18.

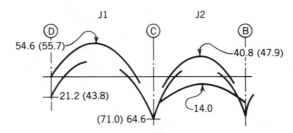

FIGURE 13.18 Maximum redistributed design moments in joists J1 and J2 (in ft-kips). (Moments in parenthesis are those obtained using moment coefficients.)

In joists J1, the total design moment

$$= \frac{21.2 + 64.1}{2} + 54.1 = 96.8'\text{k} = 1.26 \times \frac{w_u \ell_n^2}{8}$$

where $\dfrac{w_u \ell_n^2}{8} = \dfrac{0.95 \times 25.42^2}{8} = 76.6'$k

In joists J2, the total design moment

$$= 64.1 + 36.4 = 100.5'\text{k} = 1.21 \left(\frac{w_u \ell_n^2}{8} + Pa \right)$$

where $\dfrac{w_u \ell_n^2}{8} + Pa = \dfrac{0.95 \times 22.33^2}{8} + 3.5 \times 6.83 = 83.1'$k

PROBLEMS

1. What is the approximate weight (in psf) of:
   (A) ⅝-in. plate glass doors
   (B) 1-in. marble veneer
   (C) 3½-in. granite facing
   (D) rubble limestone wall, 10 to 12 in. thick
   (E) slate floor, 1 in. thick including mortar setting bed

(F) 1½-in. terrazzo

(G) a stack of round steel bars, 18 in. high, not including skids and spacer blocks?

2. How much does a fully loaded, standard 4-drawer filing cabinet weigh? What load is placed on a flat plate floor due to 4-drawer filing cabinets placed side by side, back to back, with aisles between rows just wide enough to open the drawers? Assume that the cabinets are 75% full, and arrangement allows 2′-6″ cross aisles for access at 20 ft spacing.

3. Calculate maximum design shear by statics in beams B1, B2, and B3 in Ex. 13.1 and compare to values calculated by coefficients, giving the percentage over or under.

4. Calculate moments in B4 and B5 from Ex. 13.1 in the first floor from line 1 to line 4, by moment distribution. Assume that the beam is fixed at line 4. Assume that the effect of the stair load can be estimated with sufficient accuracy by a concentrated load on beam B4 from the weight of a triple joist plus the North-South partition that it supports. At line 1, assume that B1 is supported by a basement wall similar to the one in Ex. 12.3.

Selected References

1. Xila, Liu, Wai-Fah Chen, and Mark D. Bowman. "Construction Load Analysis for Concrete Structures," *Journal of Structural Engineering*, ASCE, V.111, No.5, pp 1019–1036, May 1985.

2. Gerald B. Neville, Editor. *Simplified Design Reinforced Concrete Buildings of Moderate Size and Height*. Portland Cement Association, Skokie, Illinois, 1984.

3. *Handbook of Frame Constants*. Portland Cement Association, Skokie, Illinois.

4. *Manual of Steel Construction*. American Institute of Steel Construction, New York, 1985.

5. *Continuity in Concrete Building Frames*. Portland Cement Association, Skokie, Illinois, 1959.

CHAPTER 14

PRELIMINARY DESIGN

14.1 OBJECTIVES

Preliminary design is the selection and delineation of a framing scheme and member outlines for a reinforced concrete frame. Preliminary design for a structural frame must include a suitable system to carry all loads to adequate soil-bearing strata. The purpose of a preliminary design is to allow the final design to proceed in an orderly manner, especially when it is performed by several individuals and when the structural frame affects the work of other team members. This is the usual case.

Preliminary designs are best prepared by an engineer experienced in all framing systems. The engineer should have good conceptual ability in three dimensions and be sympathetic to the needs and problems of other people in the design and construction team. Creativity and a well-developed ability to conceive innovative solutions to unusual problems is also helpful. Engineers with these qualifications should find the discussion in this chapter helpful.

Following are the objectives in preparing a preliminary design:

1. Select a framing scheme giving the general arrangement of individual members. Locate or give the criteria for locating each primary member.

2. Select the size, that is, concrete outlines of primary structural members and members that might affect the work of other people in the design team. Minor or secondary members that do not influence the selection of a framing scheme are frequently not sized in preliminary design. Likewise, a foundation system will be selected but not the size of individual members such as spread footings.

3. Delineate features of the design that are important to the structure, to which other people in the design team must adhere. Distinguish structurally important features from those that can be changed to accommodate the work of other people.

4. Estimate the total weight of reinforcing steel when required. Such estimates are occasionally needed to prepare construction cost estimates.

14.2 PROCEDURES FOR PRELIMINARY DESIGN

In preliminary design, the engineer should consider large-scale or fundamental questions, especially those that might be overlooked in the final design. For example,

1. How does each load reach foundation soils adequate to support the load? With lateral loads especially, engineers sometimes consider the superstructure only and fail to adequately consider the substructure.
2. What could happen to threaten the safety or serviceability of the structure? Specifically, the engineer should consider the effects of the following
   A. Normal construction procedures on the partially completed structure.
   B. Construction errors, especially those that might occur due to misinterpretation of design documents relating to unusual features of construction.
   C. Overloads during construction.
   D. Loading higher or different from that specified by the controlling building code, due to normal operation of the building on completion. For example, the building code may require only 50 psf live load for offices, but many engineers recommend higher loadings to allow for concentrated areas of filing cabinets, bookshelves, computers, and the like.
   E. Changing use or function of the building requiring higher or different loading.
3. Could the building be remodeled or expanded in the future in a manner that could affect the structure adversely?

For efficiency, avoid calculating anything that is not required for the preliminary design. Likewise, omit details that do not affect other people in the design team. If a construction team has been assembled at the time preliminary design is prepared, include only such additional information as is required by construction personnel.

Preliminary Design Steps

Steps in preliminary design include the following:

1. Obtain design criteria such as those listed in Section 1.1. Some criteria will always be provided to the engineer but, frequently, they are not complete. In such cases, the engineer must supply the missing criteria, confirming it with the proper person or agency when necessary. When conflicting criteria cannot be resolved or eliminated, they must be evaluated and ranked in importance. For example, a structure cannot be least cost and also carry heavier loading or have longer spans, but one objective must take precedence over the other. Some clients will have subjective, even irrational, criteria that

must be considered by engineers and to which they must respond. Finally, the engineer must evaluate *real* criteria as opposed to *given* criteria. Clients and owners do not usually have technical training and, therefore, frequently do not state their real requirements in technical terms.

2. Prepare a preliminary design for a typical bay or panel using several different framing schemes. Rarely if ever, is a building restricted to only one or two possible schemes. Each scheme must be based on the criteria established in step 1, minimizing conflicts and compromises as much as possible. However, some compromise between conflicting criteria is always necessary. Changing the emphasis on conflicting criteria leads to different framing schemes (see Sections 14.3 and 14.4).

3. Evaluate each framing scheme. List, in writing, the advantages and disadvantages of each scheme, referring to the criteria previously established.* Select the most appropriate scheme by comparing the evaluations of all schemes, giving emphasis to the most important criteria. Evaluations can rarely be numerical or quantitative except, perhaps, on the cost of construction. Even on this point, no two contractors will estimate construction costs the same nor the same as the engineer.

4. Present all framing schemes and evaluations to the client and other people on the design team. Solicit their comments and evaluations of each scheme.

5. Using the comments received in step 4, prepare new framing schemes and revise previous framing schemes. Re-evaluate all schemes as in step 3.

6. Select one scheme to be used and complete the preliminary sizing of members in the entire structure.

14.3 SELECTING A FRAMING SCHEME

Nowhere in engineering design is intuition more appropriate, even essential, than in preliminary design. There are no formulas that lead inexorably to the most appropriate framing scheme. Rather, a brilliant scheme for an unusual problem and even good workmanlike schemes for routine problems will be the result of heavy intuitive thinking based on years of experience. Engineers do well to cultivate and respect insights to structural behavior that cannot be proved easily or quickly by normal means.

Engineers should strive for redundancy in preparing a preliminary design. Redundancy means the ability of a structural frame to carry load by more than one path. In a redundant structure, if a beam would fail, the load supported by the beam could be carried to the ground by other means, such as by catenary action of slab reinforcement. Redundant flexural members also tend to reduce deflection. It is not necessary that the secondary members carry load without distress—only that total collapse be avoided.

---

*Every scheme has both advantages and disadvantages. If you are tempted to list only one or the other, your impartiality may be questioned.

It is more difficult, but not impossible, to provide alternate load paths for columns and tension members or hangers. For such members, redundancy is defined as limiting failure to the immediate area surrounding the failed member. That is, beams and slabs can have sufficient continuity and anchorage at each end to prevent their complete collapse if support is withdrawn at one end. If a column fails, the surrounding floors and roofs might sag considerably but should not endanger life by dropping to the ground.

Engineers should favor the selection of structural frames with a high degree of redundancy because structural redundancy is essential in preventing progressive collapse. Some owners require their engineers to design the structure to minimize the risk of progressive collapse. The procedures discussed in Section 4.8 are also applicable to other floor framing methods.

If a member or a frame cannot be made redundant, more than normal care should be taken in its design. In addition, conservative engineers will use a slightly higher factor of safety for such members.

Selection of the best, most appropriate framing scheme will normally proceed in the following manner.

1. Find a Typical Bay or Panel   A typical bay is the one most often repeated in the building with the fewest modifications. In this context, a bay is the space from one column line to the next, extending across the entire width of the building, and a panel is the space between four columns (see Fig. 14.1). It will be advantageous to consider a bay rather than a panel if the column spacing has not yet been established or if the number and spacing of columns can be improved. If a typical bay does not include the most critical members, they should be considered in determining the size of corresponding members in the typical bay.

If there are no typical bays, it may mean that the structural frame (and also probably the building) will be very expensive to design and build. The engineer's first task then is to rearrange building elements so as to achieve some degree of regularity and uniformity to produce a typical bay. This need not result in compromises to the architectural design or use of the building. Nor need it result in a monotonous building facade.

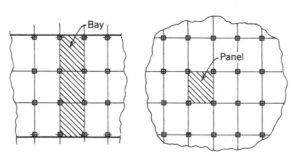

FIGURE 14.1   Typical bays and panels.

2. Determine Minimum Slab Thickness   Embedment of electrical conduit (used in almost all building slabs) requires a slab thickness of at least 3 in. in pan slabs. Other slabs should be at least 4 in. thick. If the slab has under-floor electrical distribution, greater thickness will be required. Minimum thickness may also be determined by requirements for fire protection.

The building code governing design and construction of the structural frame will specify the required fire resistance rating in hours. Some owners will want to increase the code-required fire rating, or the engineer may recommend a higher rating in some cases. Frames for small, low buildings frequently do not need a fire rating. The floors for larger and taller buildings must have a 2- or, sometimes, 3-hr fire rating. Columns and walls frequently must have a 2- to 4-hr fire rating.

The fire resistance rating of a structure is the time a structure is exposed to heat of a fire during which (a) it can support itself and superimposed loads without collapse and (b) the temperature rise on the unexposed surface of the concrete does not exceed 250°F as an average for the entire surface, nor 325°F at any one point. The purpose of the second criterion is to guard against ignition of combustible material in contact with the unexposed surface. Fire ratings are determined by testing a structural member or assembly in a furnace exposed to heat with a standard time-temperature curve.

Because concrete is more resistant to fire than steel is, a concrete frame will continue to carry load at least as long as the stress in reinforcing steel does not exceed the yield point of the steel. Neglecting load factors and capacity reduction factors (this is permissible when considering fire ratings), the stress in steel at service loads will generally be less than $0.6f_y$, where f_y is the yield strength at room temperature. The yield strength of hot rolled structural grade steel will fall below $0.6f_y$ when its temperature rises somewhat higher than 1000°F.

If sufficient concrete cover is provided to insulate the steel so that the steel temperature remains below 1000°F, the structure will continue to carry design service loads. In solid slabs, the required cover is ¾ in. for 1½-hr fire resistance rating, 1¼ in. for 3-hr fire resistance rating, and 1½ in. for 4-hr fire resistance rating. A little more cover is required for joists because the bottom steel is exposed on three sides with minimum concrete cover. A little less cover is needed when light-weight concrete is used due to its better insulating properties.

The portion of a structure exposed to fire in a limited area is stronger than an isolated, simple span structure because thermal expansion is resisted by the unexposed portions of the structure and because negative moment regions of flexural members are better protected from heat than mid-span positive moment regions. For these reasons, it is generally considered that concrete cover required by the ACI Code (Code 7.7.1, or see Section 1.8) is adequate for the fire resistance rating normally required.

The important parameters in determining whether a frame will meet the second requirement limiting heat transmission are the thickness of concrete slabs or walls and the type of aggregate used. Aggregate will usually be selected on the basis of local availability, cost, or other reasons. Thus, the required fire resistance rating will determine the minimum thickness of slabs in the structural frame. Figure A14.1

from Ref. 1 shows the effect of thickness and type of aggregate on the fire endurance of concrete slabs. A 4½-in. slab of stone concrete is usually considered adequate for a 2-hr fire rating.

Reference 1 describes standard fire test procedures, tabulates the results of many fire tests, and describes analytical methods for computing fire resistance ratings of reinforced concrete floor members and assemblies.

3. Determine Column Spacing   Column locations may be given by the architect or owner and cannot be changed. If so, proceed to the next step. When the engineer participates in or influences the decision on column locations, building use as well as structural economy must be considered. Column spacings closer than about 15 ft will not normally be economical because floor slabs spanning 15 ft will probably have minimum, or nearly minimum, thickness and reinforcement. Shorter spans do not decrease the cost of slabs as fast as the cost of columns increases. Likewise, column spacings more than about 35 to 40 ft increase the cost substantially. The cost of longer floor spans rises faster than the reduction in cost of larger but fewer columns. Because the total weight of a given building is nearly constant, the total required cross-sectional area of columns in one story is nearly constant. Considering only the columns, fewer and larger columns minimize formwork and reduce the cost.

Columns should be located in walls where they interfere the least with use of the building. In the exterior wall, the spacing and shape of columns affects tenant's views of outside scenery. Columns projecting into the room should be coordinated with interior space planning.

In most buildings there is a limited choice in column spacings. For example, in a 78-ft-wide building, columns can be spaced 2 at 39 ft, 3 at 26 ft, 4 at 19½ ft, 5 at 15.6 ft, or with minor variations of unequal spans. Considering the use of the building, only one or two of these options will be practical.

4. Sketch as Many Different Framing Schemes as Can be Conceived for the Typical Bay   The following observations will assist an engineer in arriving at appropriate framing schemes.

A. Solid slabs are economical up to about 20 ft and can span up to about 25 ft (see Fig. A14.2).

B. Pan slabs are economical up to about 35 or 40 ft. Where possible, pan slabs should span the long direction of a panel, and beams the short direction, so that the depth of beams and joists can be equal for simplicity of formwork.

C. For spans over 35 to 40 ft, prestressing or deep beams must be used to control deflections. Prestressing may unduly restrict the owner from drilling holes through the slab to provide electrical, computer, or telecommunication services at new locations when office layouts change. Deep beams may interfere with mechanical services below the floor slab and raise the cost unduly.

D. Buildings for residential occupancy, such as hotels, motels, dormitories, and apartments frequently use an exposed solid concrete slab as the finished ceiling. Surface preparation of the concrete and panting is less expensive than a separate hung ceiling. In residential buildings, spans are within the economic limits of one- or two-way solid slabs.

E. Hospitals also frequently have short spans, but mechanical services may prevent using the slab as an exposed ceiling.

F. Office buildings, banks, commercial structures, and schools generally require spans in the range of 25 to 40 ft.

G. Office buildings and, sometimes, other buildings require flexibility in providing electrical, computer, and telecommunication services at new locations when wall or room layouts change. These services can be provided by drilling into under-floor ducts an inch or two below top of the slab or by a "poke-thru" system. Electrical under-floor ducts can be installed easily in solid slabs and accommodated in pan slabs by using a pan size 2 in. shallower, as in Fig. 14.2. In a poke-thru system, holes are drilled through the slab to reach conduit just below the floor slab. Poke-thru holes in pan slabs will usually miss reinforcement. In solid slabs, poke-thru holes will sometimes cut reinforcement, and design of slabs should make allowance for this possibility.

H. Slabs exposed to the weather, especially freezing weather where deicing salts may be used (as in a parking garage), should have 2 in. cover to reinforcement, a water/cement (W/C) ratio of 0.40 or less, and should slope at least ¼ in. per foot to drains. A slope of ⅜ to ½ in. is better. Thin slabs (as in a pan slab) should be avoided as their long-term durability may not be satisfactory.

I. Flat slabs with drop panels and column capitals are generally the most economical system to support heavy live loads exceeding about 150 to 200 psf.

J. Members with a minimum size controlled by reasons other than stress should be used so that they are fully stressed. Some common examples follow:

a. If slabs must have a minimum thickness for fire protection, the structural system should use slabs that span far enough to stress minimum reinforcement in slabs to the limit. For example, a 4½-in. slab can span at least 7 or 8 ft.

b. Minimum reinforcement in flexural members should be fully stressed under load.

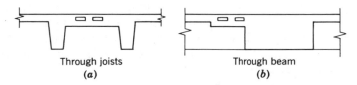

Through joists
(a)

Through beam
(b)

FIGURE 14.2   Sections through joists and under-floor electrical ducts.

    c. Concrete walls used for partitions should also be used to carry vertical loads and resist lateral loads.

    d. The spacing of lightly loaded columns of minimum size should be increased so that they are loaded to capacity.

K. Use members for more than one purpose, if possible, for example,

    a. When nonstructural, noncombustible floor finishes such as terazzo or ceramic tile are used, they may be included in the thickness required for fire protection.

    b. Walls and long, narrow columns can carry vertical loads, resist lateral loads, and serve as partitions. Basement walls can serve as grade beams to simplify foundation construction. Other walls can serve as girders to transfer column loads or other heavy loads to support points.

    c. Use temperature reinforcement in a one-way solid slab as flexural reinforcement in the long direction of a two-way slab (if the aspect ratio of the slab is two or less).

L. For economy of structural frame, consider framing systems approximately in the following order:

    a. Flat plate without drop panels or column capitals.

    b. Flat plate with wide shallow beams.

    c. Pan slabs.

    d. Flat slab with drop panels and column capitals.

    e. Waffle slabs.

    f. Slabs and deep beams.

Table 14.1 summarizes some of the advantages and disadvantages of common framing schemes.

    As indicated above, it is necessary to consider the use to which a building will be put in preparing a preliminary design. The number, size, and location of holes through the floor system for stairs, elevators, and mechanical shafts, as well as the building mechanical systems, will affect the selection of a structural frame. Underfloor electrical ducts may be installed. Laboratory equipment may be unusually sensitive to vibrations and deflections. Heavy static or moving loads should be considered in the preliminary design.

5. Design Each Proposed Framing Scheme for the Typical Bay   Each design should be complete, establishing all final concrete outlines and quantities of concrete and reinforcing steel. Loads should be as exact as possible, but approximations may be used when laborious computations are needed to determine exact loading. For example, a series of concentrated loads may be approximated as a uniform load (see Section 14.4 for methods of preliminary sizing of members). In establishing member outlines, make allowance for nontypical panels that are more highly stressed than the typical panel. However, design of the typical panel should be for loads and conditions of the typical panel.

6. Review Proposed Framing Schemes   To help ensure that all criteria have been considered and that the best schemes have been delineated, the engineer should

TABLE 14.1 Advantages and Disadvantages of Framing Schemes

Framing Scheme	Advantages	Disadvantages
Flat plate	Least cost for form work Exposed ceilings Minimum thickness Fast erection Flexible column location	Low capacity in shear Excessive concrete for longer spans
Flat slab	Economical for heavy loading	Expensive formwork
Flat banded slab (see Section 4.2.)	Extends span range of flat plate Minimum thickness	Formwork costs more than for flat plate unless many reuses of flying form are possible
Pan slabs and wide beams	Minimum concrete and steel Minimum weight, hence, reduces size of columns and foundations Long span in one direction Easy poke-thru electrical	Unattractive for a ceiling Formwork costs more than for flat plate
Skip joists and wide beams	Similar to pan slabs but use slightly less concrete Can place mechanical equipment in space between joists	Similar to pan slabs Joists must be designed as beams
Dome slab	Attractive exposed ceiling	Formwork costs more than for pan slab Uses more concrete than a pan slab
Deep one-way beams and one-way slabs	Long span in one direction	Beams interfere with mechanical work Expensive formwork
Deep two-way beams and two-way slabs	Long span in two directions Small deflection	Beams interfere with mechanical work Expensive formwork

review the preliminary designs before proceeding to step 4 (see Section 14.2). The review will be focused on important issues if the engineer asks pertinent questions about each proposed scheme. For example, the engineer should consider the following:

A. How can I make this scheme more repetitive, more economical?

B. What compromises on the part of the owner, architect, mechanical engineer, or other people will make this scheme better?

C. What compromises on my part will make this scheme more acceptable to other people?

D. What feature of this scheme costs the most? Can it be modified to reduce the cost?

E. What feature of this scheme makes it least acceptable to other people? Can that feature be modified to make it more acceptable?

Considering all schemes, the engineer should ask the following:

A. Have I considered all possible schemes?

B. Is there a better way to meet design objectives?

14.4   DETERMINING CONCRETE OUTLINES

General

Because one objective of preliminary design is to determine concrete outlines, design procedures leading directly to member sizes should be used. These will be discussed in more detail later in this section. Computations related to reinforcement can be disregarded unless the quantity of steel is needed for cost estimates.

In the following discussion, experienced engineers will recognize appropriate exceptions.

The shortest and easiest design procedures should be used so that the engineer does not lose sight of objectives in the minutiae of detailed design. If the first preliminary design reveals potentially critical members or situations, a more accurate design procedure can be used on the second iteration.

An approximate moment analysis for gravity loads using moment factors will normally suffice even for spans falling a little outside ACI Code limitations. Engineers can draw on their experience to use factors sufficiently accurate. Likewise, an approximate frame analysis by the Portal Method for lateral loads should be used. For tall or important buildings, a second iteration using more accurate methods of frame analysis may be necessary.

Do not vary the member size to fit the span and load except by increments of at least 30% to 50% or more.* For example, if a 12-in. column is inadequate, increase the size to 16 or 18 in. Make the member large enough for the most critical condition and vary the amount of reinforcement on the final design for less critical conditions.

Repeat bay sizes, panel sizes, story heights, and member sizes as often as possible. If, by inspection or by calculation, the load on a column or beam is more than 60% to 80% but less than 130% to 150% of the load on a similar member, use the same concrete outline for both members.

Do not use the maximum capacity of a concrete outline on the preliminary design. Unexpectedly higher loads, conditions slightly changed for the worse, and higher stresses by more precise analysis than those given by approximate analysis should all be accommodated by the initial concrete sizes. Only gross errors or significant changes clearly requiring a revised preliminary design should lead to changes in concrete outline on the final design. For large projects, a second iteration

*This is a "one-size-fits-all" concept, like a pair of socks.

of preliminary design may be necessary to reduce the margin of uncertainty in preliminary sizes.

Because additional cement or other expensive ingredients are required to increase the strength of concrete, specify the lowest satisfactory concrete strength needed for the project. Normally a strength of $f_c' = 3$ ksi throughout the project will be satisfactory unless there is a specific reason to use higher strength concrete. Common reasons for using higher strength concrete are as follows:

- Columns supporting more than three or four stories may be more economical if higher strength concrete is used in the lower stories.

- If concrete strength in the column is more than 40% higher than concrete strength in the floor, or $1.4 \times 3 = 4.2$ ksi, increasing the strength of concrete in the floor may be the most economical solution to passing column loads through the floor (see Section 10.2).

- Concrete exposed to freezing and thawing should have $f_c' = 4$ ksi or more in addition to air entrainment to ensure durability.

- Buildings more than 10 or 15 stories tall will usually be built on a fast schedule, requiring design strength in less than 28 days, typically 5 to 10 days. In such cases, the 28-day strength will be higher. For example, construction loads may require 3 ksi strength at 7 days, but strength at 28 days will be 3.5 to 4 ksi. If permanent service loading is heavier than construction loading, the higher concrete strength may be used in design for service loading.

- Floor framing usually does *not* require a concrete strength f_c' greater than 3 ksi because deflection usually controls the selection of concrete outlines rather than concrete strength. Increasing the strength of concrete is an inefficient method for reducing deflection or increasing shear strength.

When foundations are expensive (as in deep foundations) or the building is tall, keep member sizes as small as possible to reduce dead load. In other buildings, it is more economical to be a little generous with member sizes. When it is desirable to reduce dead load, lightweight concrete should be considered. Its higher cost may be offset by savings in columns and foundations. When slab thickness is controlled by the fire resistance rating required by the building code, the use of lightweight concrete will justify thinner slabs.

Individual Member Sizes

Considerations in establishing sizes of beams, slabs, and columns are discussed below.

Beams   The size of beams is usually controlled by flexural compression at the point of maximum negative moment. Positive moment is less critical for beams because a flange is usually present to help resist flexural compression and because positive moment is usually smaller than negative moment. Beams can be proportioned by solving Eq. 2.23 for member dimensions. Thus,

$$bd^2 = \frac{M_u}{R_c} \tag{14.1}$$

where M_u = factored moment
and R_c can be taken from Fig. A2.3.

The depth of the beam is usually established by other considerations, such as minimizing floor construction thickness, simplifying formwork, or minimizing deflection. In some cases, beam width may be established to fit within a narrow space or to allow the passage of pipes or ducts through the beam. After either width or depth is established, the other dimension is easily computed.

Use of compression steel should be avoided on preliminary design and used only on final design where necessary for strength to maintain established concrete outlines or to reduce long-term deflection.

For maximum economy, it is usually best to keep the percentage of reinforcement low so that reinforcing bars can be placed easily, especially at intersecting beams and columns. With a low percentage of reinforcement, bars can be widely spaced so that concrete can be placed easily around the bars.

Shear generally does not control the size of beams except in very short, heavily loaded members. Therefore, shear need not be computed for preliminary design.

The depth of long-span, shallow beams, especially those with little or no continuity, may be controlled by deflection. In such cases, deflection should be computed and member size revised if necessary (see Chapter 8). Often, deflection computations for the typical or most critical beam will also suffice for final computations for the entire structure.

One-Way Slabs   Thickness of solid slabs may be controlled by fire resistance requirements, as noted earlier. Alternatively, the thickness will be controlled by ACI Code requirements to limit deflection (see Table A8.3). Engineers will rarely use slabs thinner than required by Table A8.3 as the risk of unsatisfactory deflection is too great and calculations are tedious. In some cases, thicker slabs may be used to limit deflection still further.

Shear and flexural stresses rarely, if ever, control the thickness of solid one-way slabs.

Two-Way Slabs   Minimum thickness of two-way slabs is $\ell_n/32.7$, where ℓ_n is the clear span and steel f_y = 60 ksi is used (see Section 8.5). Punching shear around columns from gravity loads and from unbalanced moment transferred by shear will further control the design. Either a thicker slab, larger columns, column capitals, or shearhead reinforcement may be necessary. Engineers should be especially conservative in selecting proportions to meet shear requirements if enlargement of columns or column capitals will be difficult during final design.

Finally, deflection behavior of the slab should be considered even though the minimum thickness is exceeded. The appropriate time to compute deflection is during preliminary design rather than during final design when changes to correct unsatisfactory behavior may be difficult.

Flexure will rarely, if ever, control thickness of a flat plate or flat slab.

Pan Slabs   Shear and negative moment will control the width, spacing, and depth of joists. As discussed above for beams, positive moment will rarely, if ever, control concrete outlines of pan slabs. Increasing the width of joists near the support by using tapered pans is more effective in resisting negative moment than in resisting shear. Moment decreases more rapidly away from the support than does shear.

Deflection behavior of joists on long spans and joists with little continuity should be checked during preliminary design. If computed deflection is excessive, deeper or wider joists can be used. Such changes are more difficult to make during final design when the engineer is restricted to options (e.g., camber or more reinforcement) that do not affect other members of the design team.

Columns   In braced frames, the moment in most columns will be less than the moment at minimum eccentricity. Exceptions will generally be in columns at or near the top story where minimum reinforcement can easily be increased during final design. Thus, the required area of a column can be calculated by dividing the axial load by an average stress that includes the capacity of vertical reinforcement as well as the concrete.

$$A_g = \frac{P_u}{f_g} \tag{14.2}$$

where A_g = gross cross-sectional area of a column
$\quad P_u$ = factored ultimate load.
$\quad f_g$ = average stress on the column.

Inserting Eq. 9.5 into Eq. 14.2 where Eq. 9.5 is

$$\phi P_{n(\text{max})} = P_u = 0.8\phi[0.85f_c'(A_g - A_{st}) + f_y A_{st}],$$

using a steel strength $f_y = 60$ ksi and $\phi = 0.70$ for tied columns, and solving for f_g yields

$$f_g = 0.476f_c'(1 - \rho_T) + 33.6\rho_T \tag{14.3}$$

where $\rho_T = A_{st}/A_g$
Values for f_g are shown in Fig. A14.3.

For maximum economy, a low percentage of reinforcement should be used but not less than 0.5% to 1.0%. When reinforcement exceeds 4%, clearances should be checked to verify that column vertical bars can be placed and spliced with bars in columns above and below and that column bars passing through the floor framing will not interfere with bars in the floor.

The same size column can be maintained through 20 stories or more by increasing the concrete strength and using a high percentage of reinforcement in the lower stories. Figure 14.3 illustrates how columns might be reinforced in a 20-story build-

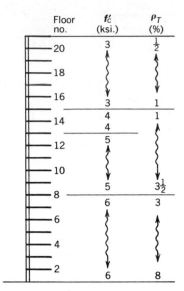

FIGURE 14.3 Example of concrete strength and reinforcement ratio in tall building columns.

ing. Only the top few stories would have columns with minimum reinforcement and excess capacity. If these columns were reduced in size, the extra cost of formwork would probably exceed the savings in concrete and steel. If a column size must be reduced, only one dimension should be changed (e.g., change from 24 × 24 in. to 24 × 18 in.).

14.5 OPTIMIZATION

General

The optimum structural frame is the best, or most favorable, in terms of an objective (see Chapter 1 and Section 14.2), from among possible satisfactory frames. "Optimization" is the process of finding the optimum frame. The term optimization is frequently used to mean the process of finding the least cost frame, but engineers must also optimize other objectives to achieve success in design practice.

Because a structural frame cannot be the optimum frame for all objectives, some balancing of conflicting objectives is required. For example, a frame cannot have long spans and also least cost. Likewise, minimum floor thickness is not consistent with small deflection. To help resolve conflicting objectives, consider the following procedure:

1. Assign priorities to objectives. Put yourself in the position of the architect, the owner, and other people in the design and construction team and visualize

how they would judge each objective, even if their judgment is subjective rather than objective.

2. Define and clarify the issues. Which objectives can be optimized and which cannot? In the range of practical solutions, which objectives will be affected most?

3. Develop options that emphasize one or more objectives. Such options are the only ones likely to be accepted.

4. Evaluate how well each option or possible structural frame meets each objective and prepare your recommendation for the frame to be used.

Before discussing optimization further, it is advisable to consider structural optimization in the context of the overall optimization hierarchy. Society and the building industry will optimize a project at several stages before completion.

The society of a community, state, or nation will influence whether to build a facility, how big it should be, for what purpose it should be built, and other such broad questions. Society makes these decisions through laws and peer pressure. In addition to government and a community's society, the owner of a proposed facility will optimize a facility in relation to these same questions.

The location of a facility will be optimized by the owner to minimize operating costs and maximize operating benefits. Retail facilities will be located near customers, schools will be located conveniently for children, and so forth. Sometimes realtors and architects assist owners in the decision on where to locate a facility. Sometimes geotechnical and structural engineers will assist in this decision when soil conditions at potential sites materially affect construction cost or constructibility, especially if the effect is adverse.

Building layout and general arrangement will be optimized, generally by the architect and the owner. The structural engineer may influence the design of the building at this stage, especially if the structural frame is an important feature of the building (as in a long-span roof or a structural frame used for its aesthetic appeal) or if the cost of the frame is a large part of the total cost (as in a warehouse, a parking garage, or a tall building).

The structural frame will be optimized by the structural engineer during preliminary design with some assistance from other people in the design and construction team.

After a structural frame has been selected, the final design should optimize, which generally means minimize, the construction cost. Because formwork costs and concrete quantities are established in preliminary design, minimizing construction costs usually means reducing the amount of reinforcing steel used. To a lesser extent, other objectives may also be optimized during detailed design. These subjects are covered in the first 13 chapters of this text. Note that structural safety is not optimized as it is an absolute requirement. Only occasionally will an owner request a building with greater-than-normal factors of safety.

The above hierarchy has been arranged in decreasing order of impact on cost and probably on other objectives as well. Societal concerns, location, and layout of a building can each affect its life cycle costs by 100% or more. Preliminary design can easily affect the cost of a structural frame by 30% to 50% whereas minimizing the amount of reinforcing steel used in the final design can affect the construction costs by only a few percent. Obviously then, engineers should concentrate their optimizing and cost-saving efforts on the preliminary design.

Owners vary in the emphasis they place on life-cycle costs. Engineers should be aware of these costs and how a structural frame affects them. Life-cycle costs include all costs in building and operating a facility for its full lifetime, such as the following:

- Construction of the frame.
- Construction of the remainder of the building.
- Financing, legal, and administrative costs during and after construction.
- Maintenance and repair.
- Remodeling.
- Demolition.
- Salvage.

Because these costs occur at different time periods, they must be compared by the use of *present value* calculations. A full discussion of life-cycle costs and *present value* calculations is beyond the scope of this text.

Cost Optimization

Even neglecting other objectives, optimum or minimum cost rarely coincides with minimum material quantities, but most technical papers on the subject make this assumption. Indeed, it is extremely difficult to reduce optimization to a mathematical process because the total cost of the frame includes the entire structure and not just the member or group of members under consideration. The structural frame is only part of a project. The costs of other parts of the project are affected by the structural frame and by individual structural members.

Minimizing the cost of construction is an important criterion in selecting a framing scheme. Engineers can usually estimate the cost of construction with satisfactory qualitative, if not quantitative, accuracy, by consulting local contractors and material suppliers on unit costs of material and labor of erection. Care must be taken to consider factors affecting unit costs by soliciting such information from contractors and suppliers. To make impartial cost comparisons, the engineer must be consistent between schemes in estimating quantities and applying unit costs.*

Include all costs in each scheme even though they may be identical from one scheme to another, such as floor finishing, and include a reasonable allowance for the contractor's overhead and profit. These will lend credibility to the engineer's cost estimate and increase the chances of its acceptance. The total cost will be easier to compare to cost estimates prepared by the architect, contractor, or other people.

Make allowance in the cost estimate for nontypical panels. For example, the average number of columns in one panel will vary from four to a little more than one,

---

*Quantities and unit costs obviously need not be identical between schemes—just consistent. A higher unit cost is justified in one scheme over another if more labor or more expensive materials are used.

depending upon the number of bays in the width and length of the building. Special conditions around the perimeter or at the stair and elevator shafts can usually be disregarded because they will affect all schemes in a similar manner.

If the thickness of the floor framing varies between schemes, allowance should be made for the cost of walls, stairs, elevators, mechanical risers, and other items of construction that vary with floor-to-floor heights.

To understand how to minimize construction costs, it is essential to know the distribution of costs for the various elements of construction and to what degree an engineer might influence these costs. This understanding comes only through the preparation of cost analyses on many structural frames. Table 14.2 has been prepared to illustrate a range of costs for typical concrete frames based on average costs in 1985. Some projects will fall below or above the ranges for both material quantities and costs. The cost of formwork is usually 50% to 60% of the total cost of the structural frame but could be as little as 35% in exceptional cases. For a particular project, engineers should prepare more detailed cost estimates than those shown in Table 14.2 to guide their optimization efforts.

The cost of reinforcing steel is minimized by using grade 60 bars, by increasing the size and reducing the number of bars to be placed, by minimizing the amount of bending required, by using standard details, and by repetition of bars exactly alike in size, length, and bending. Avoid #14 and #18 bars as they are not usually carried in stock but are rolled on special order. Use lap splices rather than welded or mechanical splices. Place bars in a single layer in beams. Use the largest practical stirrup at standard spacings. Check the fit and clearances of all bars. A well-conceived preliminary design will allow such optimization to take place. Welded wire fabric (WWF) or prefabricated mats might reduce placing costs.

The cost of steel is also minimized by reducing the quantity. This is best accomplished by selecting bar combinations as close to the required quantity as possible. Reducing the quantity of steel by cutting off bars within the span of flexural members requires much calculation effort and risks construction errors that could have serious consequences. Furthermore, the potential saving in steel is small.

Avoid spiral columns where possible, as the weight of steel is greater than that in

TABLE 14.2 Construction Cost Allocation and Opportunities for Cost Reduction

Item	Quantities (per ft^2)	Cost ($ per ft^2)	Influence of Structural Engineer percent	Amount ($ per ft^2)
Finishing	1.0 ft^2	0.30 to 0.40	0	0
Steel	2.5 to 12 lb	0.40 to 2.30	20	0.50
Concrete	0.5 to 1.0 ft^3	0.90 to 3.00	20	0.60
Formwork	1.05 to 2.0 ft^2	1.40 to 5.00	50	2.50
Total		3.00 to 9.20		3.60

This table uses approximate, average 1985 prices and approximate quantities for illustration purposes only. It is not intended as an accurate statement of probable construction costs.

tied columns except for large round columns with low concrete strength and few vertical bars (see Fig. 9.20). Also, fabrication of spirals costs more than fabrication of vertical bars.

Bundled bars in columns with more than four vertical bars may make placing concrete easier, reduce interference with intersecting beam bars, and reduce construction cost.

The cost of concrete is minimized by reducing the number of mix designs required, specifying aggregate size and mix designs appropriate for structural members with congested reinforcement, and limiting strength and other requirements to values that can be reasonably accomplished by local concrete suppliers. It is sometimes necessary and desirable to challenge concrete suppliers to furnish concrete of higher strength than they have furnished before, but the cost will also likely be higher. The quantity of concrete should be minimized on preliminary design. In final design, there is little or no flexibility in reducing concrete quantities.

The cost of concrete placement will be minimized if the engineer considers problems in placing and consolidating concrete. Because it is difficult to cast concrete through a curtain of closely spaced top reinforcement, periodic gaps of 6 to 8 in. should be provided to allow insertion of pumping hoses and vibrators. Making it easier to place concrete helps assure its high quality as well as reduce costs. Such gaps in reinforcement should be provided, even though the bars between gaps must be spaced closer, or bundled, or placed in a second layer.

Because smaller quantities are involved, because placing is more difficult and higher strength concrete is more likely to be used, the cost of concrete in columns is more than the cost of concrete in floor slabs. The cost of column concrete can be reduced by using fewer but larger columns and by specifying ties that permit an unobstructed area in the center of the column for placing concrete. Although high-strength concrete costs more than concrete of moderate strength, the increase in cost is less than the increase in strength. Thus, high-strength concrete in columns is economical if it can be fully stressed.

The cost of formwork is minimized by simplifying and repeating the shapes to be formed as much as possible and by reducing the surface area of formwork. The suggestions below will help achieve these goals.

- Space columns uniformly. Uniform loading on slabs, beams, and columns is likely to result, and sizes of these members can be made uniform. Repetition will lower construction costs even though a little extra material may be required.

- Use one column size from foundation to roof and for as many columns in each story as possible. By doing so, form building is simplified, column forms can be reused, column ties will be standardized, deck forms around the column will not need to be reworked, fewer carpenter and ironworker foremen will be required, less labor will be used, fewer supervisors and inspectors will be needed, and costs will be lower.

- Use one wall thickness from foundation to top.

- Use steel- or fiber-reinforced plastic forms where the improved surface texture will permit omission of other, more expensive finish materials.

- Use shear walls or other bracing systems to reduce the size of columns.

- Make many beams the same size even though the spans and loading differ. Use more reinforcing steel or less, as required to carry the load. Beams of uniform size are less expensive for many of the same reasons listed above for columns. Beam sizes need change only if loads and moments change by 30% to 50% or more.

- Make beams at least 2 in. wider than columns on each side, as shown in Fig. 14.4, so that column and beam bars pass each other without interference. Beams should not be narrower than columns because forms for the column-beam connection will be expensive.

- Use wide shallow beams to minimize floor-framing depth. This will simplify installation of mechanical piping, duct work, and equipment without interference with the structural frame. Coordination between trades takes everyone's time—architects, engineers, contractors, and suppliers. Lower floor-to-floor heights will reduce the cost of walls and vertical building elements, such as stairs and elevators. It is easier to place steel in a wide, shallow beam than in a deep, narrow beam. Furthermore, pipes and even duct work can pass vertically through a wide beam, whereas such penetrations may be impossible in a narrow beam. Rerouting pipes and ducts around beams is part of the cost of construction.*

- Use one deck depth, as in a flat plate or in a pan slab, where beams are the same depth as joists. The contractor can use one length of shoring, and fewer problems will be experienced in coordinating concrete construction with mechanical trades.

- Consider "flying forms" if the structural frame is large and many typical bays can be made exactly the same. A flying form is a form for an entire bay or portion of a bay built as a unit, lowered from the hardened concrete slab after the slab has gained sufficient strength, and moved, or "flown," to the next higher level. At least 8 or 10 reuses of each form are necessary so that the

*It is also a major headache and cost to the design team.

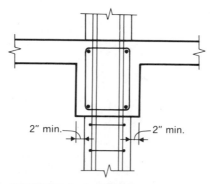

FIGURE 14.4 Avoid beam-column bar interferences.

savings of labor in reuse will exceed the extra cost of the form. For maximum efficiency, the entire building slab should be formed with flying forms.

- Consider stock lumber and plywood sizes in sizing beams and columns. Stock materials allow immediate starts, reduce worker learning curves, reduce job-site errors and lower material costs.

- For pan slabs and waffle slabs, use standard stock sizes of pans and domes.

- Use one size pan throughout a project. (Pricing of pan slab forms is based on the number of reuses and the floor area, not on the surface contact area of the pans.)

- Use no coffers or recesses in the underside of a slab unless there is a substantial advantage in doing so. The amount of concrete saved usually will not offset the additional cost of formwork.

- If depressions are required in the top of the slab for terazzo, ceramic tile, or other floor finishes, keep the bottom of the slab flush, at least in the bay in which depressions are required.

- Eliminate drop panels in a flat slab and use a flat plate instead. If column capitals are required, make all capitals the same size.

- Avoid offsets, brackets, corbels, pilasters, haunches, and all other interruptions to the smooth surface of formwork (see Fig. 14.5).

- Avoid small changes in concrete outline. Instead, make up differences in finish materials, such as masonry, partitions, and floor toppings.

- Allow for reasonable tolerances in constructing concrete so that windows, doors, walls, partitions, and other abutting construction will fit with ease.

- Plan the procedures for reshoring to minimize the amount required and simplify its placement.

Making a project simpler to build not only reduces costs but also helps ensure

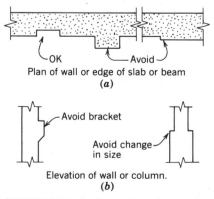

Plan of wall or edge of slab or beam
(a)

Elevation of wall or column.
(b)

FIGURE 14.5   Avoid expensive formwork.

that fewer construction mistakes will be made. Fewer mistakes mean that the project is more likely to be built according to the engineer's intent.

PROBLEMS

1. In the urban area where you live, what are the current costs of the following:
   A. One ton of fabricated reinforcing steel delivered to the job site.
   B. Placing or setting a ton of reinforcing steel.
   C. One yd^3 of concrete with design strength of f'_c = 3 ksi, delivered to the job site.
   D. Placing and consolidating 1 yd^3 of concrete, (a) in a first story column, (b) in a slab on grade, and (c) in a floor slab 10 stories above grade.
2. How much concrete is required in an interior bay 18 × 22 ft c/c of columns, when the framing system is a (A) flat plate, (B) flat slab, (C) pan slab with wide beams the same depth as the pans, and (D) waffle slab? If concrete costs $75 per yd^3, what is the difference in cost between the scheme using the least concrete and the scheme using the most concrete?
3. Given the building layout for Ex. 13.1 (Fig. 13.8), propose structural framing schemes other than the one shown. Columns may be relocated, except that the 30-ft bay in the center of the building must be free of columns. Prepare a preliminary cost estimate for a typical bay or panel for one of the schemes.

Selected References

1. The CRSI Committee on Fire Ratings. *Reinforced Concrete Fire Resistance*. Concrete Reinforcing Steel Institute, Chicago, 1980.

APPENDIXES

APPENDIX A

DESIGN TABLES AND FIGURES

TABLE A1.1 Strength Reduction Factor ϕ Applied to Nominal Strength in Flexure, Axial Load, Shear, and Torsion

Load or Stress Condition	ϕ
Flexure, without axial load	0.90
Axial tension and axial tension with flexure	0.90
Axial compression and axial compression with flexure;[a,b]	
Members with spiral reinforcement	0.75
Other members	0.70
Shear and torsion	0.85
Bearing on concrete	0.70
Flexure in plain concrete	0.65

[a] Both axial compression and flexure are multiplied by $\phi = 0.75$, or 0.70, as appropriate.

[b] When axial compression is less than 0.10 $f_c'A_g$, ϕ may be increased linearly to 0.90 as the axial compression approaches zero (see Chapter 9, Columns).

TABLE A1.2 Concrete Protection for Reinforcement

	Minimum Cover (in.)
Concrete cast against and permanently exposed to earth	3
Concrete exposed to earth or weather	
#6 through #18 bars	2
#5 bars and smaller	1½
Concrete not exposed to weather or in contact with earth	
Slabs, walls, joists	
#14 and #18 bars	1½
#11 bars and smaller	¾
Beams and columns	
primary reinforcement, ties, stirrups, and spirals	1½

TABLE A2.1 Reinforcement Ratio (%) and Maximum Depth of Rectangular Stress Block[a]

f_y (ksi)	f'_c (ksi)	3	4	5	6
	(β_1)	(0.85)	(0.85)	(0.80)	(0.75)
40	$0.75\rho_b$	2.78%	3.71%	4.37%	4.91%
	a/d	0.44	0.44	0.41	0.38
50	$0.75\rho_b$	2.06%	2.75%	3.24%	3.64%
	a/d	0.40	0.40	0.38	0.36
60	$0.75\rho_b$	1.60%	2.14%	2.52%	2.83%
	a/d	0.38	0.38	0.35	0.33

[a] For singly reinforced beams, reinforcement ratio at 75% of balanced conditions $0.75\ \rho_b$ and maximum depth of rectangular stress block a/d at $0.75\rho_b$.

TABLE A2.2 Values of d'/d at Which A'_s Will Reach f_y[a]

$f_y =$	40	50	60
$d'/d =$	0.277	0.203	0.138

[a] These values must not be exceeded.

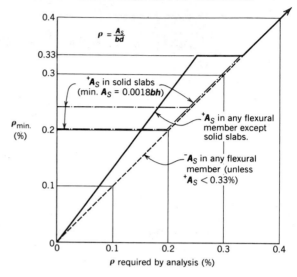

FIGURE A2.1 Minimum reinforcement in flexural members, for $f_y = 60$ ksi.

TABLE A2.3 Values of $(\phi f_y j)/12$[a]

	At $\rho = 0\%$	At 75% balanced steel ratio	
For steel strength $f_y = 40$ ksi			
	(all f'_c)	($f'_c = 3$, $0.75\ \rho_b = 2.78\%$) ($f'_c = 4$, $0.75\ \rho_b = 3.71\%$)	($f'_c = 6$, $0.75\ \rho_b = 4.91\%$)
$(\phi f_y j)/12 =$	3.0	2.34	2.42
For steel strength $f_y = 60$ ksi			
	(all f'_c)	($f'_c = 3$, $0.75\ \rho_b = 1.60\%$) ($f'_c = 4$, $0.75\ \rho_b = 2.14\%$)	($f'_c = 6$, $0.75\ \rho_b = 2.83\%$)
$(\phi f_y j)/12 =$	4.5	3.65	3.75

[a] When moment is in foot-kips and effective depth d is in inches.

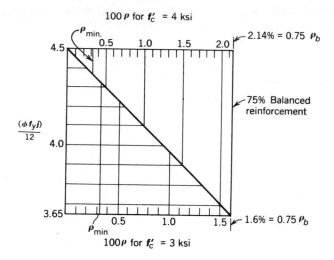

FIGURE A2.2   Values of $(\phi f_y j)/12$ for $f_y = 60$ when moment is in ft-kips and effective depth d is in inches.

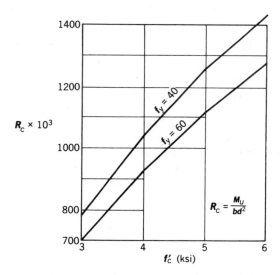

FIGURE A2.3   Resisting moment capacity of concrete R_c in ksi × 10³.

TABLE A2.4   Resisting Moment Capacity of Concrete (R_c)

	R_c [a]	
	f_y	
f'_c	60	40
3	702	783
4	936	1044
5	1116	1248
6	1273	1427

[a] R_c in ksi × 10³.

TABLE A2.5 Maximum Bar Spacing for Crack Control (in.)[a]

Bar Size (no.)	2 in. cover Exposure Conditions				1½ in. cover[b] Exposure Conditions	
	Interior No. of layers		Exterior No. of layers		Interior No. of layers	Exterior No. of layers
	1	2	1	2	1	1
3	12.0	—	6.8	—	18	11.5
4	11.3	8.5	6.4	4.8	18	10.7
5	10.7	7.9	6.1	4.5	17.5	9.9
6	10.2	7.4	5.8	4.2	16.3	d
7	9.7	7.0	5.5	4.0	15.3	d
8	9.2	6.6	5.2	3.7	14.3	d
9	8.7	6.2	5.0	3.5	13.5	d
10	8.3	5.8	4.7	3.3	12.6	d
11	7.8	5.4	4.5	3.1	11.8	d
14	7.1	4.8	4.0	c	10.4	d
18	5.9	c	c	c	8.3	d

[a] Assumptions:
- $f_y = 60$ ksi and $f_s = 0.6f_y$
- Cover to outermost layer of bars
- All bars are the same size
- When bars are in two layers, each layer has the same number of bars

[b] For slabs, spacing should be rounded to the nearest inch.

[c] Maximum spacing for crack control is less than minimum spacing for clearance between bars.

[d] For #6 bars and larger, 2 in. cover is required for exterior exposure.

TABLE A3.1 Cross-Section Properties of Pan Slabs

Joist-slab Size[a]	Wt.[b] (psf)	I_g^c (in.^{4})	Y_b^c (in.)
8 + 3, 5 @ 25	60	1104	7.48
		1582	6.75
5 @ 35	54	1223	7.88
		1914	7.07
6 @ 26	63	1254	7.32
		1709	6.67
6 @ 36	56	1393	7.72
		2051	6.98
8 + 4½, 5 @ 25	79	1630	8.49
		2340	7.74
5 @ 35	72	1813	8.88
		2825	8.08
6 @ 26	82	1852	8.33
		2528	7.64
6 @ 36	75	2058	8.73
		3028	7.99
10 + 3, 5 @ 25	67	1826	8.76
		2594	7.89
5 @ 35	58	2032	9.26
		3145	8.26
6 @ 26	70	2069	8.55
		2801	7.79
6 @ 36	61	2307	9.05
		3366	8.16
10 + 4½ 5 @ 25	85	2561	9.85
		3659	8.93
5 @ 35	77	2841	10.35
		4422	9.34
6 @ 26	89	2906	9.65
		3951	8.82
6 @ 36	80	3227	10.15
		4737	9.23
12 + 3, 5 @ 25	74	2799	9.98
		3951	9.00
5 @ 35	63	3128	10.58
		4790	9.42
6 @ 26	78	3165	9.75
		4264	8.90
6 @ 36	67	3541	10.34
		5124	9.30

Joist-slab Size[a]	Wt.[b] (psf)	I_g^c (in.^{4})	Y_b^c (in.)
12 + 4½, 5 @ 25	92	3797	11.16
		5388	10.09
5 @ 35	82	4219	11.76
		6520	10.57
6 @ 26	97	4300	10.92
		5815	9.97
6 @ 36	85	4783	11.53
		6979	10.44
14 + 3, 5 @ 25	81	4056	11.18
		5700	10.10
5 @ 35	68	4549	11.86
		6905	10.56
6 @ 26	86	4576	10.93
		6150	9.99
6 @ 36	72	5135	11.58
		7382	10.43
14 + 4½, 5 @ 25	99	5373	12.42
		7574	11.23
5 @ 35	87	5986	13.13
		9174	11.76
6 @ 26	104	6072	12.15
		8169	11.09
6 @ 36	91	6773	12.85
		9812	11.62
16 + 3, 6 @ 26	94	6342	12.08
		8512	11.07
6 @ 36	78	7127	12.80
		10197	11.54
7 @ 27	99	7029	11.85
		9127	10.96
7 @ 37	83	7890	12.54
		10844	11.42
16 + 4½, 6 @ 26	113	8259	13.36
		11069	12.21
6 @ 36	97	9238	14.14
		13295	12.77
7 @ 27	118	9160	13.11
		11861	12.08
7 @ 37	101	10246	13.88
		14137	12.64

TABLE A3.1 *(continued)*

Joist-slab Size[a]	Wt.[b] (psf)	I_g[c] (in.^{4})	Y_b[c] (in.)
20 + 3, 6 @ 26	111	11092	14.35
		14887	13.23
6 @ 36	91	12469	15.17
		17741	13.74
7 @ 27	118	12277	14.09
		15966	13.11
7 @ 37	96	13769	14.88
		18864	13.61

Joist-slab Size[a]	Wt.[b] (psf)	I_g[c] (in.^{4})	Y_b[c] (in.)
20 + 4½, 6 @ 26	130	14043	15.71
		18725	14.40
6 @ 36	109	15768	16.64
		22454	15.04
7 @ 27	136	15538	15.42
		20057	14.26
7 @ 37	115	17433	16.32
		23861	14.88

[a] Format for size designation is: depth of pan + topping slab, minimum joist width @ spacing center to center.

[b] For normal weight concretes, $w = 150$ (added weight of tapered end pans is neglected). If topping slab is different from 3 or 4½ in., the weight can be estimated by adding or subtracting 12.5 psf per inch of thickness.

[c] First value is the standard section; second value is at a tapered end.

   I_g = gross moment of inertia.

   Y_b = distance from center of gravity C.G. of gross concrete section to bottom fiber.

If topping slab is between 3 and 4½ in. thick, I_g and y_b can be estimated within 2% accuracy by straight-line interpolation between values for 3- and 4½-in. slabs.

TABLE A3.2 Average Weight of Cross Ribs[a]

Pan Depth (in.)	Weight (psf)
8	2
10	3
12	4
14	5
16	6
20	7

[a] For normal weight concrete. Ribs spaced about 12 ft c/c.

TABLE A4.1 Fraction of Unbalanced Moment Transferred by Flexure at Slab-Column Connections[a]

Y/X	0.33	0.50	1	2	3
γ_f	0.72	0.68	0.60	0.51	0.46

$$\gamma_f = \frac{1}{1 + \frac{2}{3}\sqrt{\frac{Y}{X}}}$$

$X = c_2 + d$

$Y = c_1 + d$

[a] Code 13.3.3.2.

TABLE A4.2   Distribution of Total Static Moment (M_o)[a]

For an interior span

Negative factored moment	0.65
Positive factored moment	0.35

For an end span

	1	2	3	4	5
			Slab Without Beams between Interior Supports		
	Exterior Edge Unrestrained	Slab with Beams between All Supports	Without Edge Beam	With Edge Beam	Exterior Edge Fully Restrained
Interior negative factored moment	0.75	0.70	0.70	0.70	0.65
Positive factored moment	0.63	0.57	0.52	0.50	0.35
Exterior negative factored moment	0	0.16	0.26	0.30	0.65

[a] Code 13.6.3.

TABLE A4.3 Values of α_{min} to Determine Whether Positive Factored Moments Must Be Increased[a]

	Aspect Ratio	Relative Beam Stiffness (α)				
β_a	ℓ_2/ℓ_1	0	0.5	1.0	2.0	4.0
2.0	0.5 – 2.0	0	0	0	0	0
1.0	0.5	0.6	0	0	0	0
	0.8	0.7	0	0	0	0
	1.0	0.7	0.1	0	0	0
	1.25	0.8	0.4	0	0	0
	2.0	1.2	0.5	0.2	0	0
0.5	0.5	1.3	0.3	0	0	0
	0.8	1.5	0.5	0.2	0	0
	1.0	1.6	0.6	0.2	0	0
	1.25	1.9	1.0	0.5	0	0
	2.0	4.9	1.6	0.8	0.3	0
0.33	0.5	1.8	0.5	0.1	0	0
	0.8	2.0	0.9	0.3	0	0
	1.0	2.3	0.9	0.4	0	0
	1.25	2.8	1.5	0.8	0.2	0
	2.0	13.0	2.6	1.2	0.5	0.3

Values of α_{min}

[a] Code 13.6.10.

TABLE A4.4 Increase ^+M by Multiplier δ_s[a]

β_a	2.0	1.5	1.0	0.5	0.33
δ_s	1.0	1.09	1.20	1.33	1.38

[a] Assuming columns of zero stiffness, when the ratio of dead load to live load β_a is less than 2.0 (Code 13.6.10).

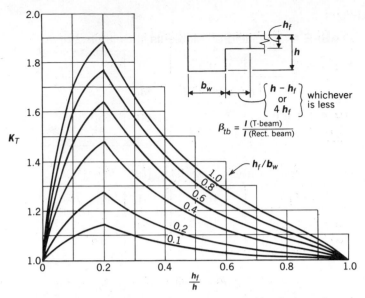

FIGURE A4.1 Ratio of moment of inertia for T-beam to moment of inertia for rectangular beam, β_{tb} for flange on one side of beam only.

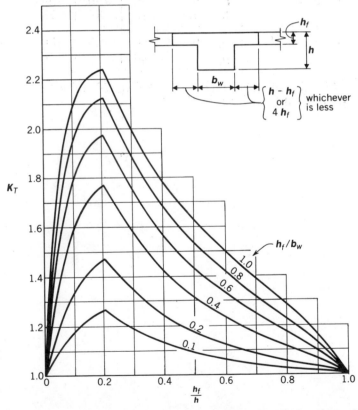

FIGURE A4.2 Ratio of moment of inertia for T-beam to moment of inertia for rectangular beam, β_{tb} for flange on both sides of beam.

471

TABLE A4.5 Factored Moments in Column Strips[a]

ℓ_2/ℓ_1	0.5	1.0	2.0

Positive moments, (^+M_u)

ℓ_2/ℓ_1	0.5	1.0	2.0
$(\alpha_1\ell_2/\ell_1) = 0$	60	60	60
$(\alpha_1\ell_2/\ell_1) \geq 1.0$	90	75	45

Interior negative moments, (*Int.* ^-M_u)

ℓ_2/ℓ_1	0.5	1.0	2.0
$(\alpha_1\ell_2/\ell_1) = 0$	75	75	75
$(\alpha_1\ell_2/\ell_1) \geq 1.0$	90	75	45

Exterior negative moments (*Ext.* ^-M_u)

ℓ_2/ℓ_1		0.5	1.0	2.0
$(\alpha_1\ell_2/\ell_1) = 0$	$\beta_t = 0$	100	100	100
	$\beta_t \geq 2.5$	75	75	75
$(\alpha_1\ell_2/\ell_1) \geq 1.0$	$\beta_t = 0$	100	100	100
	$\beta_t \geq 2.5$	90	75	45

[a] Percent of total factored negative or positive moment in panel (Code 13.6.4).

TABLE A4.6 Distribution of Total Static Moment
($M_{oo} = w_u \ell_n^2 / 8$) for DDM-Short
Procedure 2

	^+M	Interior ^-M	Exterior ^-M
Interior span			
Column strip	0.42	0.98	—
Middle strip	0.28	0.33	—
End span restrained[a]			
Column strip			
Without beams	0.62	1.05	0.52
With spandrel beams	0.60	1.05	0.45
Middle strip			
Without beams	0.42	0.35	0
With spandrel beams	0.40	0.35	0.15
End span unrestrained[b]			
Column strip	0.76	1.13	0
Middle strip	0.50	0.38	0

[a] Exterior support designed to resist slab exterior negative moment.

[b] Exterior support provides negligible restraint to slab such as slabs resting on top of masonry walls.

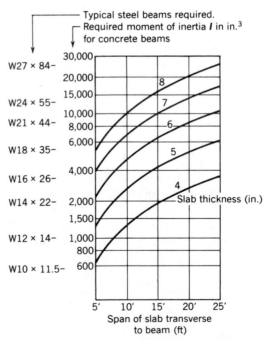

FIGURE A4.3 Required moment of inertia of supporting beams I_b for use of ACI318-63 Method 2 design of 2-way slabs.

TABLE A5.1   Basic Development Length ℓ_{db} (in.)

Bar Diam. (no.)	In Tension (Code 12.2) f'_c, (ksi)				In Compression (Code 12.3) f'_c, (ksi)		
	3	4	5	6	3	4	> 4.44
3	9 [a]	9 [a]	9 [a]	9 [a]	8.2	7.1[b]	6.8[b]
4	12	12	12	12	10.6	9.5	9
5	15	15	15	15	13.7	11.9	11.3
6	19.3	18	18	18	16.4	14.2	13.5
7	26.3	22.8	21	21	19.2	16.6	15.8
8	34.6	30.0	26.8	24.5	21.9	19	18
9	43.8	38.0	33.9	31	24.7	21.4	20.3
10	55.6	48.2	43.1	39.4	27.8	24.1	22.9
11	68.4	59.2	53	48.3	30.9	26.8	25.4
14	93.1	80.6	72.1	65.8	37.1	32.1	30.5
18	120.5	104.4	93.3	85.2	49.4	42.8	40.6

Table Assumptions	Other Conditions	Factor[c]
$f_y = 60$	$f_y < 60$	$f_y/60$
	$f_y > 60$, in tension	$(2 - 60/f_y)$
	$f_y > 60$, in compression	$f_y/60$
Bottom bars	Top bars in tension	1.4
Normal-weight concrete	All-lightweight concrete	
	Bars in tension	1.33
	Sand-lightweight concrete	
	Bars in tension	1.18
Spacing < 6 in. c/c	Spacing 6 in. c/c or more and 3 in. side clearance	
	Bars in tension	0.8
No excess steel	Excess steel	$A_s$reqd./$A_s$prov.
No spiral enclosure	With spiral enclosure	0.75

[a] Not less than 12 in., except for computation of lap splices and development of web reinforcement.

[b] Not less than 8 in.

[c] Multiply the tabulated basic development length ℓ_{db} by these factors to obtain the development length ℓ_d for the other listed conditions.

TABLE A5.2   Special Splice Requirements for Columns (Code 12.17)

Calculated Factored Load Stress in Longitudinal Bars	Splice Requirements
f_s = compression or tension $f_s \leq 0.5f_y$	Develop 200% of tension force in that face of column, with splices and continuing unspliced bars[a]
Tension $f_s > 0.5f_y$	Develop f_y in tension

[a] In any case, tensile strength in each face of the column must equal 1/4 of area of vertical steel in that face multiplied by f_y.

TABLE A5.3   Column Splices Required for Typical Conditions and f'_c = 3 ksi

Condition[a]	Splice Type[b]	Required Stagger for Alternating Bar Splices[c] (as a ratio of lap length in tension)
	Case I	
No tension	Compression lap	0
	End bearing	0.50
	Case II	
Tension ≤ 0.25 A_sf_y	Compression lap for #8 bars or smaller	0[d]
	Compression lap for #9, 10, and 11 bars	0.30
	End bearing	1.00

TABLE A5.3 *(continued)*

Condition[a]	Splice Type[b]	Required Stagger for Alternating Bar Splices[c] (as a ratio of lap length in tension)
Case III		
Tension > 0.25 $A_s f_y$	Tension lap[e]	0
	Mechanical	0

[a] The conditions are stated in terms of the calculated factored tension force in one face of the column, where A_s is the area of bars in the face of the column under consideration.

[b] *Compression lap* is a lap of a length tabulated in Table A5.4.

The effectiveness of a compression lap in tension is the ratio of required compression lap length to required tension lap length.

Tension lap is a lap of a length tabulated in Table A5.4. The effectiveness of a tension lap is 100%.

End bearing is an alignment splice transmitting compressive forces only by end bearing on the bars. Its effectiveness in tension is nil.

Mechanical is a welded splice or mechanical connection developing 125% of yield strength of the bar. Its effectiveness in tension is 100%.

[c] At any one section, no more than 50% of bars are spliced if stagger is required. In some circumstances, the tabulated amount of stagger is conservative. Computations may yield a shorter stagger distance, but the design effort may yield no construction cost savings. In no case should the stagger distance be less than 24 in. (Code 12.15.4.1).

[d] For lightweight concrete, stagger laps $0.30\ell_d$ for #7 bars and larger.

[e] When tension force > $0.50 A_s f_y$, the tension lap must be a class C splice = $1.7\ell_d$.

TABLE A5.4   Column Lap Splices[a]

Bar Size (no.)	Tension Lap Length[b] f'_c, (ksi)				Compression Lap Lengths
	3	4	5	6	($f'_c \geq 3$ ksi)
4	1'-4"	1'-4"	1'-4"	1'-4"	1'-3"
5	1'-8"	1'-8"	1'-8"	1'-8"	1'-7"
6	2'-1"	1'-11"	1'-11"	1'-11"	1'-11"
7	2'-10"	2'-6"	2'-3"	2'-3"	2'-2"
8	3'-9"	3'-3"	2'-11"	2'-8"	2'-6"
9	4'-9"	4'-1"	3'-8"	3'-4"	2'-10"
10	6'-0"	5'-3"	4'-8"	4'-3"	3'-2"
11	7'-6"	6'-5"	5'-9"	5'-3"	3'-6"
14	NP[d]	NP	NP	NP	NP
18	NP	NP	NP	NP	NP

[a] Assumptions:
- $f_y = 60$ ksi
- Tied columns
- Class B tension lap splices = $1.3\ell_d$ are required. See Table 5.1 for conditions in which longer splices are required.
- Normal-weight concrete
- Bar spacing < 6" c/c
- Vertical members

For conditions not meeting these assumptions, compute lap splice length.

[b] Tension lap length = $1.3\ell_d$. Tension lap lengths may be reduced 25% if used in spiral reinforced columns (Code 12.2.4.3).

[c] Compression lap length = $30d_b$.

[d] NP = not permitted.

TABLE A5.5   Temperature Steel Lap Splices[a]

Bar Size (no.)	Lap Length	
	Bottom[b]	Top[c]
3	1'-0"	1'-1"
4	1'-0"	1'-6"
5	1'-4"	1'-10"
6	1'-8"	2'-4"
7	2'-3"	3'-2"

[a] Assumptions:
- $f_y = 60$ ksi
- $f'_c = 3$ ksi
- Normal weight concrete
- All bars spliced at same location (class B splice)
- Bars spaced 6 in. c/c or more
- Reinforcement not required by analysis

[b] Vertical wall bars and most slab bars are considered bottom bars with 12 in. of concrete or less below the bar.

[c] Horizontal bars in walls are considered top bars.

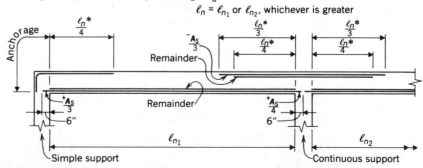

*Length of top bars $\geq$ development length ℓ_d

$\ell_n = \ell_{n_1}$ or ℓ_{n_2}, whichever is greater

FIGURE A5.1   Typical cutoff for beam and one-way slab bars. Where $\ell_n/d \geq 16$, assuming distance to point of inflection $\leq \ell_n$ 4.

TABLE A5.6 Minimum Clear Span Length of Flexural Members in Which Negative and Positive Moment Bars Can Be Developed[a]

Max. Bar Size (no.)	^-M bars[b] (ft)	^+M bars[c] (ft)
3	4½	3½
4	6	4½
5	7½	5½
6	9	7
7	10½	9½
8	13	12
9	16	15½
10	20	20
11	24	24
14	33	33
18	43	43

Assumptions:
- $f_y = 60$ ksi[d]
- $f_c' = 3$ ksi[d]
- Normal-weight aggregate concrete[e]
- All reinforcement required by analysis[d]
- > 12 in. concrete below negative moment bars[f]
- Bar spacing < 6 in. c/c[d]
- $\ell_n/d \leqslant 24$[g]
- Ends of positive moment bars are not confined by compressive reaction[h]
- At discontinuous ends of spans, positive moment bars do not extend beyond centerline of support[d]
- Uniform loads where unit live load does not exceed three times unit dead load and the longer of two adjacent spans does not exceed the shorter by more than 20% (see text for other conditions).

[a] This table has been prepared for the normal situation in which negative moment bars are in the top of the member with more than 12 in. of fresh concrete cast below the bar, and positive moment bars are in the bottom of the member with less than 12 in. of concrete below the bar. Where moment conditions are reversed, as in continuous footings and certain grade beams, divide the minimum span length for negative moment bars by 1.4 and multiply the minimum span length for positive moment bars by 1.4.

[b] Minimum span length for negative moment bars controlled by the larger of:
- $\ell_n/3 \geqslant 1.4\,\ell_d$ for #8 bars and larger in beams
- $\ell_n (\tfrac{1}{3} - \tfrac{1}{4}) \geqslant 12\,d_b$ for #3 to #7 bars in beams and for all bar sizes in slabs.

[c] Minimum span length for positive moment bars controlled by $\ell_n \geqslant 4.22\ell_d$.

[d] This table is conservative for lower strength steel, higher strength concrete, extra reinforcement not required by analysis, bar spacings wider than 6 in. c/c, and for bars extending beyond the centerline of support at discontinuous ends.

[e] For lightweight aggregate concrete, multiply the minimum span lengths by 1.33 for all-lightweight and 1.18 for sand-lightweight concrete.

[f] For negative moment bars with less than 12 in. of fresh concrete cast below the bar, divide minimum span length by 1.4.

[g] For shallower members with at least one end continuous, add to the minimum spans listed an amount equal to $4.2(d - \ell_n/24)$.

[h] If ends of positive moment bars are confined by a compressive reaction, divide minimum span length by 1.3.

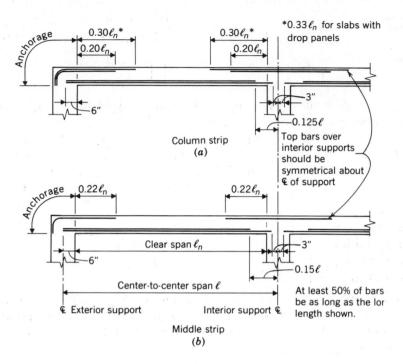

*0.33ℓ_n for slabs with drop panels

Top bars over interior supports should be symmetrical about ℄ of support

At least 50% of bars be as long as the lor length shown.

FIGURE A5.2   Minimum bar lengths for two-way slabs without beams. [Cod

TABLE A5.7   Minimum Depth of Beam for Stirrup Size and Type (in.)[a]

Bar Size (no.)	Std. Hook		135° Hook plus Long. Bars		Lapped Stirrups	
	$f_y = 60$	$f_y = 40$	$f_y = 60$	$f_y = 40$[b]	$f_y = 60$	$f_y = 40$
3	17	14	14	12	18	16
4	21	17	17	14	27	18
5	25	20	20	17	33	24
6	31	25	NP	NP[c]	42	31

[a] Assumptions:
- 1½-in. clear concrete cover over stirrups, top and bottom
- $f_c' = 3$ ksi
- Normal weight concrete
- Spacing < 6 in. c/c

[b] There are no anchorage requirements in the ACI Code for stirrups with a 135° hook and $f_y \le 40$ ksi (Code 12.13.2.3). Depths given are computed using the same procedure as that used for stirrups with $f_y = 60$ ksi.

[c] NP = Not permitted

TABLE A5.8   Width of Standard Hooks (in.)[a]

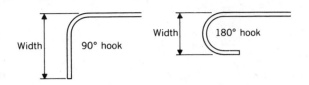

Bar Size (no.)	Width of Hook		Inside Diameter of Bend (d_b)
	90° hook	180° hook	
4	8	4	
5	10	5	
6	12	6	6
7	14	7	
8	16	8	
9	20	12	
10	22	13	8
11	24	15	
14	31	21	10
18	41	27	

[a] Rounded upward to the nearest inch.

TABLE A5.9 Development Length of Standard Hooks (in.)[a]

Bar Size (no.)	$f'_c = 3$ ksi			$f'_c = 4$ ksi		
	A	B	C	A	B	C
			ℓ_{dh}			ℓ_{dh}
	ℓ_{hb}	$0.7\ell_{hb}$[b]	$0.7 \times 0.8\ell_{hb}$[c]	ℓ_{hb}	$0.7\ell_{hb}$[b]	$0.7 \times 0.8\ell_{hb}$[c]
3	8.2	6 [d]	6 [d]	7.1	6 [d]	6 [d]
4	11	7.7	6.1	9.5	6.6	6 [d]
5	13.7	9.6	7.7	11.9	8.3	6.6
6	16.4	11.5	9.2	14.2	10.0	8.0
7	19.2	13.4	10.7	16.6	11.6	9.3
8	21.9	15.4	12.3	19.0	13.3	10.6
9	24.7	17.3	13.8	21.4	15.0	12.0
10	27.8	19.5	15.6	24.1	16.9	13.5
11	30.9	21.6	17.3	26.7	18.7	15.0
14	37.1	—	—	32.1	—	—
18	49.5	—	—	42.8	—	—

[a] ℓ_{hb} = basic development length of a standard hook in tension.

ℓ_{dh} = development length of a standard hook in tension,

= ℓ_{hb} times applicable modification factors.

Hook basic development length must also be multiplied by the following factors, where applicable:

(1) bars with yield strength f_y other than 60 ksi, $\qquad\qquad\qquad f_y/60$

(2) where anchorage or development for f_y is not specifically required and reinforcement is in excess of that required by analysis, ($A_s$reqd./$A_s$prov.)

(3) For lightweight concrete, $\qquad\qquad\qquad\qquad\qquad\qquad\qquad\qquad$ 1.3

[b] The basic development length ℓ_{hb} for #11 bars and smaller is multiplied by 0.7 when side cover (normal to plane of hook) is not less than $2\frac{1}{2}$ in., and for a 90° hook, cover on the bar extension beyond the hook is not less than 2 in.

[c] The basic development length for #11 bars and smaller is multiplied by 0.8 when the hook is enclosed vertically and horizontally within ties spaced along the full development length. Spacing of ties must not exceed 3 times the diameter of the hooked bar.

[d] Hook development length must not be less than 6 in. nor 8 bar diameters, whichever is greater. The latter requirement controls only when $f'_c > 7$ ksi.

TABLE A6.1 Maximum Permissible Shear Stress Provided by Concrete and by Reinforcement in Shallow Beams (psi)

f_c'	v_c $\phi 2\sqrt{f_c'}$	v_s $\phi 8\sqrt{f_c'}$	Total v_n $v_c + v_s$ $\phi 10\sqrt{f_c'}$
3	93	372	465
4	108	430	538
5	120	481	601

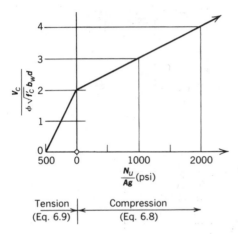

FIGURE A6.1 Minimum concrete shear capacity under axial load.

TABLE A6.2 Maximum Stirrup Spacing When $v_u \le \phi 6\sqrt{f_c'}$ (in.)[a]

	Spacing (in.)[b]					
	$f_y = 60$			$f_y = 40$		
b_w[c]	#3 264	#4 480	#5 744	#3 176	#4 320	#5 496
8				22		
10	24			18		
12	22			15	24	
14	19			13	23	
16	16			11	20	
18	15	24		10	18	
20	13	24		9	16	24
22	12	22		8	15	23
24	11	20	24	7	13	21

[a] When $v_u > \phi 6\sqrt{f_c'}$, the maximum spacing is one-half the tabulated spacing.

[b] Spacings above solid line are limited to 24 in.

[c] For other beam widths, the maximum spacing is the spacing coefficient given on this line divided by the beam width. For example, the maximum spacing **s** for a #5 stirrup in a 32-in. wide beam with $f_y = 60$ is

$$s = 744/32 = 23 \text{ in.}$$

TABLE A6.3 Shear Stress Capacity v_s for Minimum Shear Reinforcement in Deep Beams (psi)

ℓ_n/d	$f_y = 60$	$f_y = 40$
0	123	82
1	119	79
2	115	76
3	110	74
4	106	71
5	102	68

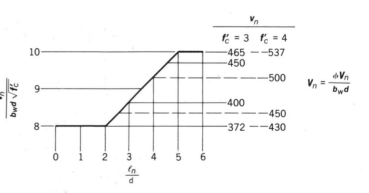

FIGURE A6.2 Max v_n in deep beams (Eq. 6.14).

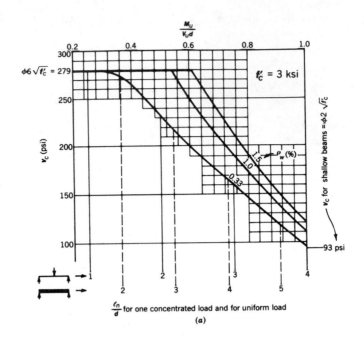

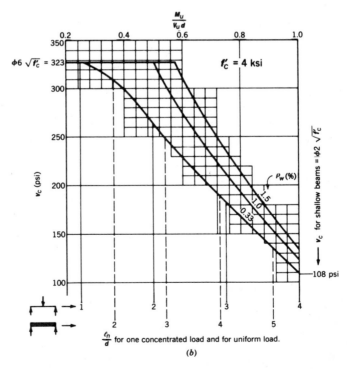

FIGURE A6.3 Concrete shear capacity v_c in deep beams as a function of the span to depth ratio ℓ_n/d (Eq. 6.15).

TABLE A6.4   Coefficient of Friction μ

Concrete placed monolithically	1.4λ[a]
Concrete placed against hardened concrete with surface intentionally roughened to a full amplitude of approximately ¼ in.	1.0λ
Concrete placed against hardened concrete not intentionally roughened.	0.6λ
Concrete anchored to as-rolled structural steel by headed studs or by reinforcing bars.	0.7λ

[a] $\lambda = 1.0$ for normal-weight concrete, 0.85 for sand-lightweight concrete, and 0.75 for all-lightweight concrete. Linear interpolation may be applied when partial sand replacement is used.

TABLE A6.5   Values of $C_t h$ for Rectangular Beams for Use in Torsion Equations

	h/b_w			
d/h	≤ 1	2	3	4
0.8	0.8	1.6	2.4	3.2
0.85	0.85	1.7	2.55	3.4
0.9	0.9	1.8	2.7	3.6
0.95	0.95	1.9	2.85	3.8
1.0	1.0	2.0	3.0	4.0

$C_t h = \dfrac{d}{h} \times \dfrac{h}{b_w}$ for $h > b_w$

$C_t h = \dfrac{d}{h}$ for $h < b_w$

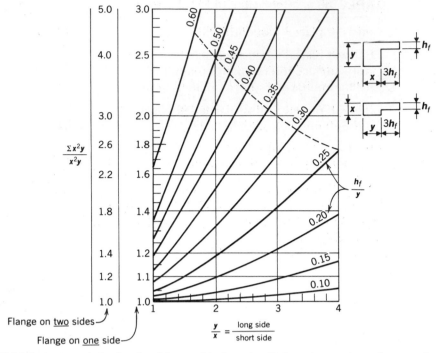

FIGURE A6.4 Increase in torsional resistance of L-shaped or T-shaped concrete section over that of a rectangular section.

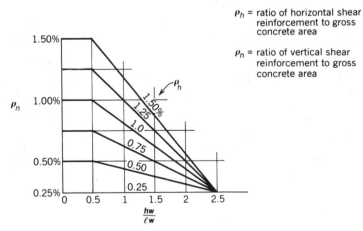

ρ_h = ratio of horizontal shear reinforcement to gross concrete area

ρ_n = ratio of vertical shear reinforcement to gross concrete area

FIGURE A6.5 Ratio of vertical shear reinforcement area to gross concrete area of horizontal section in a shear wall (Eq. 6.43).

TABLE A6.6 Torsional Moment minT_u in Rectangular Beams (ft-kips) Below Which Torsion May Be Ignored[a,b,c]

			x = short side			
y^d	12″	16″	20″	24″	28″	32″
12″	3.4					
16″	4.5	7.9				
20″	5.6	9.9	15.5			
24″	6.7	11.9	18.6	26.8		
28″	7.8	13.9	21.7	31.3	42.6	
32″	8.9	15.9	24.8	35.7	48.7	63.6

[a] For $f_c' = 3$ ksi. Multiply tabulated values by $\sqrt{f_c'/3}$ for concrete strengths other than $f_c' = 3$ ksi.

[b] Beams with flanges can accept a larger minT_u.

[c] Min$T_u = 0.5\phi\sqrt{f_c'}\Sigma x^2 y$. The maximum compatibility torsional moment for which a member must be designed is 2.67 times tabulated values (after adjusting for effect of flanges, if any).

[d] y is the long side of the beam web and may be either the width b_w or the depth h.

TABLE A6.7 Shear Strength ϕV_n of Shear Walls (k/ft)

	$f_c' = 3$			$f_c' = 4$		
Wall Thickness	Min. Reinf.[a]		Max.[b]	Min. Reinf.[a]		Max.[b]
	$f_y = 60$	$f_y = 40$		$f_y = 60$	$f_y = 40$	
8	16.9	13.7	35.8	18.0	14.8	41.3
10	21.1	17.1	44.7	22.6	18.5	51.6
12	25.4	20.5	53.6	27.1	22.2	61.9
14	29.7	23.9	62.6	31.6	25.9	72.2
16	33.9	27.4	71.5	36.1	29.6	82.6

[a] Using minimum reinforcement.

[b] Maximum shear strength with additional reinforcement.

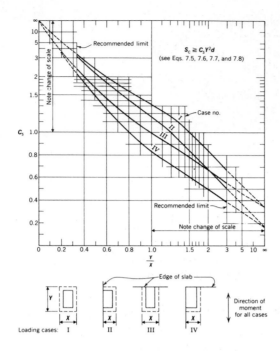

FIGURE A7.1   Analogous section modulus of slab shear perimeter for rectangular columns.

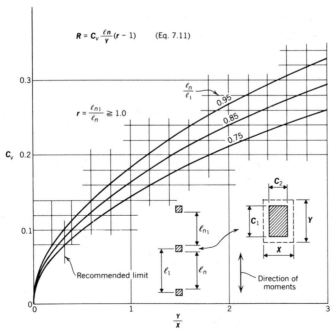

FIGURE A7.2   Ratio of punching shear stress caused by moment transfer M_t to stress caused by shear force (Eq. 7.11), for uniform loading.

TABLE A8.1 Maximum Permissible Computed Deflections

Type of Member	Deflection to Be Considered	Deflection Limitation
Flat roofs not supporting or attached to nonstructural elements likely to be damaged by large deflections	Immediate deflection due to live load **L**	$\dfrac{\ell^{\mathrm{a}}}{180}$
Floors not supporting or attached to nonstructural elements likely to be damaged by large deflections	Immediate deflection due to live load **L**	$\dfrac{\ell}{360}$
Roof or floor construction supporting or attached to nonstructural elements likely to be damaged by large deflections	That part of the total deflection occurring after attachment of nonstructural elements (sum of the long-time deflection due to all sustained loads large deflections and the immediate deflection due to any additional live load)[c]	$\dfrac{\ell^{\mathrm{b}}}{480}$
Roof or floor construction supporting or attached to nonstructural elements not likely to be damaged by large deflections		$\dfrac{\ell^{\mathrm{d}}}{240}$

[a] Limit not intended to safeguard against ponding. Ponding should be checked by suitable calculations of deflection, including added deflections due to ponded water, and considering longtime effects of all sustained loads, camber, construction tolerances, and reliability of provisions for drainage.

[b] Limit may be exceeded if adequate measures are taken to prevent damage to supported or attached elements.

[c] Longtime deflection shall be determined in accordance with ACI Code Section 9.5.2.5 or 9.5.4.2 but may be reduced by amount of deflection calculated to occur before attachment of nonstructural elements. This amount shall be determined on basis of accepted engineering data relating to time-deflection characteristics of members similar to those being considered.

[d] But not greater than tolerance provided for nonstructural elements. Limit may be exceeded if camber is provided so that total deflection minus camber does not exceed limit.

TABLE A8.2   Beam Diagrams and Deflection Formulas for Static Loading
Conditions

Case	$$\Delta = \beta_w \times \frac{w\ell^3}{EI} = \beta_m \times \frac{M\ell^2}{EI}$$	Ratio[a]
I	Simple beam—uniformly distributed load	

$$\Delta = \frac{5}{384} \times \frac{W\ell^3}{EI} = \frac{5}{48} \times \frac{M\ell^2}{EI}$$ 1.00

II   Simple beam—two equal concentrated loads at third points

$$\Delta = \frac{23}{648} \times \frac{P\ell^3}{EI} = \frac{23}{216} \times \frac{M\ell^2}{EI}$$ 1.02

III   Simple beam—concentrated load at center

$$\Delta = \frac{1}{48} \times \frac{P\ell^3}{EI} = \frac{1}{12} \times \frac{M\ell^2}{EI}$$ 0.80

IV   Beam fixed at one end, supported at other—uniformly distributed load

$$\Delta = \frac{1}{185} \times \frac{W\ell^3}{EI} = \frac{128}{1665} \times \frac{^+M\ell^2}{EI}$$ 0.74

V   Beam fixed at both ends—two equal concentrated loads at third points

$$\Delta = \frac{5}{648} \times \frac{P\ell^3}{EI} = \frac{5}{72} \times \frac{^+M\ell^2}{EI}$$ 0.67

VI   Beam fixed at both ends—uniformly distributed loads

$$\Delta = \frac{1}{384} \times \frac{W\ell^3}{EI} = \frac{1}{16} \times \frac{^+M\ell^2}{EI}$$ 0.60

TABLE A8.2   *(continued)*

Case	$$\Delta = \beta_w \times \frac{w\ell^3}{EI} = \beta_m \times \frac{M\ell^2}{EI}$$	Ratio[a]
VII	Beam fixed at one end, supported at other— concentrated load at center $$\Delta = \frac{P\ell^3}{48\sqrt{5}\,EI} = 0.00932\,\frac{P\ell^3}{EI}$$ $$\Delta = \frac{2}{5\sqrt{5}} \times \frac{M\ell^2}{EI} = 0.0596\,\frac{M_a\ell^2}{EI}$$	0.57
VIII	Beam fixed at both ends—concentrated load at center $$\Delta = \frac{1}{192} \times \frac{P\ell^3}{EI} = \frac{1}{24} \times \frac{M\ell^2}{EI}$$	0.40
IX	Cantilever beam—uniformly distributed load $$\Delta = \frac{1}{8} \times \frac{W\ell^3}{EI} = \frac{1}{4} \times \frac{^-M\ell^2}{EI}$$	2.4
X	Cantilever beam—concentrated load at free end $$\Delta = \frac{1}{3} \times \frac{P\ell^3}{EI} = \frac{1}{3} \times \frac{^-M\ell^2}{EI}$$	3.2
XI	Simple beam—moment at one end $$\Delta = \frac{1}{16} \times \frac{M\ell^2}{EI}$$	0.6

[a] Ratio of deflection for the subject case to deflection for a simple beam with uniformly distributed load causing an equivalent maximum moment.

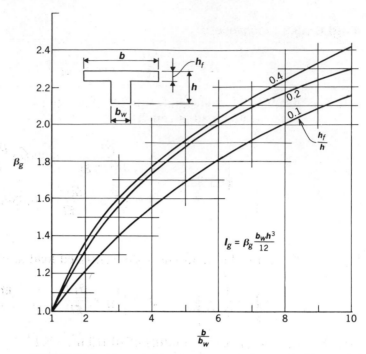

$$I_g = \beta_g \frac{b_w h^3}{12}$$

FIGURE A8.1   Moment of inertia of uncracked concrete T-beams in in⁴.

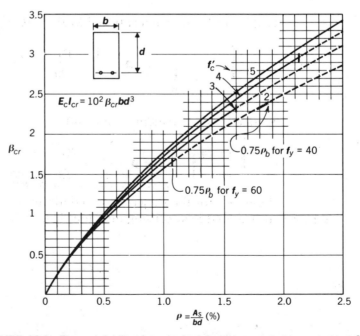

$$E_c I_{cr} = 10^2 \, \beta_{cr} b d^3$$

FIGURE A8.2   Flexural rigidity of cracked reinforced concrete beams in kip-in².

TABLE A8.3 Minimum Thickness of Nonprestressed Beams or One-way Slabs Unless Deflections Are Computed[a]

	Minimum Thickness (**h**)			
Member	Simply Supported	One End Continuous	Both Ends Continuous	Cantilever
	Members Not Supporting or Attached to Partitions or Other Construction Likely to Be Damaged by Large Deflections			
Solid one-way slabs	$\ell/20$	$\ell/24$	$\ell/28$	$\ell/10$
Beams or ribbed one-way slabs	$\ell/16$	$\ell/18.5$	$\ell/21$	$\ell/8$

[a] Span length ℓ is in inches. Values given shall be used directly for members with normal-weight concrete (w_c = 145 pcf) and grade 60 reinforcement. For other conditions, the values shall be modified as follows: (1) For structural lightweight concrete having unit weights in the range 90−120 pcf, the values shall be multiplied by ($1.65 - 0.005 w_c$) but not less than 1.09, where w_c is the unit weight in pcf. (2) For f_y other than 60,000 psi, the values shall be multiplied by ($0.4 + f_y/100,000$).

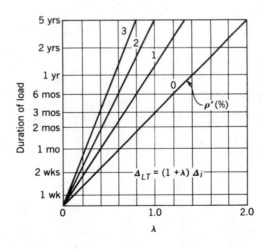

FIGURE A8.3 Values of λ for sustained load.

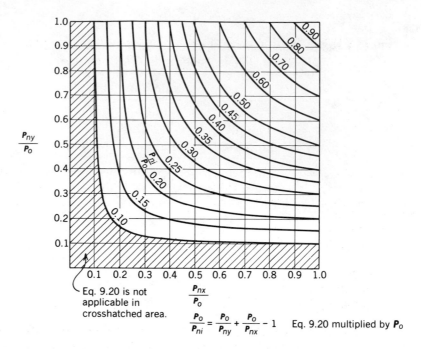

$$\frac{P_o}{P_{ni}} = \frac{P_o}{P_{ny}} + \frac{P_o}{P_{nx}} - 1 \qquad \text{Eq. 9.20 multiplied by } \mathbf{P_o}$$

FIGURE A9.1 Approximate solution for biaxial bending in columns.

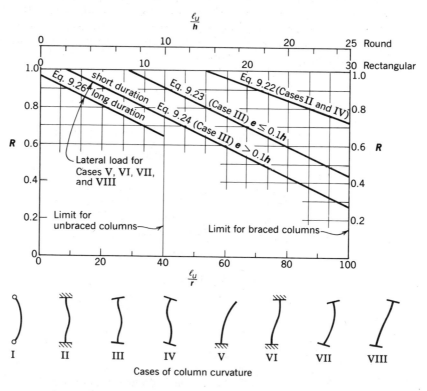

FIGURE A9.2 Strength reduction factors **R** for long columns.

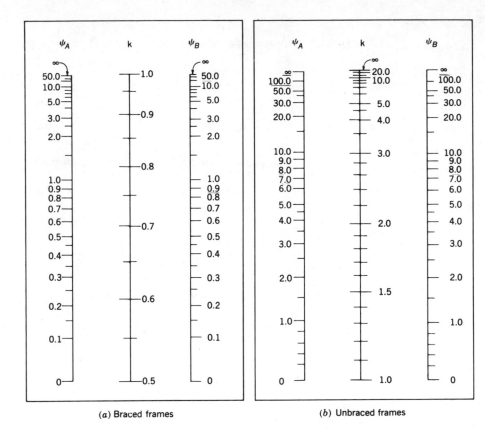

(a) Braced frames   (b) Unbraced frames

FIGURE A9.3 Increased moment in restraining beams, for unbraced columns.

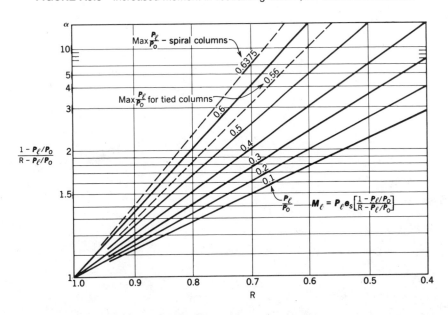

FIGURE A9.4 Effective length factor from Chapter 9, reference 5.

TABLE A9.1 Approximate Values of e_{min}/h for Symmetrical Rectangular Columns[a]

Column depth (in.)[b] d/h		55 0.95	27.5 0.90	18 0.85	14 0.80	11 0.75	9 0.70
Tied columns[c]		0.112	0.100	0.088	0.075	0.062	0.050
Tied columns[d]		0.090	0.080	0.070	0.060	0.050	0.040
Spiral columns		0.068	0.060	0.052	0.045	0.038	0.030

[a] Using Eqs. 9.6, 9.7, and 9.8. This table is usually conservative (see Table 9.1).

[b] Depth in the direction of bending for the tabulated d/h ratio, assuming concrete cover is 2 in. and #11 bar ($d' = h - d = 2\frac{3}{4}$ in.).

[c] Reinforcement in faces parallel to axis of bending.

[d] Reinforcement in faces perpendicular to axis of bending.

TABLE A13.1   Average Weight of Common Construction
Materials

Material	Weight (pcf)	Material	Weight (pcf)
Aluminum	165	Lead	710
Asphalt	80	Sand, gravel	90−105
Concrete,		Steel	490
Normal weight	145	Stone, limestone	160
Lightweight	115 or less	Stone, marble	165
Earth	75−110	Stone, granite	170
Glass	175	Timber	30−45
Gypsum board	75	Water	62.4
Paper, loose	50		

Material	Weight (psf)
Brick, 4 in. thick	40
Built-up roofing	6
Concrete block, hollow, normal weight, and lightweight	
4 in. thick	30 and 20
6 in. thick	40 and 30
8 in. thick	50 and 40
Drywall ⅝ in. thick	4
Stud and drywall partition	10

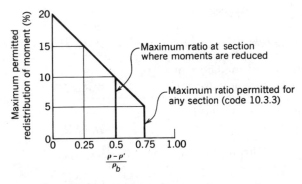

FIGURE A13.1   Redistribution of moments.

TABLE A13.2 Moment Coefficients

Positive moment	
End spans	
Discontinuous end unrestrained	$w_u \ell_n^2 / 11$
Discontinuous end integral with support	$w_u \ell_n^2 / 14$
Interior spans	$w_u \ell_n^2 / 16$
Negative moment at exterior face of first interior support	
Two spans	$w_u \ell_n^2 / 9$
More than two spans	$w_u \ell_n^2 / 10$
Negative moment at other faces of interior supports	$w_u \ell_n^2 / 11$
Negative moment at face of all supports for Slabs with spans not exceeding 10 ft and beams where ratio of sum of column stiffnesss to beam stiffness exceeds eight at each end of the span	$w_u \ell_n^2 / 12$
Negative moment at interior face of exterior support for members built integrally with supports	
Where support is a spandrel beam	$w_u \ell_n^2 / 24$
Where support is a column	$w_u \ell_n^2 / 16$
Shear in end members at face of first interior support	$1.15 \, w_u \ell_n^2 / 2$
Shear at face of all other supports	$w_u \ell_n^2 / 2$

From ACI Building Code, ACI 318-83, Section 8.3.

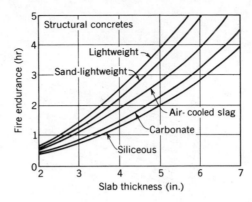

FIGURE A14.1   Fire endurance of concrete slabs—effect of thickness and type of aggregate—based on heat transmission from Chapter 14, reference 1.

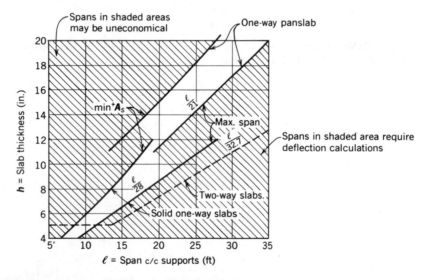

FIGURE A14.2   Preliminary selection of slab thickness for interior span.
- Maximum span based on ACI Code thickness limitations for deflection. See Table A8.3.
- Span for minimum ^+A_s based on:
  - Live load = 60 psf + superimposed dead load = 40 psf
  - Minimum reinforcement = 0.0018 bt for solid slabs
  - Minimum reinforcement $\approx$ 0.0033 $b_w d$ for pan slabs
  - f_y = 60 ksi
  - $^+M = w_u \ell^2/16$

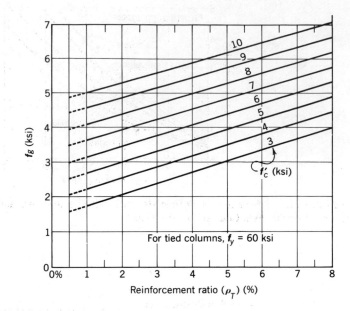

FIGURE A14.3 Average stress f_g in concentrically loaded columns. (Reproduced by courtesy of the Concrete Reinforcing Steel Institute.)

APPENDIX B

NOTATION

a = depth of equivalent rectangular stress block.

a = shear span—distance between concentrated load and face of support.

A = effective tension area of concrete surrounding the flexural tension reinforcement and having the same centroid as that reinforcement, divided by the number of bars or wires (in.^{2}). When the flexural reinforcement consists of different bar or wire sizes, the numbers of bars or wires shall be computed as the total area of reinforcement divided by the area of the largest bar or wire used.

A = area of footing.

A_b = area of an individual bar (in.^{2}).

A_c = area of core of spirally reinforced compression member measured to outside diameter of spiral (in.^{2}).

A_c = area of concrete section resisting shear transfer.

A_f = area of reinforcement in bracket or corbel resisting factored moment, $[V_{ua} + N_{uc}(h - d)]$ (in.^{2}).

A_g = gross area of section (in.^{2}).

A_h = area of shear reinforcement parallel to flexural tension reinforcement (in.^{2}).

A_n = area of reinforcement in bracket or corbel resisting tensile force N_{uc} (in.^{2}).

A_ℓ = total area of longitudinal reinforcement to resist torsion (in.^{2}).

A_s = area of nonprestressed tension reinforcement (in.^{2}).

A_s' = area of compression reinforcement (in.^{2}).

A_{st} = total area of longitudinal reinforcement, [bars or steel shapes (in.^{2})].

A_t = area of structural steel shape, pipe, or tubing in a composite section (in.^{2}).

A_t = area of one leg of a closed stirrup resisting torsion within a distance s (in.^{2}).

A_{vf} = area of shear-friction reinforcement (in.^{2}).

A_{vh} = area of shear reinforcement parallel to flexural tension reinforcement within a distance s_2 (in.^{2}).

501

A_w = area of an individual wire to be developed or spliced (in.^{2}).

A_1 = loaded area.

A_2 = maximum area of the portion of the supporting surface that is geometrically similar to and concentric with the loaded area.

b = width of compression face of member (in.).

b = effective compressive flange width of a structural member (in.).

b_o = perimeter of critical section for slabs and footings (in.).

b_t = width of that part of cross section containing the closed stirrups resisting torsion.

b_w = web width, or diameter of circular section (in.).

c = distance from extreme compression fiber to neutral axis (in.).

c_1 = size of rectangular or equivalent rectangular column, capital, or bracket measured in the direction of the span for which moments are being determined (in.).

c_2 = size of rectangular or equivalent rectangular column, capital, or bracket measured transverse to the direction of the span for which moments are being determined (in.).

C = cross-sectional constant to define torsional properties.

C_m = a factor relating actual moment diagram to an equivalent uni-uniform moment diagram.

C_s = constant used in determining analogous section modulus of critical section for shear in a slab.

C_t = factor relating shear and torsional stress properties.

   $= \dfrac{b_w d}{\Sigma x^2 y}$

C_v = constant used in determining

ratio of punching shear stress caused by moment transfer M_t to stress caused by shear force (Eq. 7.11), for uniform loading.

d = distance from extreme compression fiber to centroid of tension reinforcement (in.).

d' = distance from extreme compression fiber to centroid of compression reinforcement (in.).

d_b = nominal diameter of bar or wire.

d_c = thickness of concrete cover measured from extreme tension fiber to center of bar or wire located closest thereto (in.).

d_p = diameter of pile at footing base.

d_s = distance from extreme tension fiber to centroid of tension reinforcement (in.).

D = dead loads, or related internal moments and forces.

e = eccentricity of column axial load measured from centroid of section.

e' = eccentricity of column axial load measured from centroid of tension reinforcement.

e_x, e_y = eccentricity of load from center of gravity of footing, in the x direction and y direction, respectively.

E = load effects of earthquake, or related internal moments and forces.

E_1 = effectiveness in tension of splices in group 1 bars.

E_2 = effectiveness in tension of splices in group 2 bars.

E_c = modulus of elasticity of concrete (psi).

E_{cb} = modulus of elasticity of beam concrete.

E_{cc} = modulus of elasticity of column concrete.

E_{cs} = modulus of elasticity of slab concrete.

EI = flexural stiffness of a member.

E_s = modulus of elasticity of reinforcement, (psi).

f_b = flexural stress in a plain concrete footing.

f_{ba} = permissible flexural stress in a plain concrete footing.

f'_c = specified compressive strength of concrete (psi).

$\sqrt{f'_c}$ = square root of specified compressive strength of concrete, (psi).

f_{ct} = average splitting tensile strength of lightweight aggregate concrete (psi).

f_g = average stress on a column.

f_r = modulus of rupture of concrete (psi).

f_s = calculated stress in reinforcement at service loads (ksi).

f_y = specified yield strength of nonprestressed reinforcement (ksi).

F = loads due to weight and pressures of fluids with well-defined densities and controllable maximum heights, or related internal moments and forces.

GK = torsional rigidity of a member.

GK_{cr} = cracked torsional rigidity of a member.

GK_g = uncracked torsional rigidity of a member.

h = overall thickness of member (in.).

h_v = total depth of shearhead cross section (in.).

h_w = total height of wall from base to top (in.).

H = loads due to weight and pressure of soil, water in soil, or other materials, or related internal moments and forces.

I = moment of inertia of section resisting externally applied factored loads.

I_b = moment of inertia about centroidal axis of gross section of beam.

I_c = moment of inertia of gross section of column.

I_{cr} = moment of inertia of cracked section transformed to concrete.

I_e = effective moment of inertia for computation of deflection.

I_g = moment of inertia of gross concrete section about centroidal axis, neglecting reinforcement.

I_s = moment of inertia about centroidal axis of gross section of slab.

= $h^3/12$ times width of slab defined in notations α and β_t.

I_{se} = moment of inertia of reinforcement about centroidal axis of member cross section.

I_t = moment of inertia of structural steel shape, pipe, or tubing about centroidal axis of composite member cross section.

k = effective length factor for compression members.

k = ratio of height of compressive stress block by elastic theory to effective depth (see Fig. 8.5).

K_3, K_4 = ratio of distance from neutral axis to perimeter of critical section to the length of critical section for shear in a two-way slab, for Cases III and IV, respectively.

K_b = flexural stiffness of beam; moment per unit rotation.

K_c = flexural stiffness of column; moment per unit rotation.

K_s = flexural stiffness of slab; moment per unit rotation.

K_t = torsional stiffness of torsional member; moment per unit rotation.

ℓ = span length of beam or one-way slab; clear projection of cantilever (in.).

ℓ_1 = length of span in direction that moments are being determined, measured center to center of supports.

ℓ_2 = length of span transverse to ℓ_1, measured center to center of supports.

ℓ_a = additional embedment length at support or at point of inflection (in.).

ℓ_c = vertical distance between supports (in.).

ℓ_d = development length (in.).

ℓ_{dh} = development length of standard hook in tension, measured from critical section to outside end of hook [straight embedment length between critical section and start of hook (point of tangency) plus radius of bend and one bar diameter] (in.).

= $\ell_{hb} \times$ applicable modification factors.

ℓ_{hb} = basic development length of standard hook in tension (in.).

ℓ_n = clear span for positive moment or shear and average of adjacent clear spans for negative moment.

ℓ_s = stagger distance between splices in group 1 and group 2 bars.

ℓ_u = unsupported length of compression member.

ℓ_v = length of shearhead arm from centroid of concentrated load or reaction (in.).

ℓ_w = horizontal length of wall (in.).

ℓ_x, ℓ_y = length of rectangular footing in the x direction and y direction, respectively.

L = live loads, or related internal moments and forces.

M = design moment.

M_s = maximum moment in member at stage deflection is computed.

M_c = factored moment to be used for design of compression member.

M_{cr} = moment causing flexural cracking at section due to externally applied loads.

M_ℓ, M_s = moment capacity of long columns and short columns, respectively.

M_{max} = maximum factored moment at section due to externally applied loads.

M_o = total factored static moment.

M_n = nominal moment strength at section.

= $A_s f_y (d - a/2)$.

M_p = required plastic moment strength of shearhead cross section.

M_u = factored moment at section.

M_v = moment resistance contributed by shearhead reinforcement.

M_{1b} = value of smaller factored end moment on compression member due to the loads that result in no appreciable sidesway, calculated by conventional elastic frame analysis, positive if member is bent in single curvature, negative if bent in double curvature.

M_{2b} = value of larger factored end moment on compression member due to loads that result in no appreciable sidesway, calculated by conventional elastic frame analysis.

M_{2s} = value of larger factored end moment on compression member due to loads that result in appreciable sidesway calculated by conventional elastic frame analysis.

n = modular ratio of elasticity.

= E_s / E_c.

N_u = factored axial load normal to cross section occurring simultaneously with V_u, to be taken as positive for compression, negative for tension, and to include effects of tension due to creep and shrinkage.

N_{uc} = factored tensile force applied at

P top of bracket or corbel acting simultaneously with V_u, to be taken as positive for tension.

P = axial load on footing.

p = allowable soil-bearing pressure.

P_b = nominal axial load strength at balanced strain conditions.

P_c = critical axial load.

P_n = nominal axial load strength at given eccentricity.

P_o = nominal axial load strength at zero eccentricity.

P_ℓ, P_s = axial load capacity of long columns and short columns, respectively.

P_u = factored axial load at given eccentricity $\leq \phi P_n$.

P_{nw} = nominal axial load strength of wall.

r = radius of gyration of cross section of a compression member.

r = ratio of longest clear span to shortest clear span, each adjacent to the column, in the direction for which moments are being determined.

R = capacity reduction factor, applied to both axial load and moment (Chapter 9).

R = proportion of negative moments calculated by elastic theory that may be increased or decreased at supports of continuous flexural members (Chapter 13).

R_t = ratio of tensile strength of all bars in connection with splices that are less than 100% effective to strength of all bars where none are spliced.

R_1 = ratio of area of group 1 bars to area of all bars.

R_2 = ratio of area of group 2 bars to area of all bars.

s = spacing of shear or torsion reinforcement in direction parallel to longitudinal reinforcement (in.).

s = spacing of stirrups or ties (in.)

s_w = spacing of wire to be developed or spliced (in.).

s_1 = spacing of vertical reinforcement in wall (in.).

s_2 = spacing of shear or torsion reinforcement in direction perpendicular to longitudinal reinforcement—or spacing of horizontal reinforcement in wall (in.).

S = section modulus.

T = tensile force.

T = cumulative effects of temperature, creep, shrinkage, and differential settlement.

T_c = nominal torsional moment strength provided by concrete.

T_n = nominal torsional moment strength.

T_r = splitting force at bend in flexural member.

T_s = nominal torsional moment strength provided by torsion reinforcement.

T_u = factored torsional moment at section.

u = equivalent unit bond stress.

U = required strength to resist factored loads or related internal moments and forces.

v = design shear stress.

v_c = permissible shear stress carried by concrete (psi).

v_h = permissible horizontal shear stress (psi).

V = design shear force at section.

V_c = nominal shear strength provided by concrete.

V_n = nominal shear strength.

V_s = nominal shear strength provided by shear reinforcement.

V_u = factored shear force at section.

w_c = unit weight of plain concrete.

w_c = weight of concrete (lb/ft^3).

w_d = factored dead load per unit area.

w_{rc} = unit weight of reinforced concrete.

w_s = unit weight of steel.

w_t = factored live load per unit area.

w_u = factored load per unit length of beam or per unit area of slab.

W = wind load, or related internal moments and forces.

x = shorter overall dimension of rectangular part of cross section.

x_1 = shorter center-to-center dimension of closed rectangular stirrup.

x_o = shorter dimension between longitudinal corner bars.

X,Y = dimensions of critical section for shear in two-way slab.

y = longer overall dimension of rectangular part of cross section.

y_t = distance from centroidal axis of gross section, neglecting reinforcement, to extreme fiber in tension.

y_1 = longer center-to-center dimension of closed rectangular stirrup.

y_c = distance from centroidal axis of gross section, neglecting reinforcement, to extreme fiber in compression.

y_o = longer dimension between longitudinal corner bars.

z = quantity limiting distribution of flexural reinforcement.

α (alpha) = angle between inclined stirrups and longitudinal axis of member.

α = ratio of flexural stiffness of beam section to flexural stiffness of a width of slab bounded laterally by centerlines of adjacent panels (if any) on each side of the beam.
$$= \frac{E_{cb}I_b}{E_{cs}I_s}$$

α_c = ratio of flexural stiffness of columns above and below the slab to combined flexural stiffness of the slabs and beams at a joint taken in the direction of the span for which moments are being determined.
$$= \frac{\Sigma K_c}{\Sigma(K_s + K_b)}$$

α_f = angle between shear-friction reinforcement and shear plane.

α_m = average value of α for all beams on edges of a panel.

α_m = average ratio of stiffness of columns to floor members at the two ends of a column.

α_{min} = minimum α_c.

α_t = coefficient as a function of y_1/x_1.

α_v = ratio of stiffness of shearhead arm to surrounding composite slab section.

α_1 = α in direction of ℓ_1.

α_2 = α in direction of ℓ_2.

β (beta) = ratio of clear spans in long to short direction of two-way slabs.

β = ratio of long side to short side of footing.

β_a = ratio of dead load per unit area to live load per unit area (in each case without load factors).

β_b = ratio of area of reinforcement cut off to total area of tension reinforcement at section.

β_c = A coefficient for computing cracked moment of inertia of beams.

β_c = ratio of long side to short side of concentrated load or reaction area.

β_d = absolute value of ratio of maximum factored dead load moment to maximum factored total load moment, always positive.

β_g = a coefficient for computing gross moment of inertia of T-beams (see Fig. A8.1)

β_m = a deflection coefficient (see Table A8.2).

β_s = ratio of length of continuous edges to total perimeter of a slab panel.

β_t = ratio of torsional stiffness of edge beam section to flexural stiffness of a width of slab equal to span length of beam, center-to-center of supports.

$$= \frac{E_{cb}C}{2E_{cs}I_s}$$

β_{tb} = ratio of moment of inertia of a T-beam to moment of inertia of a rectangular beam.

β_w = a deflection coefficient (see Table A8.2).

β_1 = factor defining rectangular compression block in flexural members.

γ_f
(gamma) = fraction of unbalanced moment transferred by flexure at slab-column connections.

γ_v = fraction of unbalanced moment transferred by eccentricity of shear at slab-column connections.
$= 1 - \gamma_f$.

δ_b
(delta) = moment magnification factor for frames braced against sidesway, to reflect effects of member curvature between ends of compression member.

δ_s = moment magnification factor for frames not braced against sidesway, to reflect lateral drift resulting from lateral and gravity loads.

δ_s = factor defined by Eq. 4.4.

Δ_1
(delta) = initial deflection.

Δ_{LT} = long-term deflection.

ϵ
(epsilon) = strain.

ϵ_{cu} = ultimate strain in concrete = 0.003.

ϵ_y = yield strain in steel.

η
(eta) = number of identical arms of shearhead.

θ
(theta) = rotation of a member at a support (Chapter 8).

θ = twist of a member due to torsion (Chapter 8).

θ = angle of bend in a bent flexural member (Chapter 10).

λ
(lambda) = multiplier for additional long-time deflection.

λ = correction factor related to unit weight of concrete for shear friction.

μ
(mu) = coefficient of friction.

ξ
(xi) = time-dependent factor for sustained load.

ρ
(rho) = ratio of nonprestressed tension reinforcement.
$= A_s/bd$.

ρ' = ratio of nonprestressed compression reinforcement.
$= A_s'/bd$.

ρ_b = reinforcement ratio producing balanced strain conditions.

ρ_g = ratio of total longitudinal reinforcement area to cross-sectional area of column.

ρ_h = ratio of horizontal shear reinforcement area to gross concrete area of vertical section.

ρ_n = ratio of vertical shear reinforcement area to gross concrete area of horizontal section.

ρ_s = ratio of volume of spiral reinforcement to total volume of core (out-to-out spirals) of a spirally reinforced compression member.

ρ_T = total area of longitudinal reinforcement in a column
$= A_{st}/A_g$.

ρ_w = A_s/b_wd.

ϕ
(phi) = strength reduction factor.

$\Sigma x^2 y$ = torsional section properties.

ψ
(phi) = ratio of $\Sigma(EI/\ell_u)$ for columns to $\Sigma(EI/\ell)$ for flexural members in a plane at one end of a compression member.

APPENDIX C

ABBREVIATIONS

Text	Tables, Figures, and Examples	Definition
c/c	c/c	center to center
ft	′	foot or feet
ft-kip	′k	foot-kip
FRP	FRP	fiber-reinforced plastic (the fibers are usually glass)
in.	″	inch or inches
in.-kip	″k	inch kip
kip	k	kilo pound = 1000 pounds
klf	k/′	kips per lineal foot
ksf	ksf	kips per square foot
ksi	ksi	kips per square inch
lbs	lbs	pounds
pcf	pcf	pounds per cubic foot
plf	plf	pounds per lineal foot
psf	psf	pounds per square foot
psi	psi	pounds per square inch
ft^2	□′	square foot
sq. in.	□″	square inch
WWF	WWF	welded wire fabric, or fabric made of cold-drawn steel wires welded at all points of intersection

APPENDIX D

SI CONVERSION FACTORS

Length
1 in. = 25.4 mm (millimeter)
1 ft = 0.3048 m (meter)

Area
1 in.^{2} = 645.2 mm^2
1 ft^2 = 0.0929 m^2
1 yd^2 = 0.8361 m^2

Volume
1 ft^3 = 0.0283 m^3
1 yd^3 = 0.7645 m^3

Load
1 lb = 4.448 N (Newton)
1 kip = 4.448 kN (kiloNewton)
1 klf = 14.59 kN/m
1 psf = 0.0479 kN/m^2
1 ksf = 47.9 kN/m^2

Structural Properties
1 in.^{3} = 16.39 × 10^3 mm^3
1 in.^{4} = 0.4162 × 10^6 mm^4

Material Properties and Stress
1 pcf = 16.03 kg/m^3
1 psi = 0.006895 MPa (Mega Pascal)
1 ksi = 6.895 MPa

Moment
1 ft-kip = 1.356 kNm (kiloNewton-meter)

APPENDIX E

BAR AND WIRE PROPERTIES

Properties of Wire used in Welded Wire Fabric

Wire Size Number		Diameter	Cross-sectional	Weight
Smooth	Deformed	(in.)	Area (in.^{2})	(plf)
W1.4		0.134	0.014	0.049
W2		0.160	0.02	0.068
W2.9		0.192	0.029	0.098
W4	D4	0.226	0.04	0.136
W5	D5	0.252	0.05	0.170
W5.5		0.265	0.055	0.187

ASTM Standard Reinforcing Bars

Bar Size (no.)	Nominal Dimensions—Round Sections			Weight (plf)
	Diameter (in.)	Cross-sectional area (in.²)	Perimeter (in.)	
3	0.375	0.11	1.178	0.376
4	0.500	0.20	1.571	0.668
5	0.625	0.31	1.963	1.043
6	0.750	0.44	2.356	1.502
7	0.875	0.60	2.749	2.044
8	1.000	0.79	3.142	2.670
9	1.128	1.00	3.544	3.400
10	1.270	1.27	3.990	4.303
11	1.410	1.56	4.430	5.313
14	1.693	2.25	5.32	7.65
18	2.257	4.00	7.09	13.60

INDEX